INTRODUCTION TO CMOS OP-AMPS AND COMPARATORS

INTRODUCTION TO CMOS OP-AMPS AND COMPARATORS

ROUBIK GREGORIAN

A Wiley-Interscience Publication

JOHN WILEY & SONS, INC.

New York / Chichester / Weinheim / Brisbane / Singapore / Toronto

This book is printed on acid-free paper. ♾

Published simultaneously in Canada.

For ordering and customer information call, 1-800-CALL WILEY.

Library of Congress Cataloging-in-Publication Data

Gregorian, Roubik.
Introduction to CMOS OP-AMPs and comparators / Roubik Gregorian.
p. cm.
"A Wiley-Interscience publication."
Includes Index.
ISBN 0-471-31778-0 (hardcover : alk. paper)
1. Operational amplifiers—Design and construction. 2. Comparator circuits—Design and construction. 3. Metal oxide semiconductors, Complementary. I. Title.
TK7871.58.06G74 1999
621.39′5—dc21 98-23233
CIP

Printed in the United States of America.
10 9 8 7 6 5 4 3 2 1

To my wife, Agnes
And our children, Aris and Talin

CONTENTS

PREFACE

Operational amplifiers (op-amps) and comparators are two of the most intricate, and in many ways the most important, building blocks of an analog circuit. These components are used in such devices as switched-capacitor filters, analog-to-digital (A/D) and digital-to-analog (D/A) converters, amplifiers, modulators, rectifiers, peak detectors, and so on. The performance of op-amps and comparators usually limits the high-frequency application and dynamic range of the overall circuit. Without a thorough understanding of the operation and basic limitation of these components, the circuit designer cannot determine or even predict the actual response of the overall system. Hence this book gives a fairly detailed explanation of the overall configurations and performance limitations of op-amps and comparators exclusively in CMOS technology. While the scaling properties of the very large scale integration (VLSI) processes have resulted in denser and higher-performance digital circuits, they have also changed the design techniques used for CMOS analog circuits. Therefore, the main purpose of these discussions is to illustrate the most important principles underlying the specific circuits and design procedures. Nevertheless, the treatment is detailed enough to enable the reader to design high-performance CMOS op-amps and comparators suitable for most analog circuit applications.

The main emphasis of this book is on physical operation and design process. It has been written as a unified text dealing with the analysis and design of CMOS op-amps and comparators. It is intended for classroom adoption to be used as a senior or graduate-level text in the electrical engineering curriculum of universities and also as training and reference material for industrial circuit designers. To increase the usefulness of the book as a text for classroom teaching, numerous problems are included at the end of each chapter; these problems may be used for homework assignments. To enhance its value as a design reference, tables and numerical design examples are included to clarify the step-by-step processes involved. The first two chapters provide a concise, basic-level, and (I hope) clear description of analog MOS integrated circuits and the necessary background in semiconductor device physics. The remainder of the book is devoted to the design of CMOS op-amps and comparators and to the practical problems encountered and their solutions. The book also includes two introductory chapters on the applications of op-amps and comparators in A/D and D/A converters. For a more detailed discussion on the important subject of data converters, readers are referred to the *Principles of Data Conversion System Design* by Behzad Rezavi, and *Delta-Sigma Data Converters: Theory, Design and Simulation* by Steven R. Norsworthy, Richard Schreier, and Gabor C. Temes.

This book is based in part on a previous book I coauthored with Gabor C. Temes, titled *Analog MOS Integrated Circuits for Signal Processing.* The original material has been augmented by the latest developments in the area of analog MOS integrated circuits, in particular op-amps and comparators. Most of the material and concepts originated from the publications cited at the end of each chapter as well as from many practicing engineers who worked with me over the years.

Since the original book evolved from a set of lecture notes written for short courses, the organization of the material was therefore influenced by the need to make the presentation suitable for audiences of widely varying backgrounds. Hence I tried to make the book reasonably self-contained, and the presentation is at the simplest level afforded by the topics discussed. Only a limited amount of preparation was assumed on the part of the reader: mathematics on the junior level, and one or two introductory-level courses in electronics and semiconductor physics are the minimum requirements.

The book contains eight chapters. Chapter 1 provides a basic introduction to digital and analog signal processing, followed by several representative examples of circuits and systems utilizing CMOS op-amps and comparators. This material can be covered in one lecture (two-hour lectures are assumed here and throughout the preface).

In Chapter 2 the physics of MOS devices is described briefly and linearized models of MOSFETs, as well as MOS capacitors and switches are discussed. The technology used to fabricate CMOS devices is also discussed briefly. Once again, depending on the background of the audience, two or three lectures should suffice to cover the content of this chapter.

Chapter 3 covers some of the basic subcircuits commonly utilized in analog MOS integrated circuits. These subcircuits are typically combined to synthesize a more complex circuit function. Complete coverage of all topics of this chapter requires about three lectures.

In Chapter 4 circuit design techniques for realizing CMOS operational amplifiers are discussed. The most common circuit configurations, as well as their design and limitations, are included. Full coverage of all topics in this chapter requires about four lectures.

In Chapter 5 the principles of CMOS comparator design are discussed. First the single-ended auto-zeroing comparator is examined, followed by simple and multistage differential comparators, regenerative comparators, and fully differential comparators. Two lectures should be sufficient for complete coverage of this chapter.

Chapters 6 and 7, which cover CMOS digital-to-analog and analog-to-digital converters, serve as practical application examples of op-amps and comparators. The fundamentals and performance metrics of the data converters are presented first, followed by a discussion of popular architectures of Nyquist-rate converters. Digital-to-analog converters are divided into voltage, charge, and current scaling types. Analog-to-digital converters include high-speed flash, medium-speed successive-approximation, and low-speed serial converters. Complete coverage of all topics may require three to four lectures.

In Chapter 8 the design principles presented in Chapter 4 and 5 are employed to work out several design examples to acquaint the reader with the problems and trade-offs involved in op-amp and comparator designs. Practical considerations such as dc biasing, systematic offset voltage, and power supply noise are discussed in some detail. All topics in this chapter can be covered in three lectures; if the detailed discussion in Sections 8.2 and 8.3 is condensed, the material can be presented in two lectures.

Thus, depending on the depth of the presentation, full coverage of all material in the book may require as many as 20 two-hour lectures or as few as 16.

I am grateful to many people who have helped me directly or indirectly in the elaborate and sometimes overwhelming task of publishing this book. In particular, I would like to thank my colleagues Drs. S. C. Fan, B. Fotouhi, B. Ghaderi, and G. C. Temes, who read and criticized versions of the manuscript. Their comments have been most helpful and are greatly appreciated. Most of the difficult typing task was done by Ms. W. Irwin and D. Baker. I am grateful for their excellent and painstaking help. Last, but not least, I would like to express my gratitude to my family for graciously suffering neglect during the writing of this book. Without their understanding and support this work would not have been possible.

ROUBIK GREGORIAN

Saratoga, California
January 1999

INTRODUCTION TO CMOS OP-AMPS AND COMPARATORS

CHAPTER 1

INTRODUCTION

Operational amplifiers (op-amps) and comparators are two of the most important building blocks for analog signal processing. Op-amps and a few passive components can be used to realize such important functions as summing and inverting amplifiers, integrators, and buffers. The combination of these functions and comparators can result in many complex functions, such as high-order filters, signal amplifiers, analog-to-digital (A/D) and digital-to-analog (D/A) converters, input and output signal buffers, and many more. Making the op-amp and comparator faster has always been one of the goals of analog designers. In this chapter the basic concept of digital and analog signal processing is introduced. Then a third category of signal processing, the sampled-data analog technique, which is in between the two main classifications, is described. Finally, a few representative examples are given of circuits and systems utilizing CMOS op-amps and comparators, to illustrate the great potential of these components as part of an MOS-LSI chip.

1.1 CLASSIFICATION OF SIGNAL PROCESSING TECHNIQUES [1–4]

Electrical signal processors are usually divided into two categories: analog and digital systems. An *analog system* carries signals in the form of voltages, currents, charges, and so on, which are *continuous* functions of the *continuous*-time variable. Some typical examples of analog signal processors are audio amplifiers, passive- or active-*RC* filters, and so on. By contrast, in a *digital system* each signal is represented by a sequence of numbers. Since these numbers can contain only a finite number of digits (typically, coded in the form of binary digits, or bits) they can only take on discrete values. Also, these numbers are the sampled values of the signal, taken at discrete time instances. Thus both the dependent and independent variables of a digital signal are discrete. Since the processing of the digital bits is usually performed synchronously, a timing or clock circuit is an important part of the digital system. The timing provides one or more clock signals, each containing accurately timed pulses that operate or synchronize the operation of the components of the system. Typical examples of digital systems are a general-purpose digital computer or a special-purpose digital signal processor dedicated to (say) calculating the Fourier transform of a signal via the fast Fourier transform (FFT), or a digital filter used in speech analysis, and so on.

By contrast, analog signal processing circuits utilize op-amps, comparators, resistors, capacitors, and switches to perform such functions as filters, amplifiers, rectifiers, and many more. To understand the basic concepts of the most commonly used configurations of an analog circuit, consider the simple analog transfer function

$$\frac{V_{\text{out}}(s)}{V_{\text{in}}(s)} = \frac{b}{s^2 + as + b}. \tag{1.1}$$

It is easy to verify that the *RLC* circuit shown in Fig. 1.1*a* can realize this function (Problem 1.1). Although this circuit is easy to design, build, and test, the presence of the inductor in the circuit makes its fabrication in integrated form impractical. In fact, for low-frequency applications, this circuit may well require a very large valued, and hence bulky, inductor and capacitor. To overcome this problem, the designer may decide to realize the desired transfer function using an active-*RC* circuit. It can readily be shown that the circuit in Fig. 1.1*b*, which utilizes three *operational amplifiers,* is capable of providing the transfer function specified in

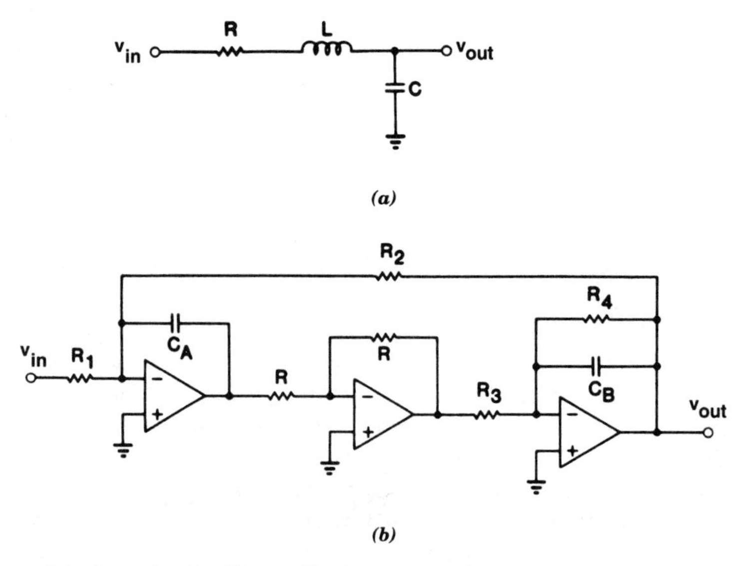

Figure 1.1. Second-order filter realization: (*a*) passive circuit; (*b*) active-*RC* circuit.

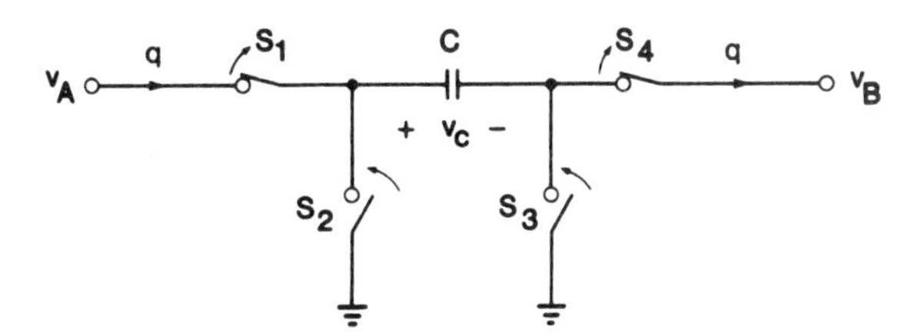

Figure 1.2. Switched-capacitor realization of a resistive branch.

Eq. (1.1). This circuit needs no inductors and may be realized with small discrete components for a wide variety of specifications (Problem 1.2). It turns out, however, that while integration of this circuit on a bipolar chip is, in principle, feasible (since the amplifiers, resistors, and capacitors needed can all be integrated), there are some major practical obstacles to integration. These include the very large chip area needed by the RC components, as well as the stringent accuracy and stability requirements for these elements. These requirements cannot readily be satisfied by integrated components, since neither the fabricated values nor the temperature-induced variations of the resistive and capacitive elements track each other. The resulting pole–zero variations are too large for most applications.

Prior to mid-1970s, analog circuits such as the one shown in Fig. 1.1 were implemented using integrated bipolar op-amps and discrete passive components. In the 1970s two developments made it possible to fully integrate analog circuits in metal-oxide semiconductor (MOS) technology. The first development was the emergence of a technique called switched-capacitor (SC) circuits [6], which is an effective strategy for solving both the area and the matching problems by replacing each resistor in the circuit by the combination of a capacitor and a few switches. Consider the branches shown in Fig. 1.2. Here, the four switches S_1, S_2, S_3, and S_4 open and close periodically, at a rate which is much faster than that of the variations of the terminal voltage v_A and v_B. Switches S_1 and S_4 operate synchronously with each other but in opposite phase with S_2 and S_3. Thus when S_2 and S_3 are closed, S_1 and S_4 are open, and vice versa. Now when S_2 and S_3 close, C is discharged. When S_2 and S_3 open, S_1 and S_4 close, and C is recharged to the voltage $v_C = v_A - v_B$. This causes a charge $q = C(v_A - v_B)$ to flow through the branch of Fig. 1.2. Next, C is again discharged by S_2 and S_3, and so on. If this cycle is repeated every T seconds (where T is the *switching period or clock period*), the average current through the branch is then

$$i_{\text{av}} = \frac{q}{T} = \frac{C}{T}(v_A - v_B). \tag{1.2}$$

Thus i_{av} is *proportional* to the branch voltage $v_A - v_B$. Similarly, for a branch containing a resistor R, the branch current is $i = (1/R)(v_A - v_B)$. Thus the average current flowing in these two branches are the same if the relation $R = T/C$ holds.

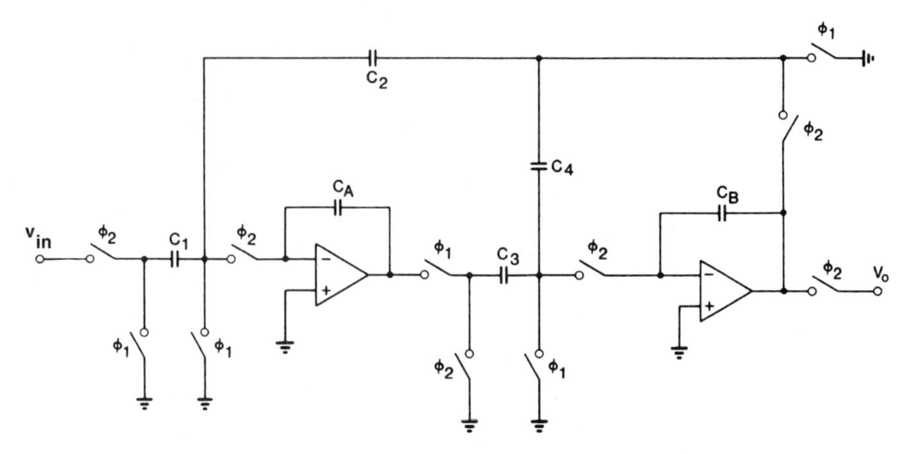

Figure 1.3. Second-order switched-capacitor filter section.

Physically, the switches transform the capacitor *C*, a nondissipative memoried element, into a dissipative memoryless (i.e., resistive) one.

It is plausible therefore that the branch of Fig. 1.2 can be used to replace all resistors in the circuit of Fig. 1.1*b*. The resulting stage [3] is shown in Fig. 1.3. In this circuit, switches that belong to different "resistors" but perform identical tasks have been combined. Furthermore, the second operational amplifier (op-amp) in Fig. 1.1*b*, which acted merely as a phase inverter, has been eliminated. This was possible since by simply changing the phasing of two of the switches associated with capacitor C_3, the required phase inversion could be accomplished without an op-amp.

As Fig. 1.3 illustrates, the transformed circuit contains only capacitors, switches, and op-amps. A major advantage of this new arrangement is that now all time constants, previously determined by the poorly controlled *RC* products, will be given by expressions of the form $(T/C_1)C_2 = T(C_2/C_1)$. Here the clock period T is usually determined by a quartz-crystal-controlled clock circuit and hence is very accurate and stable. The other factor of the time constant is C_2/C_1, that is, the ratio of two on-chip MOS capacitances. Using some simple rules in the layout of these elements, it is possible to obtain an accuracy and stability on the order of 0.1% for this ratio. The resulting overall accuracy is at least 100 times better than what can be achieved with an on-chip resistor and capacitor for the *RC* time constant.

A dramatic improvement is also achievable for the area required by the passive elements. To achieve a time constant in the audio-frequency range (say 10 krad/s), even with a large (10-pF) capacitor, a resistance of 10-MΩ is required. Such a resistor will occupy an area of about 10^6 μm^2, which is prohibitively large; it is nearly 10% of the area of an average chip. By contrast, for a typical clock period of 10 μs, the capacitance of the switched capacitor realizing a 10-MΩ resistor is $C = T/R = 10^{-5}/10^7 = 10^{-12}\,\mathrm{F} = 1\,\mathrm{pF}$. The area required realizing this capacitance is about 2500 μm^2, only 0.25% of that needed by the resistor that it replaces.

The second development that made the realization of the fully integrated analog MOS circuits possible was the design of the MOS op-amp. Perhaps the most generally useful analog circuit function is that of the operational amplifier. Prior to about 1977, there existed a clear separation of the bipolar and MOS technologies, according to the function required [1,5]. MOS technology, with its superior device density, was used mostly for digital logic and memory applications, while all required analog functions (such as amplification, filtering, and data conversion) were performed using bipolar integrated circuits, such as bipolar op-amps. Since that time, however, rapid progress made in MOS fabrication techniques made it possible to manufacture much more complex and flexible chips. In addition, new developments occurred in communication technology (such as digital telephony, data transmission via telephone lines, adaptive communication channels, etc.) which required analog and digital signal processing circuitry in the same functional blocks. The analog functions most often needed are filtering (for antialiasing, smoothing, band separation, etc.), amplification, sample-and-hold operations, voltage comparison, and the generation as well as precise scaling of voltages and currents for data conversion. The separation of these analog functions from the digital ones merely because of the different fabrication technologies used is undesirable, since it increases both the packaging costs and the space requirements and also, due to the additional interconnections required, degrades the performance. Hence there was strong motivation to develop novel MOS circuits, which can perform these analog functions and which can also share the area on the same chip with the digital circuitry.

Compared with bipolar technology, MOS technology has both advantages and disadvantages. MOS device has extremely high impedance at its input (gate) terminal, which enables it to sense the voltage across a capacitor without discharging it. Also, there is no inherent offset voltage across the MOS device when it is used as a charge switch. Furthermore, high-quality capacitors can be fabricated reliably on an MOS chip. These features make the realization of such circuits as precision sample-and-hold stages feasible on an MOS chip [1]. This is usually not possible in bipolar technology.

On the negative side, the transconductance of MOS transistors is inherently lower than that of bipolar transistors. A typical transconductance value for a moderate-sized MOS device is around 2.5 mA/V; for a bipolar transistor, it may be about 50 times larger. This leads to a higher offset voltage for an MOS amplifier than for a bipolar amplifier. (At the same time, however, the input capacitance of the MOS transistor is typically much smaller than that of a bipolar transistor.) Also, the noise generated in an MOS device is much higher, especially at low frequencies, than in a bipolar transistor. The conclusion is that the behavior of an amplifier realized on an MOS chip tends to be inferior to an equivalent bipolar realization in terms of offset voltage, noise, and dynamic range. However, it can have much higher input impedance than that of its bipolar counterpart.

As a result of these properties, the largest use of the MOS op-amp is expected to be as part of an MOS-LSI (large-scale integration) chip. Here the design of the op-amp can take advantage of the important performance specifications that are needed. The loading of the op-amp is often very light and usually only a small-valued capacitor has to be driven by these op-amps. Switched-capacitor circuits fall especially into this category, where element-value accuracy is important but the signal frequency is not too high and the dynamic range required is not excessive. Voice- and audio-frequency filtering and data conversion are in this category and represent the bulk of the past applications.

In addition to frequency-selective switched-capacitor filtering introduced in Fig. 1.3, which has been the most common application of MOS op-amps, there are many other functions for which op-amps and comparators can be used. These include analog-to-digital (A/D) and digital-to-analog (D/A) data conversion, programmable-gain amplification for AGC and other applications, peak-detection, rectification, zero-crossing detection, and so on. They have also been used extensively in large mixed-signal analog/digital systems such as voice codecs, high-speed data communication modems, audio codecs, and speech processors. This range will expand continuously as the quality (bandwidth, dynamic range, power consumption, etc.) of the components, especially op-amps and comparators, improves.

1.2. EXAMPLES OF APPLICATIONS OF OP-AMPS AND COMPARATORS IN ANALOG MOS CIRCUITS

In this section, a few selected examples of practical analog MOS circuits are given where CMOS op-amps and comparators are used extensively. Of course, the reader should not expect to understand the details of these systems at this stage. However, the diagrams may give an idea of the potentials of these components in analog signal processing.

As mentioned earlier, one of the most important applications of CMOS op-amps is in switched-capacitor filters. Figure 1.4*a* shows the circuit diagram of a seventh-order switched-capacitor filter. Its measured frequency response is shown in Fig. 1.4*b*. The measured *passband* variation for the device is less than 0.06 dB. This represents a superior performance, which could not have been achieved without extensive trimming using any other filter technology.

An obvious application of a CMOS op-amp is the realization of charge-mode digital-to-analog converters (DAC). It can be obtained by combining a programmable capacitor array and an offset-free switched-capacitor gain stage. An example of an N-bit charge-mode DAC is shown in Fig. 1.5, where V_{ref} is a temperature-stabilized constant reference voltage. The output of the DAC is the product of the reference voltage and the binary-coded digital signal ($b_1, b_2, b_3, \ldots, b_N$). In Chapter 6 the design of such circuits is discussed in some detail.

Modulators, rectifiers, and peak detectors [6] belong to an important class of nonlinear circuits, which can be implemented with a combination of op-amps and comparators. In an amplitude modulator the amplitude of a signal $x(t)$ (usually called the *carrier*) is varied (modulated) by $m(t)$, the modulating signal. Hence the output signal $y(t)$ is the product of $x(t)$ and $m(t)$, or $y(t) = x(t)m(t)$. A periodic carrier signal, which is readily generated from a stable clock source, is a square wave alternating between two equal values $\pm V$. An easy way to perform modulation with

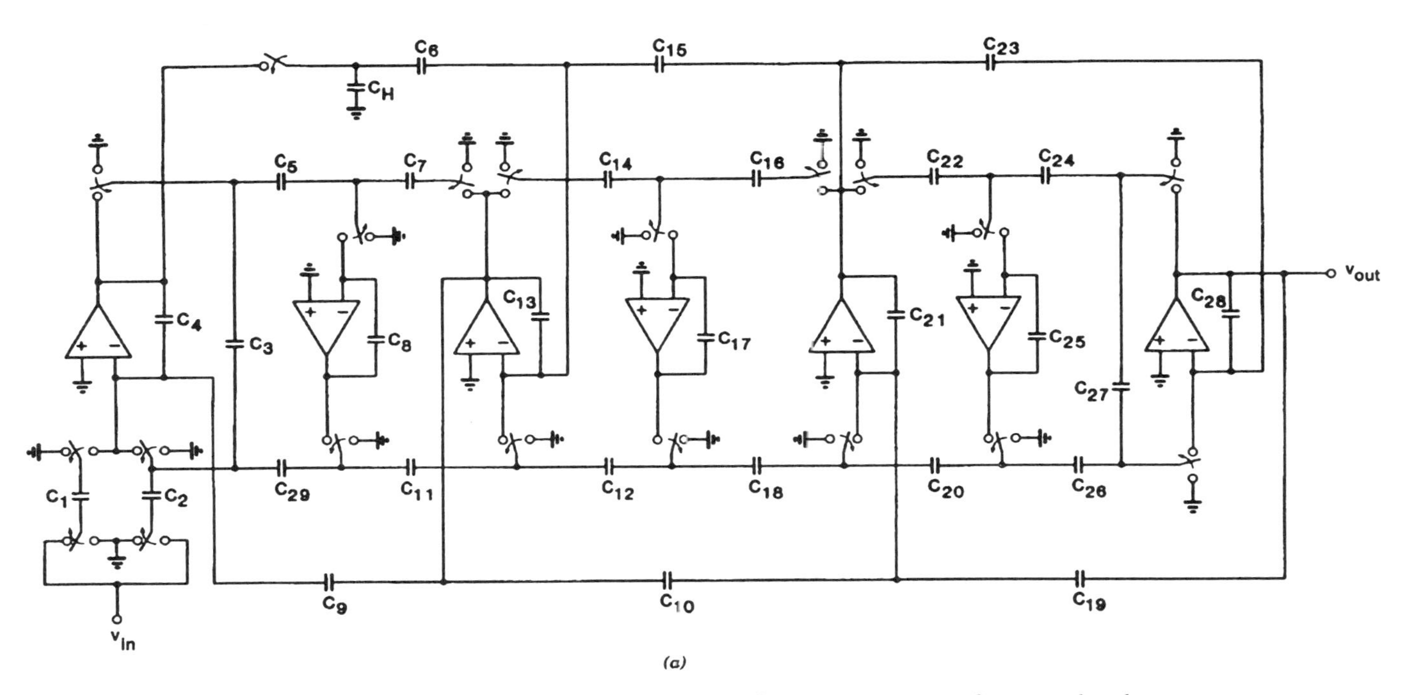

Figure 1.4. (*a*) Circuit diagram; (*b*) measured frequency response of a seventh-order switched-capacitor low-pass filter.

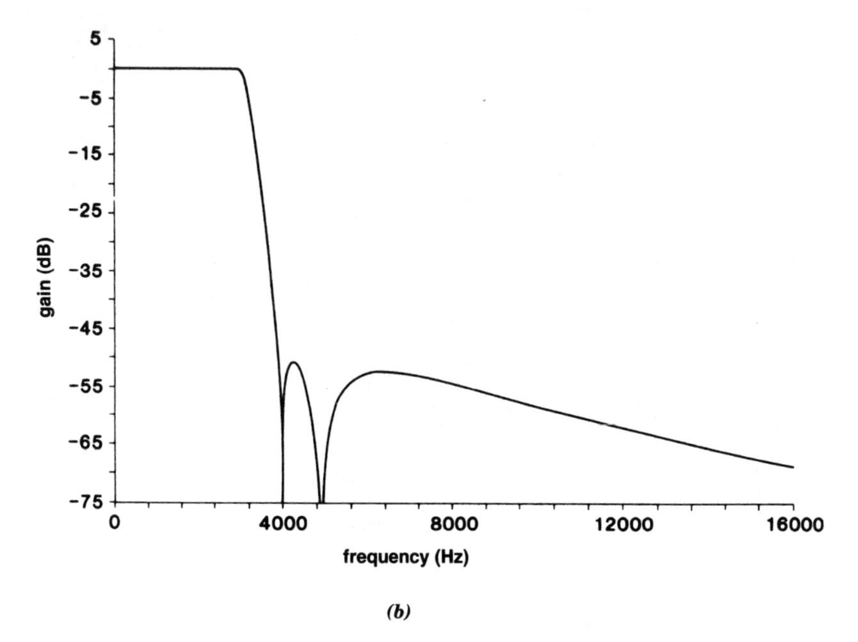

(b)

Figure 1.4. *Continued*

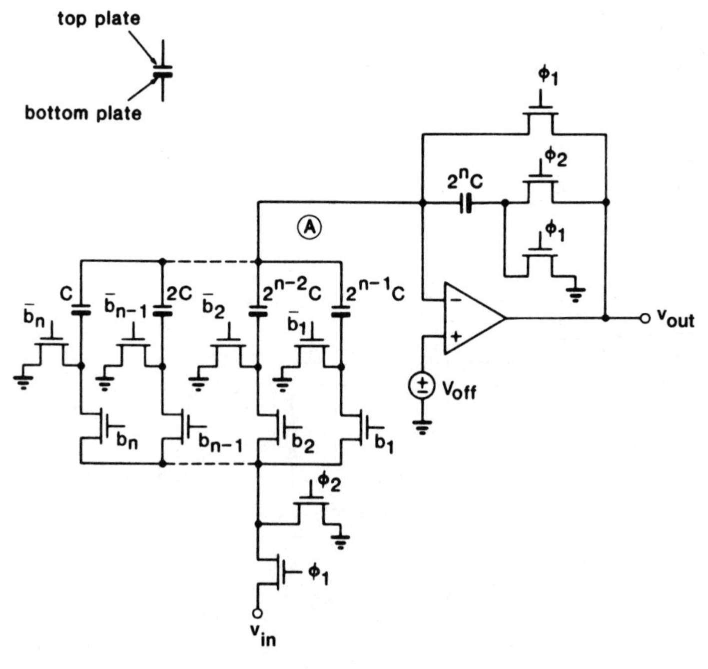

Figure 1.5. Multiplying digital-to-analog converter.

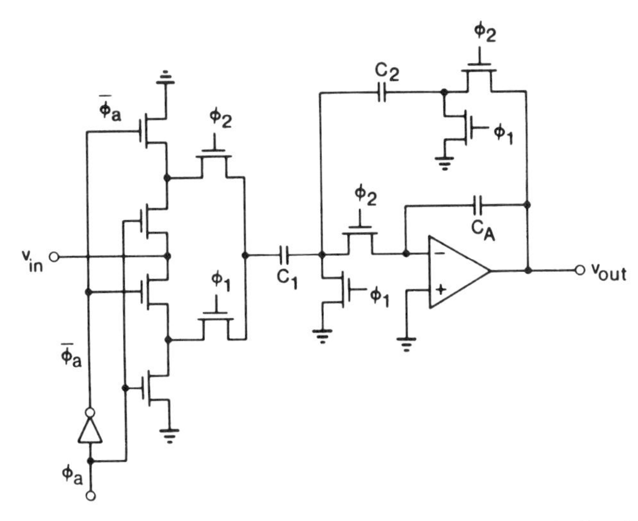

Figure 1.6. Switched-capacitor modulator with two clock signals.

a square-wave carrier is to switch the polarity of the input signal $m(t)$ periodically. A stray-insensitive switched-capacitor modulator circuit which performs according to this principle is shown in Fig. 1.6. The clock phases ϕ_1 and ϕ_2 are operated at the fast clock rate ω_c, while the phase ϕ_a changes at the slow carrier-frequency rate ω_{ca}. Normally, ω_c is much larger (by a factor of 30 or more) than ω_{ca}.

Another nonlinear circuit is a full-wave rectifier that converts an input signal $v_{in}(t)$ to its absolute value $|v_{in}(t)|$. A simple way of implementing a switched-capacitor full-wave rectifier is to add a comparator to an amplitude modulator. The circuit of a switched-capacitor full-wave rectifier based on the modulator of Fig. 1.6 is shown in Fig. 1.7*a*. Here A is set to "1" if $v_{in} > 0$ and to "0" if $v_{in} < 0$, while B is set to $\bar{A}$ by the comparator and the latch that follows it each time ϕ_1, goes high. The signals A and B then set the polarity of the transfer function so that it inverts the negative input signals, but not positive ones. Figure 1.7*b* shows an auto-zeroing comparator, which is discussed in detail in Chapter 5.

A peak detector is a circuit whose output holds the largest positive (or, if so specified, negative) voltage earlier attained by the input signal. An MOS peak detector is shown in Fig. 1.8. The op-amp acts as a comparator, with $v_{out} = V_{max}$ and v_{in} as its inputs. If $v_{in} > V_{max}$, the op-amp output goes high and M_1 conducts, charging C until $v_{out} \approx v_{in}$ is reached. If $v_{in} < V_{max}$, the op-amp output is low, M_1 is cut off, and $v_{out} = V_{max}$ is held by C.

One of the most important applications of the comparators is in A/D converters. A successive-approximation A/D converter is one type of medium-speed Nyquist-rate converter that can be realized using a programmable capacitor array (PCA) and a voltage comparator. A 5-bit converter is shown in Fig. 1.9. For high-speed operation, flash A/D converters can be used. In this configuration an array of 2^N comparators are used for an N-bit A/D converter. A conceptual diagram of an N-bit flash

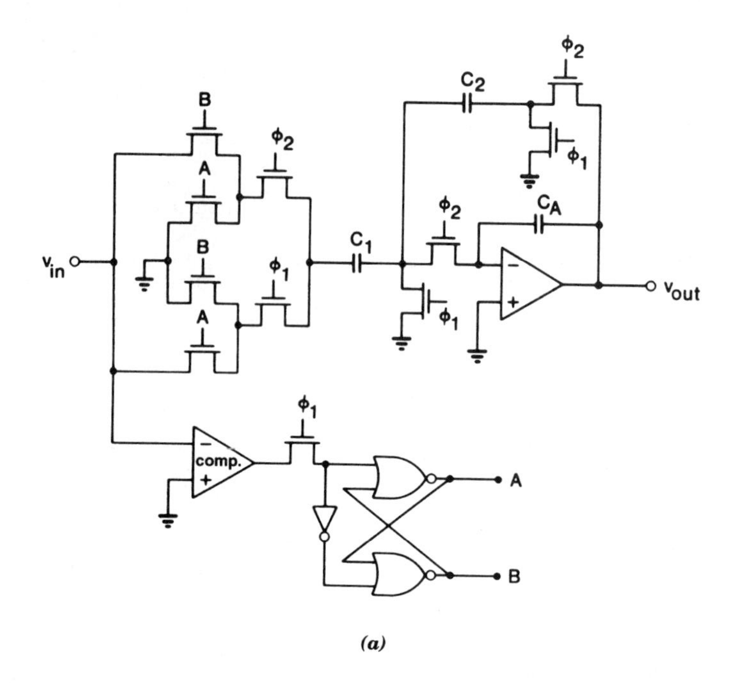

(a)

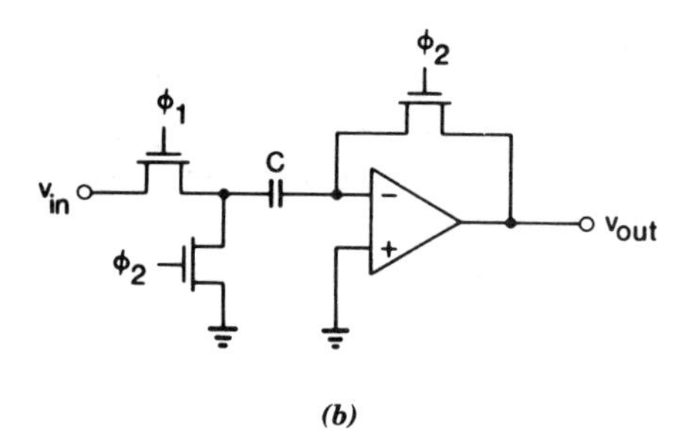

(b)

Figure 1.7. Switched-capacitor full-wave rectifier: (*a*) complete circuit; (*b*) offset-compensated comparator.

A/D converter is shown in Fig. 1.10. Analog-to-digital converters are discussed in detail in Chapter 7.

With the recent rapid progress made in MOS fabrication techniques and the emergence of the submicron CMOS technology, many intricate systems containing analog and digital functions have been combined in a fully integrated form. One drawback of the submicron CMOS technology is the reduction in the power supply voltage, which results in a reduced signal swing and hence a lower dynamic range. To improve the performance of the system and reduce the effects of noise injection from the power, ground, and clock lines, most modern high-performance mixed-

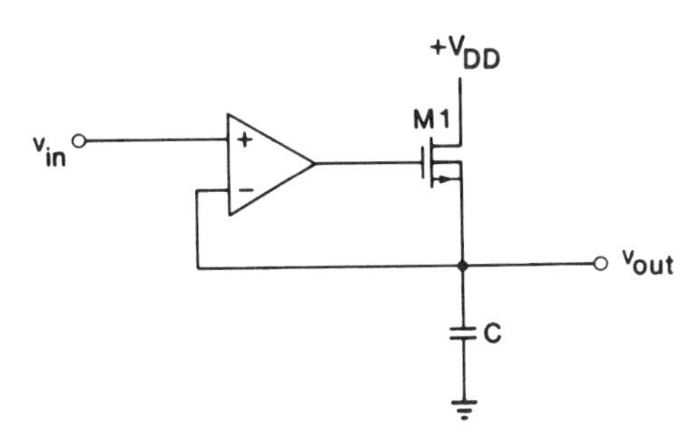

Figure 1.8. Continuous-time peak detector.

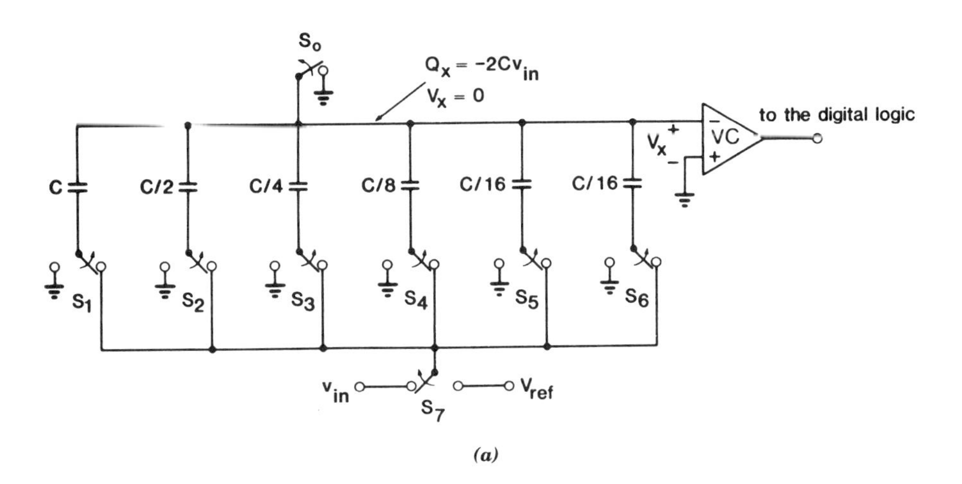

(a)

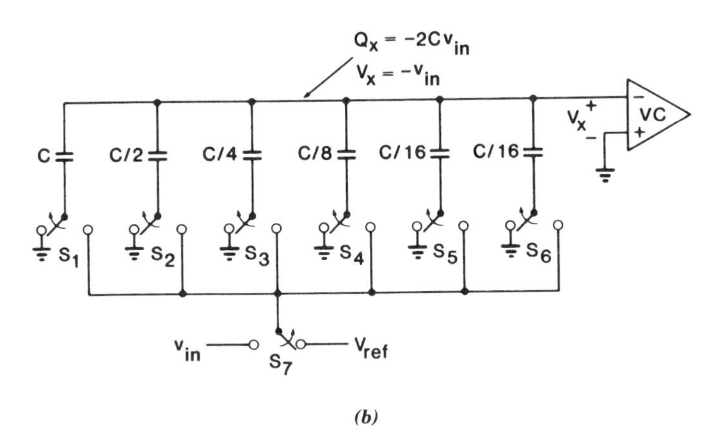

(b)

Figure 1.9. Five-bit successive-approximation A/D converter.

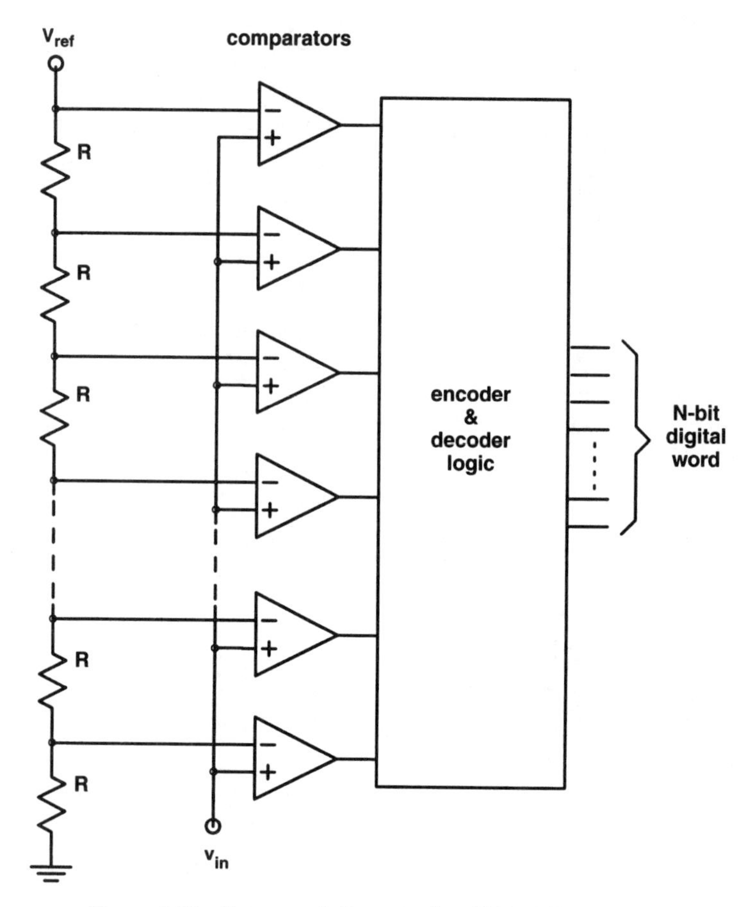

Figure 1.10. Conceptual diagram of an N-bit A/D converter.

signal integrated circuits make use of fully differential signal paths. With op-amps and comparators, the fully differential signal paths require fully differential outputs as well as inputs, and they are known as fully differential op-amps and comparators. Since this technique uses symmetrical layout, many of the noise voltages (power supply noise, clock-feedthrough noise, offset voltages) appear as common-mode signals. They are to a considerable extent canceled in the differential output voltage v_{out} at all frequencies. A high-frequency high-Q switched-capacitor bandpass filter that uses a fully differential signal path is shown in Fig. 1.11. This filter is typically used in a radio-frequency (RF) receiver system, which requires high selectivity at high frequencies [7]. The two complementary switch blocks (X_1 and X_2) are shown in Fig. 1.12. The filter uses fully differential single-pole transconductance folded-cascode op-amps with source-follower common-mode feedback as illustrated in Fig. 1.13 [8]. This op-amp achieves 100-MHz unity-gain bandwidth and 60 dB of gain with 1 mA of total current consumption. Fully differential op-amps are discussed in detail in Chapter 4.

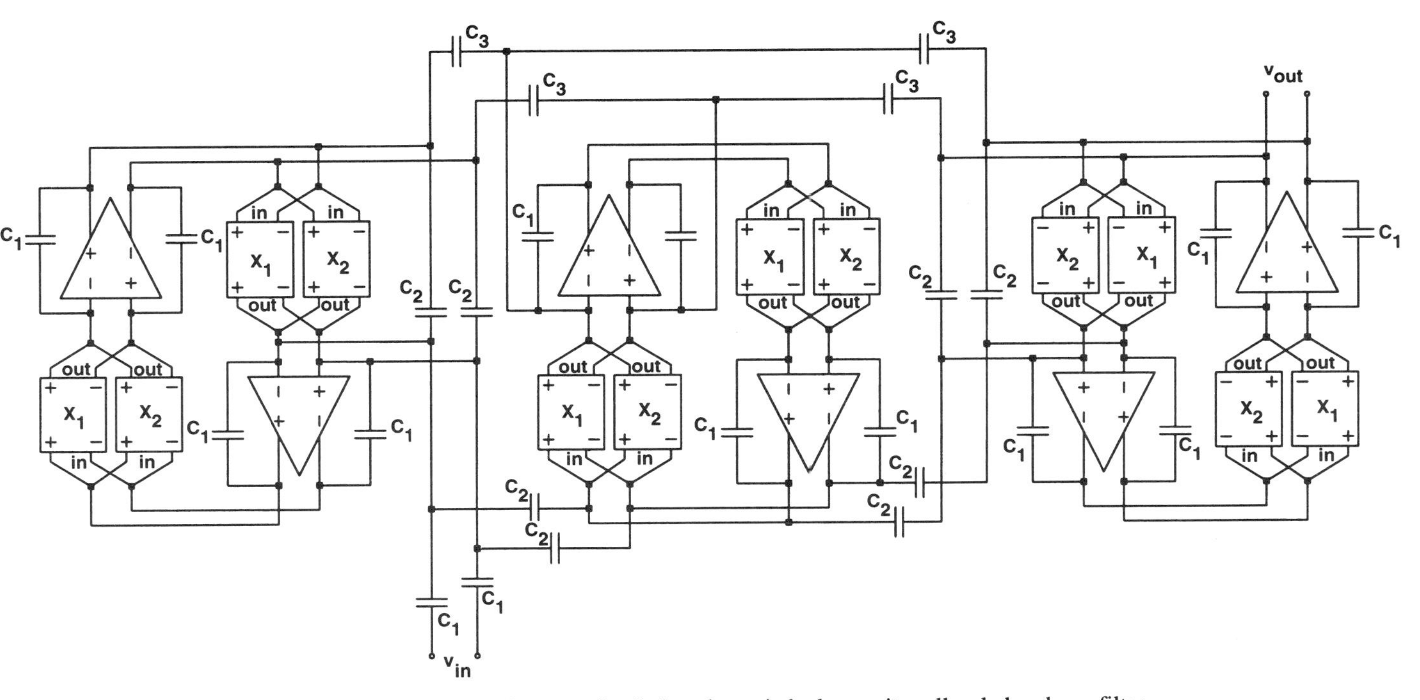

Figure 1.11. Schematic diagram of a sixth-order switched-capacitor all-pole bandpass filter.

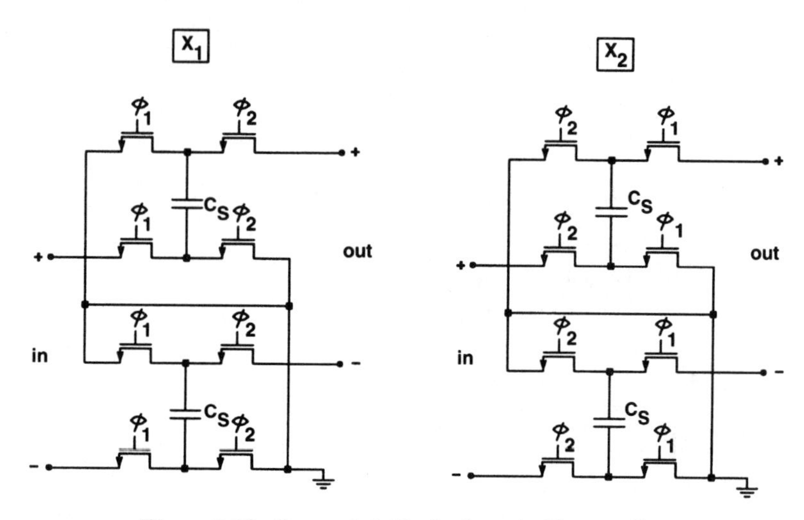

Figure 1.12. Two switch blocks for a double sampling.

Another application of the fully differential op-amps is in oversampling, or delta-sigma A/D converters. The oversampling converters operate at sampling rates of 16 to 512 times the Nyquist rate and increase the signal-to-noise ratio by subsequent filtering. The oversampling techniques lend themselves most favorably to applications that require a relatively low frequency (<1 MHz) and high resolution (>12 bits). The most obvious application of delta-sigma converters is in digital telephony

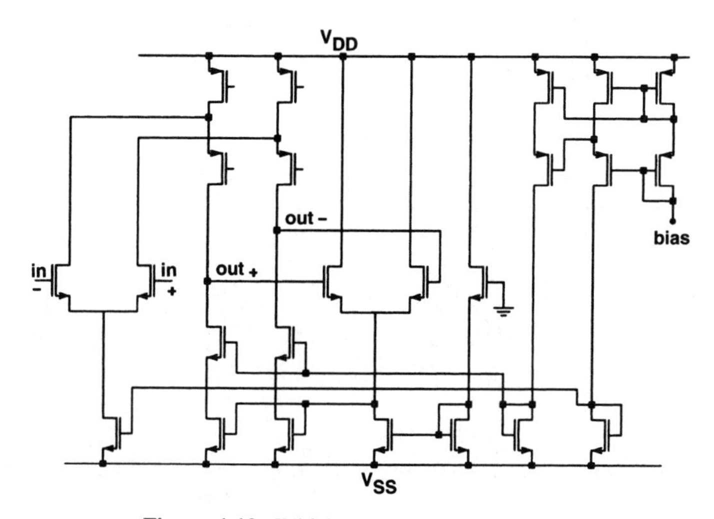

Figure 1.13. Wideband op-amp for the filter.

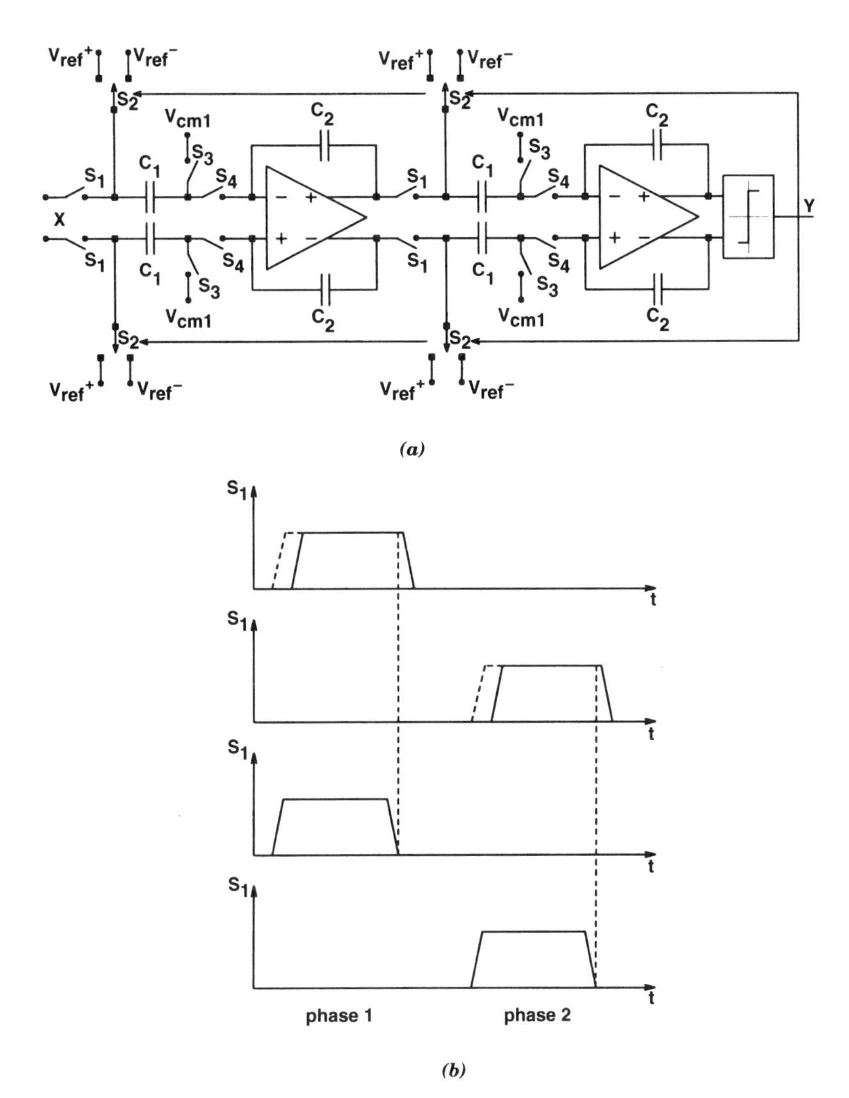

Figure 1.14. (*a*) Fully differential CMOS implementation of a second-order delta-sigma modulator; (*b*) two-phase clock scheme.

and digital audio. Figure 1.14 shows a fully differential, switched-capacitor CMOS implementation of a second-order delta-sigma modulator [9]. It consists of two parasitic-insensitive integrators, a comparator that serves as a 1-bit A/D converter, and a two-level (1-bit) D/A converter. Use of a fully differential configuration attenuates power supply noise, clock feedthrough, and even-order harmonic distortion. The modulator operates on two-phase nonoverlapping clocks consisting of a sampling phase and an integration phase. It achieves 16-bit dynamic range with an oversampling ratio of 256 and a signal bandwidth of 20 kHz.

As the examples above illustrate, present-day CMOS op-amps and comparators and their use in analog MOS circuits have reached a certain level of maturity. Already, almost any analog signal processing task in the voice- or audio-frequency range has a possible solution using such circuits. As fabrication technology and circuits design techniques continue to advance, the speed and dynamic range of these circuits will increase, allowing their use in such large-volume applications as video and radio systems, image processing, high-speed transmission circuits, and so on.

PROBLEMS

1.1. Show that the circuit of Fig. 1.1*a* can realize the transfer function of Eq. (1.1). What should be the element values R, L, and C?

1.2. Calculate the transfer function of the active-RC circuit of Fig. 1.1*b*. Assume that the circuit is to realize the transfer function of Eq. (1.1). Write the available equations for the element values. How many element values can be chosen arbitrarily?

REFERENCES

1. R. W. Brodersen, P. R. Gray, and D. A. Hodges, *Proc. IEEE, 67,* 61–75 (1979).
2. Y. Tsividis, *Proc. IEEE, 71,* 926–940 (1983).
3. R. Gregorian, K. W. Martin, and G. C. Temes, *Proc. IEEE, 71,* 941–966 (1983).
4. D. J. Allstot and W. C. Black, Jr., *Proc. IEEE, 71,* 967–986 (1983).
5. P. R. Gray and R. G. Meyer, *Analysis and Design of Analog Integrated Circuits,* 2nd ed. Wiley, New York, 1984.
6. R. Gregorian and G. C. Temes, *Analog MOS Integrated Circuits for Signal Processing,* Wiley, New York, 1986.
7. Bang-Sup Song and P. R. Gray, *IEEE, J. Solid-State Circuits, SC-21*(6), 924–933 (1986).
8. T. C. Choi, R. T. Kaneshira, R. W. Broderson, P. R. Gray, W. B. Jett, and M. Wilcox, *IEEE J. Solid-State Circuits, SC-18*(6), 652–664 (1983).
9. B. P. Brandt, D. E. Wingard, and B. A. Wooley, *IEEE J. Solid-State Circuits, SC-26*(4), 618–627 (1991).

CHAPTER 2

MOS DEVICES AS CIRCUIT ELEMENTS

In this chapter the physics of MOS (metal-oxide semiconductor) devices is discussed briefly. The most important and simplest current–voltage relations are given, and simple models introduced for MOS transistors in linear operation. The discussion here is in the simplest possible terms, aimed at providing some physical understanding of the highly complex device operation for the circuit designer. Precision and depth have regretfully been sacrificed in the process. The ambitious reader is referred to the excellent specialized works listed as references at the end of the chapter.

2.1. SEMICONDUCTORS

In metals (e.g., aluminum, copper, silver) that are good electrical conductors, the atoms are arranged in a regular crystal array. The electrons from the outer (valence) shell of the atoms are free to move within the material. Since the number of atoms, and thus the number of free electrons, is very large (on the order of 10^{23} cm^{-3}), even a small electric field results in a large electron current—hence the high conductivity observed for these metals.

The picture is quite different for an insulator such as silicon dioxide (SiO_2). Here the valence electrons form the bonds between adjacent atoms and hence are themselves tied to these atoms. Thus no free electrons are available for conduction and the conductivity is very low.

Semiconductors (such as silicon or germanium) are in between conductors and insulators in their electrical properties. At very low temperatures, the valence electrons are bound to their atoms, which again form a regular lattice. However, as the temperature is raised, due to the thermal vibrations of the atoms, some bonds will be broken, and an electron escapes from each of these bonds. Such electrons are capable of conducting electricity. Furthermore, each fugitive electron leaves a charge deficit (called a *hole*) behind in the bond. A valence electron in a bond close to a hole can easily move over, filling the hole and leaving a new hole in its own bond. The effect is the same as if the hole had moved from one bond to the next. Since the hole "moves" in a direction opposite that of the moving valence electron, in an electronic field it behaves like a positively charged particle.

Electrical conduction is thus possible for a semiconductor at room temperature. The density of thermally generated electrons and holes is, however, much smaller than that of the free electrons in metal. Typical numbers are 10^{10} charge carriers per cubic centimeter for silicon and 10^{13} in germanium. In what follows, the currently dominant material, silicon, is discussed exclusively.

Adding foreign elements (dopants) to the pure silicon can raise the number of free charge carriers in a semiconductor. Silicon (and germanium) has *four* valence electrons. If an atom of an element with *five* valence electrons (such as arsenic, phosphorus, or antimony) is added to the semiconductor, it may take the place of a silicon atom in the crystal lattice. Thus four of its valence electrons will participate in the four bonds tying the atom to adjacent semiconductor atoms in the lattice. The fifth valence electron of the foreign atom, however, will not have a place in any bond and will thus be free to move away from its parent atom. Hence such a dopant element (called a *donor,* since it contributes free electrons to the semiconductor) enhances the conductivity of the material.

Adding atoms of an element with *three* valence electrons will also contribute to the conductivity. Now there will be a bond *lacking* a valence electron for each dopant atom. Thus each such atom creates one hole. These dopants (e.g., boron, aluminum, and gallium) are called *acceptors,* since the holes will propagate by accepting bound valence electrons from adjacent semiconductor atoms.

In doped semiconductors there will be carriers due to thermal effects as well as to the donor (or acceptor) atoms. Materials containing *donors* will thus have both free electrons and holes, but there will be more electrons than holes. Such semiconductors will be called *n-type,* where *n* stands for *negative.* Materials containing acceptors will have a majority of holes; they are called *p-type* semiconductors, where *p* stands for *positive.*

A semiconductor structure can also be fabricated that contains two adjacent regions of different types (Fig. 2.1). The surface joining the two regions is called *a pn junction.* When the junction is fabricated, the random thermal motion of the

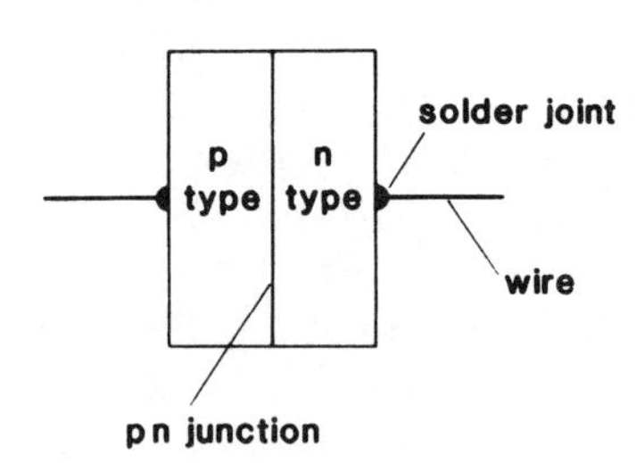

Figure 2.1. A *pn* junction diode.

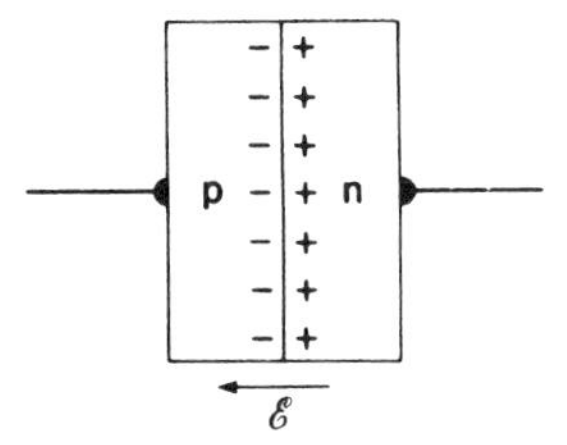

Figure 2.2. Ion layers in a *pn* junction.

majority carriers (electrons in the *n*-type region, holes in the *p*-type region) will cause electrons to spill over from the *n*-type region to the *p*-type region. Vice versa, holes will move from the *p*-type region to the *n*-type semiconductor. Thus this random motion (called *diffusion*) results in the *p*-type semiconductor being charged negatively while the *n*-type region is charged positively. The effect will be strongest near the junction: There, in the *p*-type region, the negatively charged acceptor atoms will no longer be neutralized by holes, and (in the *n*-type region) free electrons will no longer surround the positively charged donor ions. Hence in this area a dipole layer of fixed ions will be formed (Fig. 2.2). The electric field $\mathscr{E}$ created by the dipole opposes further majority-carrier diffusion, but it helps the thermally generated minority carriers (electrons in *p*-type regions, holes in *n*-type regions) to migrate from one region to another. Thus, after a short transient, equilibrium will be obtained. Four different carrier currents will flow: Majority carriers will move by diffusion from region to region despite $\mathscr{E}$, and minority carriers will flow aided by $\mathscr{E}$. These currents cancel each other in equilibrium, since the effects of $\mathscr{E}$ compensate for the larger number of available majority carriers.

The equilibrium will be upset if a voltage source is connected to the wires soldered to the semiconductor (Fig. 2.3). Assume first that the polarity of the source is such that it makes the *p*-region more positive with respect to the *n*-region; that is, $v > 0$ in Fig. 2.3. Then v will reduce $\mathscr{E}$ and thus increase the current of majority carriers held back by $\mathscr{E}$ from spilling over the boundary. Even a small reduction of $\mathscr{E}$ caused by, say, a battery of $v = 0.8$ V can result in a large majority-carrier current (say,

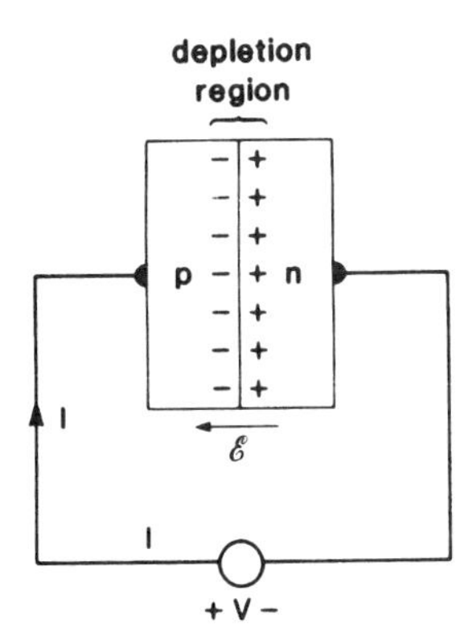

Figure 2.3. Circuit for testing a *pn* diode.

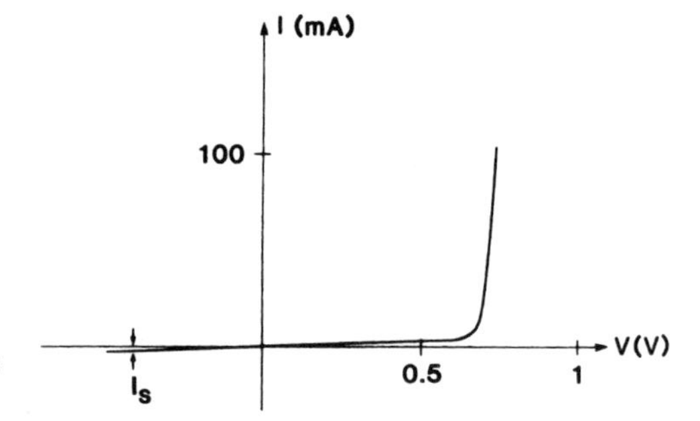

Figure 2.4. Current versus voltage characteristics of a *pn* junction diode.

$i = 1$ A) in the circuit. Hence v with the polarity indicated will be called *forward voltage* and i *forward current.*

Let us now reverse the polarity of the voltage source so that $v < 0$ in Fig. 2.3. Now v will aid $\mathscr{E}$ in obstructing the flow of majority carriers from region to region. If v is large enough, the majority current is essentially eliminated, and only the flow of minority carriers (electrons moving from the p region to the n region, and holes moving in the opposite direction) remains. Since the number of minority carriers is, however, small and nearly independent of v, the resulting net current will be small and nearly constant. This is the case of *reverse voltage* and *current.* With the reference directions used in Fig. 2.3, now $i < 0$. Figure 2.4 illustrates the overall behavior of i as a function of v. A detailed theoretical analysis [1, Sec. 4.3; 2, Sec. 6.6] reveals that the describing equation is, to a good approximation,

$$i = I_S(e^{qv/kT} - 1). \tag{2.1}$$

Here I_S is the *saturation current,* determined by the geometry and the material properties of the device, $q \approx 1.6 \times 10^{-19}$ C is the electron charge, and $k \approx 1.38 \times 10^{-23}$ J/K is Boltzmann's constant. T is the temperature of the semiconductor, in Kelvin. At room temperature ($T = 300$ K), $kT/q \approx 26$ mV. I_S is usually very small, on the order of 10^{-9} A or less. Thus i increases exponentially with v for $v > 0$, while $i \approx -I_S$ and is very small if $v < 0$ (Fig. 2.4).

The behavior of the region directly adjacent to the boundary between the p and n regions is of prime importance. As mentioned earlier, the majority carriers are very sparse in this area; some have immigrated into the other region, and the others have been pushed back into the inside of their native region by the field $\mathscr{E}$. Hence the border area contains only the fixed ions, charged negatively in the p-type region and positively in the n-type material (Fig. 2.3). This area is hence called the *depletion region.* Its width increases with increasing $\mathscr{E}$; hence it will be greater (smaller) for reverse (forward) voltage v.

Due to the field $\mathscr{E}$, a voltage ϕ_i (often called the *built-in voltage*) appears across the depletion region for $v = 0$. The total potential across the junction, for $v \neq 0$, is thus $\phi_i - v$. Typically, $\phi_i = 0.5$ to 1 V.

For $v < 0$, the *pn* junction can be regarded as a capacitor, since only a small saturation current I_S flows for a dc voltage v, and since adjacent positive $(+Q)$ and negative $(-Q)$ charges are stored in the depletion region (Fig. 2.3). Since the charge stored is a nonlinear function of v, the capacitance is nonlinear. We shall define the capacitance C by the incremental relation $C = dQ/dv$. It can then be shown [1, Sec. 3.3; 2, Sec. 6.5] that for the device illustrated in Fig. 2.3,

$$C = \left\{ \frac{q\epsilon_s/[2(1/N_a + 1/N_d)]}{\phi_i + |v|} \right\}^{1/2} \tag{2.2}$$

holds. Here $\epsilon_S \simeq 1.04$ pF/cm is the permittivity of the silicon; $\epsilon_S = \epsilon_0 K_S$, where ϵ_0 is the permittivity of free space ($\epsilon_0 \approx 8.86 \times 10^{-14}$ F/cm); and $K_S \approx 11.7$ is the *dielectric constant* (also called the relative permittivity) of silicon. A is the area of the junction in square centimeters and N_a (N_d) is the number of acceptor (donor) atoms per cubic centimeter. Note that C decreases with $|v|$. It can be shown [1, Sec. 3.3] that the quantity under the square-root sign is $(\epsilon_S/x_d)^2$, where x_d (in centimeters) is the width of the depletion region; hence $C = \epsilon_S A/x_d$ holds.

2.2. MOS TRANSISTORS [1, 2, Sec. 9.2]

Consider next the structure shown in Fig. 2.5. It is a sandwich of several layers: From top to bottom, it contains layers of metal, silicon dioxide (SiO_2, an excellent insulator), p-type silicon, and a second metal layer connected to ground. It is called a *metal-oxide semiconductor* (MOS) structure. Let v be *negative*; then an electric field will be created across the dioxide layer, which will attract positive charges (holes) to the region R under the top metal electrode (Fig. 2.5). Thus negative charges will be stored in the top metal electrode and positive charges in R. The device will thus behave as a capacitor C of magnitude

$$C = \epsilon_{ox} \frac{A}{l}, \tag{2.3}$$

where ϵ_{ox} is the permittivity of the SiO_2, and $\epsilon_{ox} = \epsilon_0 K_{ox} \simeq 0.35$ pF/cm, where K_{ox} is the dielectric constant of SiO_2 ($K_{ox} \approx 3.9$). A is the area of the top electrode,

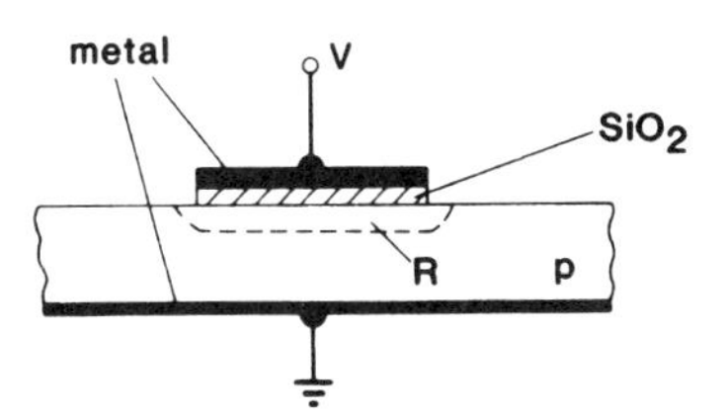

Figure 2.5. Metal-oxide-semiconductor (MOS) structure.

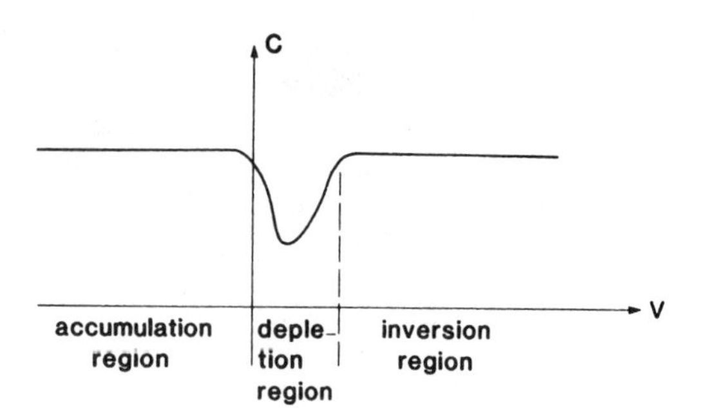

Figure 2.6. Capacitance versus voltage characteristics of an MOS structure.

and l is the thickness of the SiO_2 layer.* The p-type Si layer between R and the bottom metal layer behaves as a resistor; hence the overall structure simulates a lossy capacitor.

Next, let v be a small *positive* voltage in Fig. 2.5. The electric field will now repel holes. As a result, the fixed negatively charged ions in R will be abandoned by the mobile holes, and a net negative space charge will appear in R, which is now a depletion layer. Thus charge is again stored in the top electrode and a capacitor is created. For very small values of v ($v \ll 1$ V), Eq. (2.3) will remain valid for the magnitude of the capacitance. As the value of v is increased, however, the charge in R becomes greater since the depletion region widens. Since the average ion is now farther away from the surface, the effective value of l in Eq. (2.3) increases and C decreases.

If v is increased even further, a new effect appears. Since the thermal generation of holes and electrons occurs continuously in the semiconductor, if the field created by a positive v is strong enough, it can attract thermal electrons to R; these will then move to the surface. When this occurs, the capacitor stores positive charges in the top electrode, while negative ones (electrons) are stored in the surface layer. Thus, in Eq. (2.3), l again becomes the SiO_2 thickness, and hence C has the same value as it had for negative voltage v. The overall behavior of C as a function of v is illustrated schematically in Fig. 2.6, which also gives the names of the three operating regions. The names of the first two are evident; the third one is called the *inversion region,* since (due to the high voltage v) mobile electrons are attracted into R, which thus behaves as an n (rather than a p)-type material. It should be noted that since thermal electrons are generated at a slow rate, the voltage v must be present for some time before the inversion layer is formed; hence it will not appear if v is a high-frequency (say, $f > 1$ kHz) signal rather than a constant voltage.

Consider next the structure shown in Fig. 2.7. A new feature is the presence of two n^+ (i.e., heavily doped n-type) regions in the p-type material. The one on the left will be called the *source*; a voltage v_S is connected to it. The n^+ region on the

* Often, the oxide thickness l is measured in angstroms (1 Å $= 10^{-8}$ cm). Usual values of l are between 50 and 200 Å.

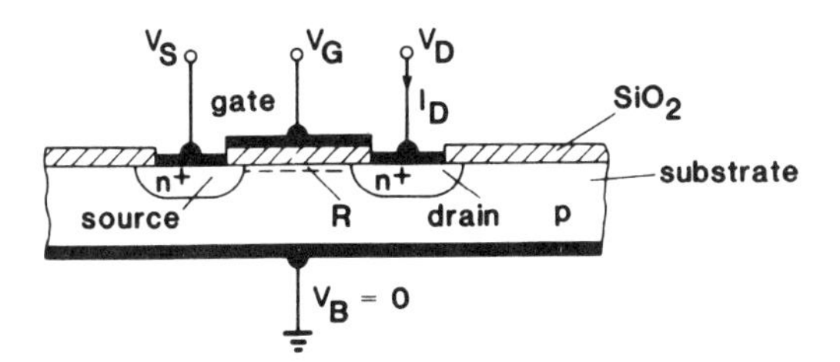

Figure 2.7. MOS transistor.

right will be called the *drain*; its voltage is denoted by v_D. The top metal electrode will be called the *gate;* its voltage is v_G. The body of the semiconductor is usually called the *substrate* or *bulk.* The overall device is the MOS *transistor.* Its operation is discussed briefly next.

Let the source be grounded, so that $v_S = 0$. Also, let v_D have a small positive value, say 0.5 V. We will consider the behavior of the drain current i_D as v_G is raised from zero to higher positive values. Since the gate is insulated from the rest of the device by the oxide layer, it will not conduct any current. The n^+ drain region and the surrounding *p*-type substrate form a *pn* junction. Since the substrate is grounded, while $v_D > 0$, this junction is reverse biased. Hence for $v_G = 0$, $i_D \approx 0$.

As v_G is increased, the region R under the gate will first be depleted, then inverted, as discussed earlier in connection with Figs. 2.5 and 2.6. When R is depleted, i_D remains zero, since the area around the drain is still reverse biased. However, the situation changes when v_G is so large that inversion occurs, so that R is filled with electrons. Now, a layer containing mobile electrons, called an *inversion layer* or *channel,* connects the drain to the source. Since the drain is positive with respect to the source, electrons will flow from the source to the drain and a positive current $i_D > 0$ will be observed. The smallest voltage v_G necessary to produce a channel is called the *threshold voltage* and is denoted by V_T. Usually, V_T is given as the v_G value needed for $i_D = 1$ μA; it may range from a fraction of a volt to several volts.

It should be noted that for the structure of Fig. 2.7, most of the electrons in the channel do *not* originate from thermal effects in the bulk; instead, they are drawn by the electric field due to v_G out of the source. Some electrons are also drawn from the drain; however, since $v_D > 0$, the drain–substrate junction is more reverse biased, and hence it is harder for electrons to escape from the drain.

Since a potential difference v_D exists between the two ends of the channel, the electrons in the channel will be attracted to the drain. Therefore, in addition to the random thermal motion of the electrons, a steady motion (called *drift*) will occur, which causes the current flow. For small v_D, the channel will therefore behave as a resistor, and hence $i_D \approx v_D/R$, where the channel resistance R is given by

$$R = \frac{L}{W\mu_n|Q_n|}\,. \tag{2.4}$$

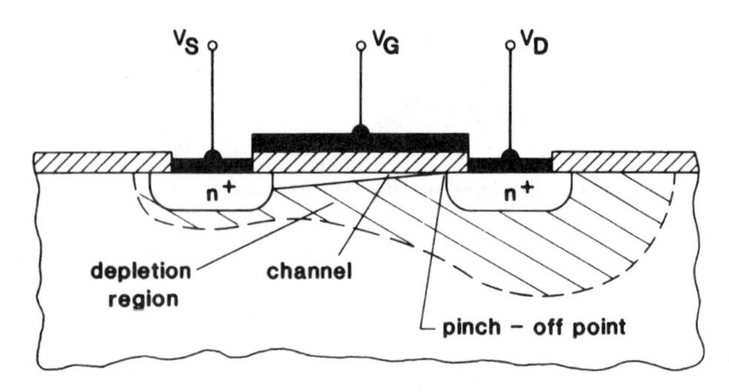

Figure 2.8. Pinch-off in an MOS transistor.

Here L is the length and W the width of the channel, and μ_n is the mobility of the electrons in the channel,* defined by the relation (electron drift velocity) = (mobility) × (electric field). Finally, Q_n is the charge density (in C/cm^2) of the electrons in the channel. Since v_G can be considered as the sum of two terms, V_T (necessary to maintain the depletion region under the channel) and $v_G - V_T$ (necessary to maintain the channel), we have

$$Q_n = -C_{ox}(v_G - V_T), \tag{2.5}$$

where $C_{ox} = \epsilon_{ox}/l$ is the capacitance (per unit area) of the oxide layer separating the gate from the channel. Hence, for small v_D (i.e., $v_D \ll v_G - V_T$), the relation

$$i_D = \mu_n C_{ox} \frac{W}{L}(v_G - V_T)v_D \tag{2.6}$$

holds. Thus the transistor acts as a resistor, with resistance $R = [\mu_n C_{ox} W/L(v_G - V_T)]^{-1}$ controlled by v_G.

If v_D is increased so that it is no longer negligible compared to v_G, Eq. (2.6) will become inaccurate. Since the potential of the channel at the grounded source is zero while at the drain it is v_D, we can assume that its average potential is $v_D/2$. Hence the average voltage between the gate and channel is $(v_G - v_D/2)$. Replacing v_G by $(v_G - v_D/2)$ in Eq. (2.6) gives

$$i_D = \mu_n C_{ox} \frac{W}{L}\left(v_G - V_T - \frac{v_D}{2}\right)v_D. \tag{2.7}$$

Equation (2.7) remains a good approximation for $v_D < v_G - V_T$. This range is called the *linear region* (or *triode region*) of operation of the MOS transistor.

When $v_D \geq v_G - V_T$, a new phenomenon occurs. Consider the situation illustrated in Fig. 2.8, where only the structure near the semiconductor surface is shown, magni-

* The mobility *in the bulk* of the semiconductor is higher, since μ_n decreases with the concentration of ionized impurities. Typical values are $\mu_n \approx 1000$ cm^2/V · s for $N_D = 10^{16}$ cm^{-3}, while $\mu_n \approx 100$ cm^2/V · s for $N_D = 10^{19}$ cm^{-3}.

fied. As the figure indicates, due to the variation of the potential along the channel, the charge density Q_n decreases near the drain. If $v_D = v_G - V_T$, at the drain the gate-to-channel voltage is no longer sufficient to maintain the channel. Thus the depletion region surrounding the source, the channel, and the drain extends all the way to the surface. This phenomenon is sometimes called *pinch-off,* and the region where it occurs is the *pinch-off point* (Fig. 2.8). If v_D is increased further, the pinch-off point will move toward the source, since the area where $v_G - v_D \leq V_T$ will increase. Hence the channel will now extend only from the source to the pinch-off point, the latter being somewhere under the gate. The region between the pinch-off point and the drain is depleted. Electrons from the channel are injected into this depletion region at the pinch-off point and are swept to the drain by the field created by the potential difference between the drain and the pinch-off point. The voltage $v_{DS} \triangleq v_D - v_S$ is thus divided between the two series-connected regions: the channel between the source and the pinch-off point, and the depletion region between the pinch-off point and the drain. Clearly, the latter has a higher resistance, and hence most of v_{DS} in fact appears across it. Any increase of v_D will, to a good approximation, result in an equal voltage increase across the depletion region and will hardly change i_D. Thus, for v_D . $v_G - V_T$, from Eq. (2.7),

$$i_D(v_D) \approx i_{D\text{sat}} \triangleq i_D(v_{D\text{sat}}) \qquad (2.8)$$

$$= \frac{\mathrm{m}_n C_{\text{ox}}}{2} \frac{W}{L} (v_G - V_T)^2.$$

This phenomenon is called *saturation;* $v_{D\text{sat}} = v_G - V_T$ is the *drain saturation voltage* and $i_{D\text{sat}}$ as given by Eq. (2.8), is the *drain saturation current.*

The drain current does, in reality, increase somewhat with increasing v_D. This can be attributed to the move of the pinch-off point toward the source for increasing v_D and hence to the shortened channel; as Eq. (2.8) indicates, i_D increases as L is reduced. As an approximation, this effect (often called *channel-length modulation*) can be included in the formula for $i_D\,(v_D)$ in the form of an added factor $(1 + 1 v_D)$. Here 1 is a device constant that depends on L, on the doping concentration of the substrate, and on the substrate bias (discussed in the next section). For $L \approx 10$ mm, typically $1 \approx 0.03\ \text{V}^{21}$; generally, $1 \sim 1/L$.

It is usual to introduce the abbreviations $k' \triangleq \mathrm{m}_n C_{\text{ox}}/2$ and $k \triangleq k'\,(W/L)$. Then the saturation current given by Eq. (2.8) becomes

$$i_D = k(v_G - V_T)^2(1 + 1 v_D), \qquad v_G \geq V_T, \qquad (2.9)$$

which incorporates channel-length modulation. Figure 2.9 shows the variation of i_D with v_G for constant v_D. Figure 2.10 illustrates its dependence on v_D for various v_G values, where v_{G1} , v_{G2} , $v_{G3} \cdots$.

All derivations of this section were performed for the structure shown in Fig. 2.7, whose source, drain, and channel were all n type. This device is called an *n-channel* MOS, or NMOS *transistor.* A similar arrangement can be constructed by creating p^+ drain and source diffusions in an n-type substrate. Now a negative v_G is needed to create a p-type channel under the gate, and a negative v_D is used to

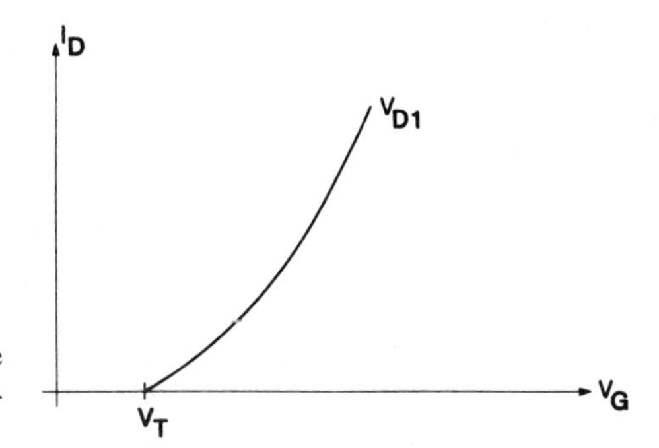

Figure 2.9. Drain current versus gate voltage characteristics of an MOS transistor.

attract the holes in the channel to the drain. Also, i_D will be negative if the reference direction of Fig. 2.7 is used. The resulting device is called a *p*-channel MOS or PMOS transistor. Formulas (2.3) to (2.9) remain valid if some small changes are made. The mobility μ_n of electrons must be replaced by μ_p, the hole mobility in the channel. As would be expected from the more elaborate mechanism of hole conduction, $\mu_p < \mu_n$. Typical mobility values in the channel region for an impurity concentration of 10^{16} cm^{-3} are $\mu_n = 1000$ cm^2/V · s and $\mu_p = 400$ cm^2/V · s. The electron charge density Q_n in the channel is to be replaced by Q_p, the hole charge density; also, a negative sign must be included in Eqs. (2.6) to (2.9) to account for the change in the charge of the carriers. Finally, v_D must be replaced by $|v_D|$ in Eq. (2.9), since now $v_D < 0$. In conclusion, Eq. (2.7) becomes

$$i_D = -2k\left(v_G - V_T - \frac{v_D}{2}\right)v_D. \tag{2.10}$$

Here $k \triangleq \mu_p C_{ox} W/2L$ and $V_T < 0$. Equation (2.10) describes the drain current characteristics in the linear range. The behavior of i_D in the saturation region can be obtained by modifying (2.9):

$$i_D = -k(v_G - V_T)^2(1 + \lambda\,|v_D|). \tag{2.11}$$

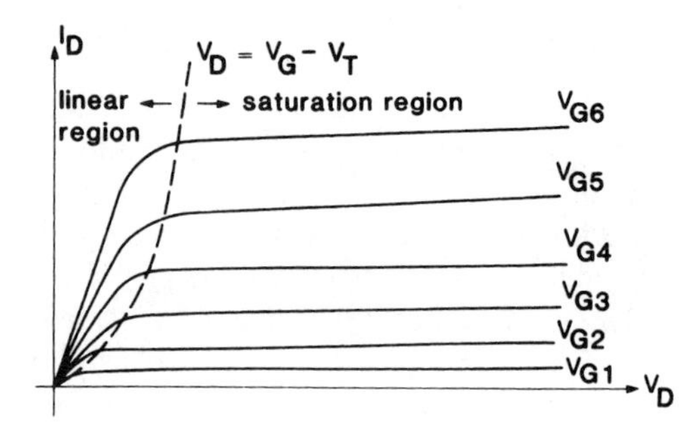

Figure 2.10. Drain current versus drain-to-source voltage characteristics of an MOS transistor.

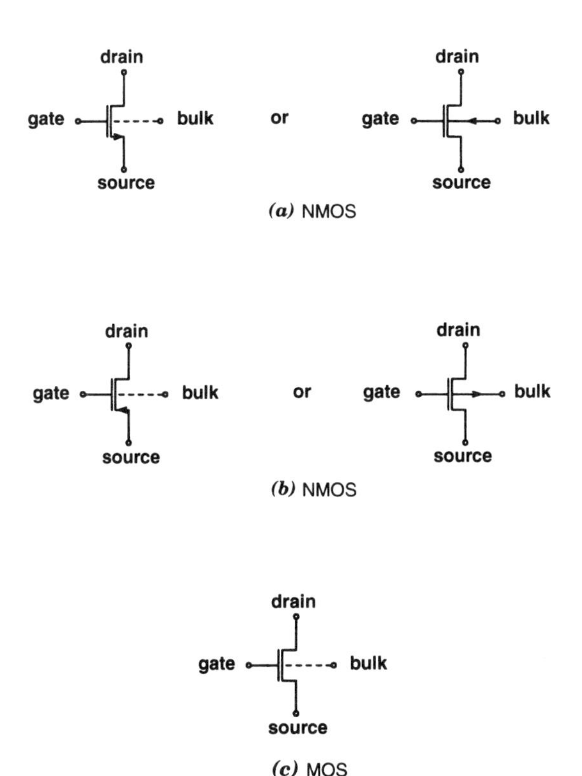

Figure 2.11. Transistor symbols.

The circuit symbols used for NMOS and PMOS transistors are shown in Fig. 2.11*a* and *b,* respectively. If the type is unimportant, the simplified symbol of Fig. 2.11*c* may be used for both NMOS and PMOS devices.

Since the operation of the devices described in this section is dependent on the electric field induced by the gate voltage, they are called field-effect devices (FETs), or MOSFETs.*

Since PMOS transistors are more easily fabricated than NMOS transistors, they were initially predominant. However, later, when the techniques for the reliable production of NMOS devices were developed, the latter became standard. The main reason for this is the higher mobility of electrons, which makes the NMOS transistors faster than PMOS transistors.

2.3. MOS TRANSISTOR TYPES: BODY EFFECT

The MOS transistors described in Section 2.2, both NMOS and PMOS types, share several features. In the structure, the gate is insulated electrically from the rest of

* Since the charge carriers here are either electrons *or* holes (not both), FETs are also sometimes called *unipolar* devices, to contrast them with bipolar transistors, in which *both* electron and hole currents exist.

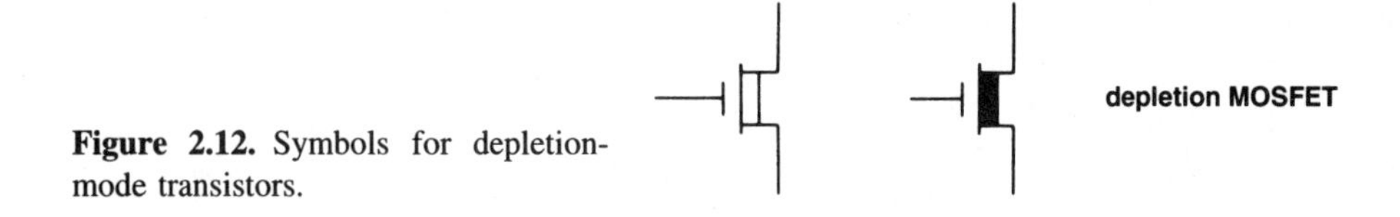

Figure 2.12. Symbols for depletion-mode transistors.

the device by the SiO_2 layer under it. Hence it is often called an *insulated-gate field effect transistor* (IGFET). Also, the voltage v_G induces and enhances the drain current. Thus the devices described operate in the *enhancement mode.*

It is also possible to fabricate an MOS transistor that conducts drain current when $v_G = 0$. For example, an n-type layer can be introduced by doping, which connects the source and drain of an NMOS device. With such a doped channel, the field of the gate is not needed to produce an inversion layer; the region R (Fig. 2.7) now has a ''built-in'' conducting n-type channel.

However, if a *negative* gate voltage is applied, the field thus created will repel electrons and create a depletion layer in the channel adjacent to the SiO_2 surface, thereby reducing the conductivity and thus the drain current. If the magnitude of the negative gate voltage is sufficiently large, the channel becomes completely depleted and $i_D \approx 0$ results. The value v_G at which this occurs is again called threshold voltage and is denoted by V_T. Now, however, $V_T < 0$. Such a device is called a *depletion-mode* FET.

It should be noted that even without a doped layer, an NMOS transistor can conduct for $v_G = 0$ due to oxide charges [1, Sec. 7.4] if the bulk is very lightly doped. It is also possible to create a depletion-mode PMOS device, with $V_T > 0$, by establishing a p-type doped channel.

The relations (2.6) to (2.11) remain valid for depletion-mode devices if the value and sign of V_T is chosen appropriately, as described above. Two symbols often used to denote depletion-mode MOSFETs are shown in Fig. 2.12.

A totally different structure can also be used to produce a depletion-mode field-effect transistor (Fig. 2.13). Here, a lightly doped n-type layer (channel) connects the n^+ source and drain regions and the gate is a p^+ region implanted in this layer. Hence, for $v_S = v_G = 0$ and $v_GD > 0$, a drain current will flow. If v_G is made negative, the p^+ implant acting as the gate will be surrounded by a depletion layer; the greater $|v_G|$, the deeper the layer. The mobile electrons in the channel cannot

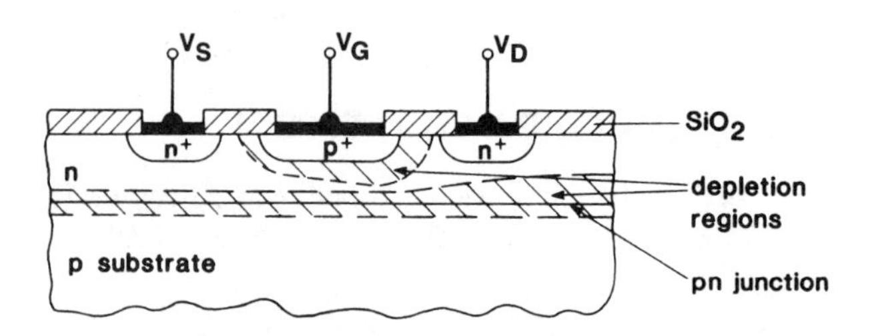

Figure 2.13. Junction field-effect transistor.

enter this depletion layer, or the one along the *pn* junction between the channel and the substrate. Hence the effective cross section of the channel is reduced as $|v_G|$ is increased. At some value $v_G = V_P$ ($V_P < 0$), i_D becomes zero (in practice, < 1 μA). Thus V_P plays the same role as V_T for a depletion-mode MOSFET; it is called the *pinch-off voltage*. It can be visualized as the gate voltage, which causes the two depletion regions in the channel to merge, leaving no conductive path between source and drain.

The device described and shown in Fig. 2.13 is called an *n*-channel *junction field-effect transistor* (JFET), since its gate is separated from the rest of the device by a reverse-biased *pn* junction rather than by an SiO_2 layer as for the MOSFET (IGFET). Since the JFET is hardly ever used in analog MOS integrated circuits, it is not discussed in detail here. A clear description of its physics and current–voltage characteristics is given in Sec. 2.5 of Ref. 1.

Next, a key limitation (called the *body effect*) of MOSFETs used as analog circuit elements is described. In the discussion in Section 2.2 it was always assumed that both the bulk and source are grounded, so that $v_B = v_S = 0$ held. Often, circuit considerations make this convenient arrangement impossible and $v_S \neq v_B$ must be used. Obviously, the voltage $v_S - v_B$ must be such that the source–bulk junction is *reverse biased;* otherwise, a large junction current will flow inside the transistor. This current may damage the device, and in any case will impede its proper operation. Thus, say, in an NMOS transistor, the bulk must be biased to make it negative with respect to both source and drain.

If the source potential is not zero, the voltages v_G and v_D must be replaced in all equations by $v_{GS} = v_G - v_S$ and $v_{DS} = v_D - v_S$, respectively. In addition, the depletion region around the channel (Fig. 2.8) will become wider if the reverse voltage between the bulk and the source (and hence the channel) is increased. Since the voltage $v_G = V_T$ is the gate voltage necessary to maintain the depletion region (without creating a channel), V_T will increase in magnitude. The dependence of V_T on the voltage $v_{SB} \triangleq v_S - v_B$ can be shown [1, Sec. 8.2] to be in the form

$$|V_T| = |V_{T0}| + \gamma\left(\sqrt{2|\phi_p| + |v_{SB}|} - \sqrt{2|\phi_p|}\right). \tag{2.12}$$

Here, V_{T0} is the threshold voltage for $v_{SB} = 0$ and γ is a device constant given by

$$\gamma = \frac{\sqrt{2\epsilon_s q N_{\text{imp}}}}{C_{\text{ox}}}. \tag{2.13}$$

In Eq. (2.13), ϵ_S is the permittivity of silicon: $\epsilon_S = \epsilon_0 K_S$, $K_S \approx 11.7$. Also, N_{imp} is the density of the impurity ions in the bulk. For NMOS, $N_{\text{imp}} = N_A$, the acceptor ion density; for PMOS, $N_{\text{imp}} = N_D$, the donor ion density. For example, for $N_{\text{imp}} = 10^{15}$ cm^{23} and 800 Å oxide thickness (i.e., $C_{\text{ox}} \approx 4.4 \times 10^{28}$ F/cm^2), $\gamma \approx 0.423 V^{1/2}$. Finally, ϕ_p is a material constant* of the bulk; its value is around 0.3 V.

* $\phi_p \triangleq (E_i - E_f)/q$, where E_i is the intrinsic Fermi energy and E_f the Fermi energy of the semiconductor [1, p. 318].

TABLE 2.1. Key Units and Constants for MOS Transistors

$1\ \mu m = 10^{-4}$ cm $= 10^4$ Å
1 mil $= 25.4\ \mu m = 0.0254$ mm
Electron charge (magnitude): $q = 1.6 \times 10^{-19}$ C
Permittivity of free space: $\epsilon_0 = 8.86 \times 10^{-14}$ F/cm
Permittivity of silicon: $\epsilon_{si} = \epsilon_0 K_{si} = 1.04 \times 10^{-12}$ F/cm; $K_{si} = 11.7$
Permittivity of silicon dioxide: $\epsilon_{ox} = \epsilon_0 K_{ox} = 3.5 \times 10^{-13}$ F/cm; $K_{ox} = 3.9$
Oxide capacitance: $C_{ox} = \epsilon_{ox}/t_{ox} = 3.5 \times 10^{-13}/t_{ox}^{cm}$ F/cm^2
Intrinsic carrier concentration: $n_i = 1.5 \times 10^{10}$ cm^{-3}, $T = 300$ K
Boltzmann's constant: $k = 1.38 \times 10^{-23}$ J/K; kT/q (at $T = 300$ K) $= 0.026$ V
Electron mobility in Si ($N_{imp} = 10^{17}$ cm^{-3}, $T = 300$ K): 670 cm^2/V · s
Hole mobility in Si ($N_{imp} = 10^{17}$cm^{-3}, $T = 300$ K): 220 cm^2/V · s
Body-effect coefficient: $\gamma = \sqrt{\dfrac{2qK_{si}N_{imp}}{\epsilon_0}}\dfrac{t_{ox}}{K_{ox}} \approx 1.67 \times 10^{-3}\ t_{ox}^{cm}\sqrt{N_{imp}^{cm^{-3}}}\ V^{1/2}$
Bulk potential: $\phi_p = -\dfrac{kT}{q}\ln\dfrac{N_{imp}}{n_i} = 0.026 \ln(0.67 \times 10^{-10}\ N_{imp}^{cm^{-3}})$

This phenomenon, the body effect, is a major limitation of MOS devices operated with $v_S \neq v_B$; its evil influence will be lamented repeatedly later in the book. As Eqs. (2.12) and (2.13) show, to reduce the body effect, N_{imp} should be made small. However, for very small N_A values (say, $N_A < 10^{13}$ cm^{-3}), an NMOS may behave as a depletion-mode device, as explained earlier. Thus the body effect cannot be eliminated completely. Some key constants and formulas on MOSFETs are summarized in Tables 2.1 and 2.2.

2.4. SMALL-SIGNAL OPERATION AND EQUIVALENT CIRCUIT OF MOSFET TRANSISTORS

Earlier the physical principles and basic operation of MOS transistors were discussed briefly, and formulas derived that gave the drain current as a function of the voltages and/or currents at the various terminals of the device. In these earlier discussions it was assumed that all voltages and currents were constant or that they varied sufficiently slowly so that all capacitive currents (and hence all capacitances themselves) could be neglected in the discussions. On the other hand, the formulas derived were valid for large as well as small voltage and current variations.

In many important linear applications (such as operational amplifiers, discussed in Chapter 4) the voltages and currents of the transistor vary so rapidly that capacitive effects cannot be ignored, and hence the capacitances of the device must be included in the analysis. At the same time, in such circuits the signals are sufficiently small, so that linear approximations may be used in all nonlinear relations. This simplifies the equations and permits the use of linear models (simple linear equivalent circuits) for these nonlinear devices.

TABLE 2.2. Drain–Current Relations for MOSFETs in Large-Signal Low-Frequency Operation

Region of Operation	NMOS	PMOS
Triode region:	$i_D = \mu_n C_{ox} \frac{W}{L}\left(v_{GS} - V_T - \frac{v_{DS}}{2}\right)v_{DS}$	$-i_D = \mu_p C_{ox} \frac{W}{L}\left(v_{GS} - V_T - \frac{v_{DS}}{2}\right)v_{DS}$
$\lvert v_{GS}\rvert > \lvert V_T\rvert$;		
$\lvert v_{DS}\rvert < \lvert v_{GS}\rvert - \lvert V_T\rvert$		
Saturation region	$i_D = \frac{\mu_n C_{ox} W}{2L}(v_{GS} - V_T)^2 (1 + \lambda v_{DS})$,	$-i_D = \frac{\mu_p C_{ox} W}{2L}(v_{GS} - V_T)^2 (1 - \lambda v_{DS})$,
$\lvert v_{GS}\rvert > \lvert V_T\rvert$;	where $\lambda \propto L^{-1}(v_{DG} + V_T)^{-1/2}N_{imp}^{-1/2}$,	where $\lambda \propto L^{-1}(-v_{DG} - V_T)^{-1/2}N_{imp}^{-1/2}$,
$\lvert v_{DS}\rvert > \lvert v_{GS}\rvert - \lvert V_T\rvert$	$V_T = (V_T)_{v_{SB}=0} + \gamma(\sqrt{2\lvert\phi_p\rvert + v_{SB}} - \sqrt{2\lvert\phi_p\rvert})$	$V_T = (V_T)_{v_{SB}=0} - \gamma(\sqrt{2\lvert\phi_p\rvert - v_{SB}} - \sqrt{2\lvert\phi_p\rvert})$
	(see Table 2.1 for the values of γ and ϕ_p)	

In the following discussion, we concentrate on the linearized approximation and modeling process for MOSFETs operating in their saturation regions, which is the usual condition for linear (analog) operation. Afterward, we give a brief overview of the linearization and modeling for devices that operate in their triode (nonsaturated) or cutoff region. Assuming an NMOS transistor, and combining Eqs. (2.9) and (2.12), the relation

$$i_D = k(v_{GS} - V_{T0} - \gamma\sqrt{2\phi_p + v_{SB}} + \gamma\sqrt{2\phi_p})^2(1 + \lambda v_{DS}) \tag{2.14}$$

results. Here we used $v_{GS} \triangleq v_G - v_S$ and $v_{DS} \triangleq v_D - v_S$ to replace v_G and v_D, since in general $v_S \neq 0$. For small variations of i_D, v_{GS}, v_{DS}, and v_{SB}, the nonlinear expression (2.14) can be replaced by a first-order Taylor approximation. Specifically, near a constant bias point $i_D^0 = f(v_{GS}^0, v_{DS}^0, v_{SB}^0)$ we can write

$$i_D^0 + \Delta i_D \approx i_D^0 + \left(\frac{\partial i_D}{\partial v_{GS}}\right)^0 \Delta v_{GS} + \left(\frac{\partial i_D}{\partial v_{DS}}\right)^0 \Delta v_{DS} + \left(\frac{\partial i_D}{\partial v_{SB}}\right)^0 \Delta v_{SB}. \tag{2.15}$$

Here $(\partial i_D/\partial v_{GS})^0$ and so on denote the partial derivatives evaluated at the bias point. Δi_D is the deviation (increment) of i_D from its bias value; Δv_{GS}, Δv_{DS}, and Δv_{SB} are the increments of v_{GS}, v_{DS}, and v_{SB}. All deviations must be small for Eq. (2.15) to hold. If only the incremental (small-signal ac) components are of interest, Eq. (2.15) can be written as

$$\Delta i_D \approx g_m \ \Delta v_{GS} + g_d \ \Delta v_{DS} + g_{mb} \ \Delta v_{SB}, \tag{2.16}$$

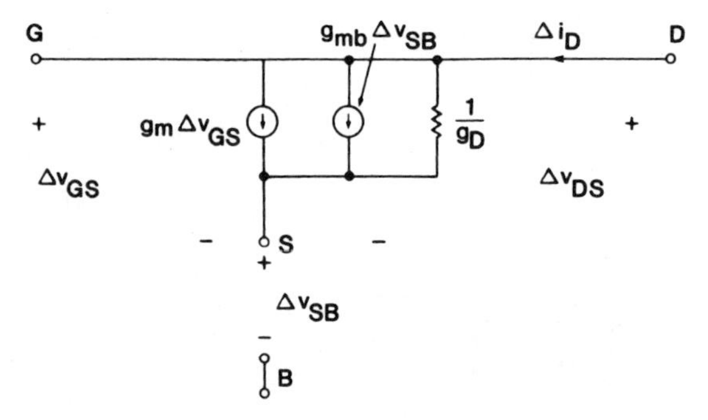

Figure 2.14. Low-frequency equivalent circuit of a MOSFET.

where

$$g_m \triangleq \left(\frac{\partial i_D}{\partial v_{GS}}\right)^0, \qquad g_d \triangleq \left(\frac{\partial i_D}{\partial v_{DS}}\right)^0, \qquad g_{mb} \triangleq \left(\frac{\partial i_D}{\partial v_{SB}}\right)^0. \tag{2.17}$$

Here g_d is the (incremental) drain conductance; g_m and g_{mb} are transconductances that can be represented by voltage-controlled current sources (VCCSs). Hence an equivalent-circuit model, shown in Fig. 2.14, can be constructed. The values of g_m, g_{mb}, and g_d can be found from Eq. (2.14):

$$\begin{aligned} g_m \triangleq \left(\frac{\partial i_D}{\partial v_{GS}}\right)^0 &= 2k\left(v_{GS}^0 - V_{T0} - \gamma\sqrt{2\phi_p + v_{SB}^0} + \gamma\sqrt{2\phi_p}\right)(1 + \lambda v_{DS}^0) \\ &= 2\sqrt{k(1 + \lambda v_{DS}^0)i_D^0}, \end{aligned} \tag{2.18}$$

$$\begin{aligned} g_{mb} \triangleq \left(\frac{\partial i_D}{\partial v_{SB}}\right)^0 &= -k\left(v_{GS}^0 - V_{T0} - \gamma\sqrt{2\phi_p + v_{SB}^0} + \gamma\sqrt{2\phi_p}\right) \\ &\quad \times (1 + \lambda v_{DS}^0)\frac{\gamma}{\sqrt{2\phi_p + v_{SB}^0}} \\ &= \frac{-\gamma g_m/2}{\sqrt{2\phi_p + v_{SB}^0}}, \end{aligned} \tag{2.19}$$

$$\begin{aligned} g_d \triangleq \left(\frac{\partial i_D}{\partial v_{DS}}\right)^0 &= k\left(v_{GS}^0 - V_{T0} - \gamma\sqrt{2\phi_p + v_{SB}^0} + \gamma\sqrt{2\phi_p}\right)^2 \lambda \\ &= \frac{\lambda}{1 + \lambda v_{DS}^0} i_D^0 . \end{aligned} \tag{2.20}$$

Hence, to a good approximation, g_m and g_{mb} are proportional to $\sqrt{i_D^0}$, while g_d is proportional to i_D^0.

The other important components of the complete small-signal model of the MOSFET are the capacitors representing the incremental variations of stored charges with changing electrode voltages. These play an important role in the high-frequency operation of the device. The *intrinsic* components of the terminal capacitances of the MOSFET devices (associated with reverse-biased *pn* junctions and with channel and depletion regions) are strongly dependent on the region of operation, while the *extrinsic* components (due to layout parasitics, overlapping regions, etc.) are relatively constant. Assuming again that the transistor operates in the saturation region, it can be assumed that the channel begins at the source and extends over two-thirds of the distance to the drain. In this region of operation, the most important capacitances are the following:

1. C_{gd}: *Gate-to-Drain Capacitance.* This is due to the overlap of the gate and the drain diffusion. It is a thin-oxide capacitance, and hence to a good approximation can be regarded as being voltage independent.
2. C_{gs}: *Gate-to-Source Capacitance.* This capacitance has two components: C_{gsov}, the gate-to-source thin-oxide overlap capacitance, and C'_{gs}, the gate-to-channel capacitance. The latter (in the saturation region) is around $\frac{2}{3}C_{ox}$, where C_{ox} is the total thin-oxide capacitance between the gate and the surface of the substrate. In the triode region, $C'_{gs} = C_{ox}$. C_{gs} is nearly voltage independent in the saturation region.
3. C_{sb}: *Source-to-Substrate Capacitance.* This capacitance also has two components: C_{sbpn}, the *pn* junction capacitance between the source diffusion and the substrate, and C'_{sb}, which can be estimated as two-thirds of the capacitance of the depletion region under the channel. The overall capacitance C_{sb} has a voltage dependence which is similar to that of an abrupt *pn* junction.
4. C_{db}: *Drain-to-Substrate Capacitance.* This is a *pn* junction capacitance and is thus voltage dependent.
5. C_{gb}: *Gate-to-Substrate Capacitance.* This capacitance is usually small in the saturation region; its value is around $0.1C_{ox}$.

Figure 2.15 illustrates the physical structure of an NMOS transistor and the locations of the capacitances in the cutoff (Fig. 2.15*a*), saturation (Fig. 2.15*b*), and nonsaturation or triode (Fig. 2.15*c*) regions. Table 2.3 lists the terminal capacitors of the NMOS device and their estimated values in the three regions of operation. The notations used are those shown in Fig. 2.15*a* to *c*. Figure 2.16 depicts the complete high-frequency (ac) small-signal model of the MOSFET. In analyzing the small-signal behavior of MOSFETs, the model of Fig. 2.14 can be used if only low-frequency signals are present; if the capacitive currents are also of interest, the circuit of Fig. 2.16 must be applied.

From the models of Figs. 2.14 and 2.16 and accompanying discussions, a number of general statements can be made about the desirable construction of a MOSFET:

1. For high ac gain, g_m should be large. This will be the case, by Eq. (2.18), if $k \triangleq (\mathrm{m}_n C_{ox} W)/2L$ is large. Thus the oxide should be thin to maximize C_{ox} (which is the oxide capacitance per unit area); also, W/L should be as large as possible. These measures, however, tend to increase the size and thus the cost of the integrated circuit. Also, by Eq. (2.18), the quiescent (bias) current i_D^0 should be as large as the allowable dc power dissipation permits.
2. As the negative sign in Eq. (2.19) indicates, the body effect reduces the gain. To minimize g_{mb}, by Eqs. (2.19) and (2.13), we need large C_{ox}, small N_{imp} (i.e., lightly doped substrate), and a large bias voltage v_{SB}^0 for the source.

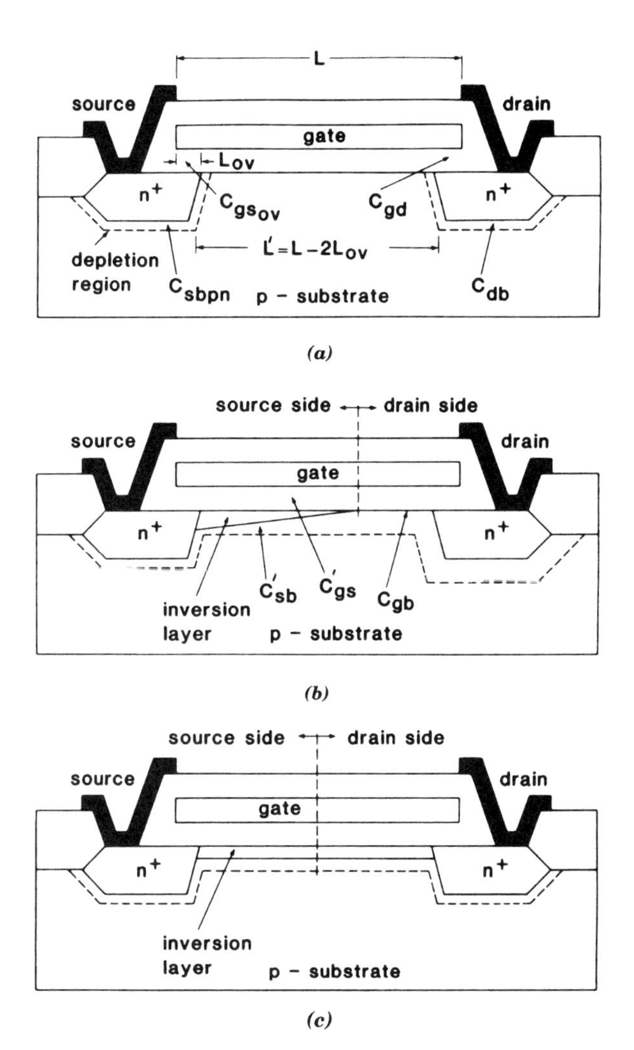

Figure 2.15. Parasitic capacitances in MOSFET in the (*a*) cutoff region, (*b*) saturation region, and (*c*) triode region.

(Of course, if v_{SB} is constant, no incremental body effect occurs and these requirements are irrelevant.)

3. Ideally, the MOSFET in saturation should behave as a pure current source. Hence, as Fig. 2.14 illustrates, g_d should be small. By Eq. (2.20) this requires a small bias current i_D^0, a large bias voltage v_{DS}^0, and a small λ. Since λ is introduced by channel-length modulation, it can be reduced by increasing L and also [1, Sec. 8.4] by increasing N_{imp}. A summary of the formulas derived in this section is given in Table 2.4.

TABLE 2.3. Terminal Capacitances of a MOSFET in the Three Main Regions of Operation[a]

			Capacitance		
Region of Operation	C_{gs}	C_{gd}	C_{gb}	C_{sb}	C_{db}
Cutoff region	$WL_{ov}C_{ox}$	$WL_{ov}C_{ox}$	$WL'C_{ox}$	$A_sC_{pn}(V_{sb})$	$A_dC_{pn}(V_{db})$
Saturation region	$WC_{ox}(L_{ov} + \frac{2}{3}L')$	$WL_{ov}C_{ox}$	$\frac{\frac{1}{3}WL'C_{ox}C_{pn}(V_{db})}{C_{ox} + C_{pn}(V_{cb})}$	$A_sC_{pn}(V_{sb}) + \frac{2}{3}WL'C_{pn}(V_{sb})$	$A_dC_{pn}(V_{db})$
Nonsaturated (triode) region	$WL_{ov}C_{ox} + \frac{1}{2}WL'C_{ox}$	$WC_{ox}(L_{ov} + \frac{1}{2}L')$	0	$A_sC_{pn}(V_{sb}) + \frac{1}{2}WL'C_{pn}(V_{sb})$	$A_dC_{pn}(V_{db}) + \frac{1}{2}WL'C_{pn}(V_{db})$

[a] L' denotes $L - 2L_{ov}$.

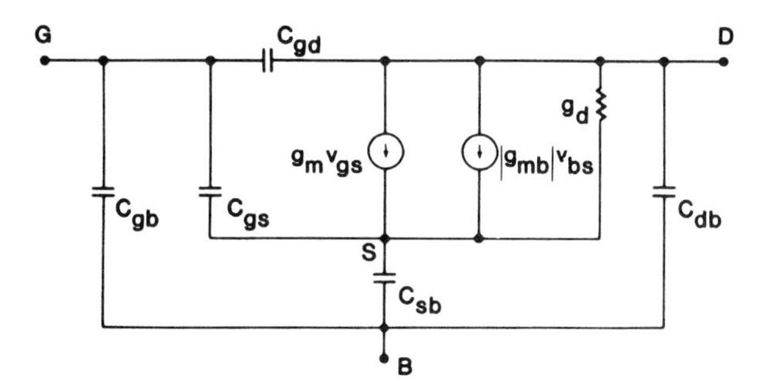

Figure 2.16. High-frequency equivalent circuit of a MOSFET.

Next, we discuss briefly the linear model of a MOSFET that is biased in its nonsaturated (triode) region. Usually, such a device is used as a switch that is opened or closed (turned on or off) by a large gate voltage or as a fairly linear large-valued variable resistor. Hence, here we derive its model only to analyze its behavior in such applications. We assume that v_{GS} is constant and that $g_{mb} \ll g_{ds}$ and hence negligible. Under these conditions it can be seen from Eq. (2.7) that when the device conducts drain current, it behaves like a resistor connected between the drain and source terminals. The equivalent small-signal drain-to-source resistance for the case of v_{DS} near zero is given by

$$r_{ds} = \frac{1}{g_{ds}} = \frac{1}{2k(v_{GS}^0 - V_{T0})} \tag{2.21}$$

and is thus controlled by the gate voltage overdrive $v_{GS}^0 - V_{T0}$.

In high-frequency application, the device capacitances must also be included in the model. A simple equivalent circuit is shown in Fig. 2.17. Since now a continuous channel extends from the source to the drain, the gate-to-channel capacitance C'_{gs} is connected to both the drain and the source. An accurate high-frequency representation of the channel should include a distributed resistive line extending from the source to the drain and capacitively coupled to both the gate and the substrate. However, as a first approximation, C'_{gs} can be treated as a lumped capacitance partitioned equally between C_{gs} and C_{gd}, as indicated in the last row of Table 2.3.

Finally, if the device is cut off, no channel exists and the model contains only

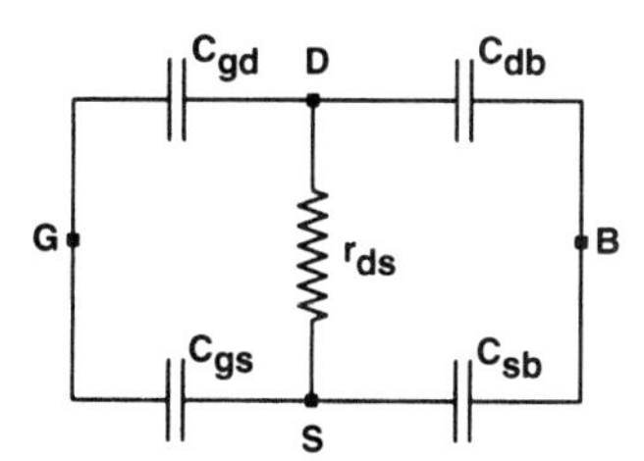

Figure 2.17. High-frequency model of a MOSFET in its triode region.

TABLE 2.4. Small-Signal Parameters of MOSFETS in Saturation[a]

Parameter	NMOS	PMOS
Transconductance: $g_m \triangleq \dfrac{\partial i_D}{\partial v_{GS}}$	$\dfrac{\mu_n C_{ox} W}{L}(v_{GS}^0 - V_T) = \sqrt{\dfrac{2\mu_n C_{ox} W i_D^0}{L}}$	$-\dfrac{\mu_p C_{ox} W}{L}(v_{GS}^0 - V_T) = -\sqrt{\dfrac{2\mu_p C_{ox} W(-i_D^0)}{L}}$
Body-effect transconductance: $g_{mb} \triangleq \dfrac{\partial i_D}{\partial v_{SB}}$	$-\dfrac{\gamma/2}{\sqrt{2\phi_p + v_{SB}^0}}\, g_m$	$-\dfrac{\gamma/2}{\sqrt{2\phi_p - v_{SB}^0}}\, g_m$
Drain conductance: $g_d \triangleq \dfrac{\partial i_D}{\partial v_{DS}}$	$\dfrac{\lambda i_D^0}{1 + \lambda v_{DS}^0}$	$\dfrac{\lambda i_D^0}{1 - \lambda v_{DS}^0}$
Gate-to-source capacitance C_{gs}	$\frac{2}{3} WLC_{ox}$	$\frac{2}{3} WLC_{ox}$
Gate-to-drain capacitance C_{gd}	C_{gd} overlap	C_{gd} overlap
Source (or drain)-to-bulk capacitance $C_{sb}(C_{db})$	$\dfrac{C_{sb0}}{\sqrt{1 + v_{SB}^0/2\phi_p}}; \dfrac{C_{db0}}{\sqrt{1 + v_{DB}^0/2\phi_p}}$	$\dfrac{C_{sb0}}{\sqrt{1 - v_{SB}^0/2\phi_p}}; \dfrac{C_{db0}}{\sqrt{1 - v_{DB}^0/2\phi_p}}$

[a] See Fig. 2.16. $|\lambda v_{DS}| \ll 1$ is assumed in all formulas.

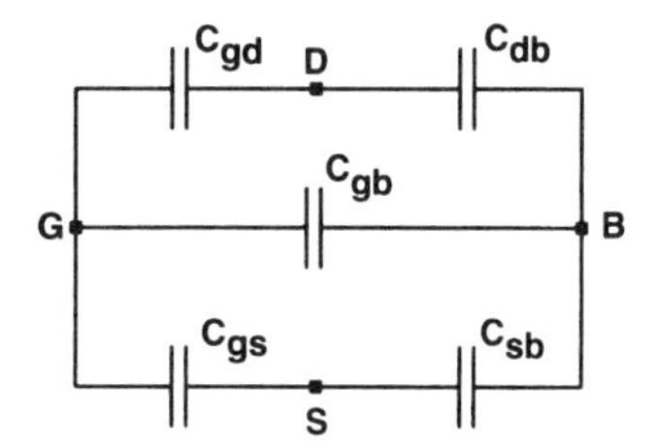

Figure 2.18. High-frequency model of a MOSFET in its cutoff region.

the capacitances, with the values listed in the first row of Table 2.3. A simplified model of a MOSFET in the cutoff region is shown in Fig. 2.18, where the drain–source resistance is infinity and C_{gs} and C_{ds} capacitors are due to gate overlap and fringing capacitances. The gate-to-substrate capacitance C_{gd} in the cutoff region is, however, highly nonlinear, and for the gate-to-source voltage around zero its value is approximately equal to $WL'C_{\text{ox}}$, where $L' = L - 2L_{\text{ov}}$ and L_{ov} is the length of the overlap between the gate and the source–drain diffusion regions.

2.5. WEAK INVERSION

The triode and saturation regions of operation discussed earlier in this chapter assume that the device is operated in strong inversion and $V_{GS} - V_T \geq 100$ mV (for an NMOS transistor). If $V_{GS} - V_T < 100$ mV, the device is in the weak inversion region (also called the *subthreshold region*) and operation of the *n*-channel MOS transistor with the source connected to substrate is more accurately described by the following exponential relationship between the gate-to-source control voltage and drain current:

$$I_D \approx \frac{W}{L} I_{D0} e^{qV_{GS}/nkT}, \tag{2.22}$$

where

$$I_{D0} = \mu_0 C_{\text{ox}} \frac{1}{m} \left(\frac{nkT}{q} \right)^2 e^{-(qV_T/nkT + 1)} \tag{2.23}$$

and the parameters m and n are defined in terms of the various capacitances in the device [3]. Equation (2.22) assumes that the device is operated with $V_{DS} \gg kT/q$.

The MOS transistors operating in the weak inversion region, similar to bipolar devices, have an exponential relationship. However, since I_{D0} is very small (on the order of 10 to 20 nA), the available current to charge and discharge capacitances is also small, resulting in poor frequency performance. In practice, MOS transistors are operated in weak inversion only in low-frequency applications when low power consumption is desired.

2.6. IMPACT IONIZATION [4]

One of the severe problems in submicron MOS technologies operating at supply voltages around 5 V is impact ionization. Figure 2.19 illustrates an n-channel MOS device cross section showing the impact ionization current flow and the $I-V$ characteristic as a result of impact ionization. As depicted in the figure, when the drain-to-source voltage is increased, the strength of the electric field at the drain end of the channel eventually becomes high enough to induce significant impact ionization current which originates from the drain depletion region and flows into the substrate. Once this happens the current that flows into the drain terminal has two components. One component is the MOS transistor channel current that flows from the drain to the source, and the other is the impact ionization current that flows from the drain to the substrate. The impact ionization current is not a function of the transistor channel length, and the magnitude of the current is not reduced dramatically simply by making the length longer. The current is largely determined by the peak electric field, which in turn is a function of the gate oxide thickness, drain junction depth, doping concentration in the substrate, the voltage between the drain terminal and the drain end of the channel region, and the gate-to-drain voltage. In technologies with feature sizes in the range of 2 μm, for an n-channel MOS device, the impact ionization current equals 1% of the drain current when the voltage between the drain and the drain end of the channel is in the range 4 to 9 V, and the device is biased

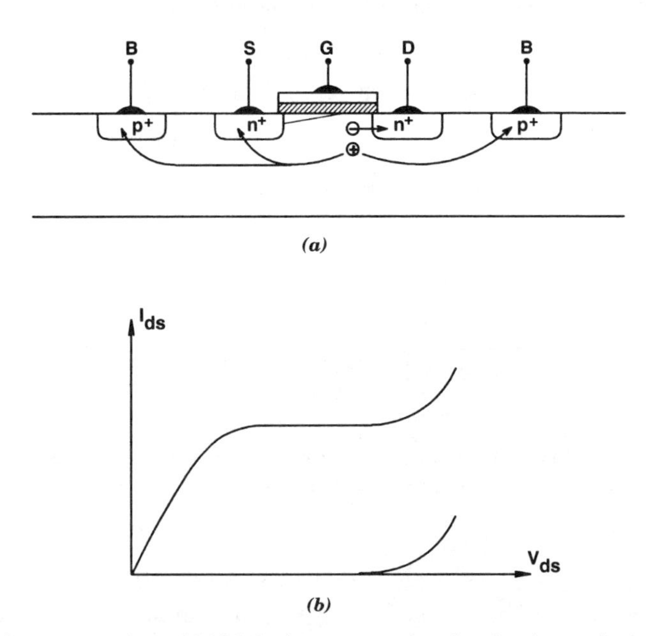

Figure 2.19. (a) An n-channel MOS device cross section showing impact ionization current flow; (b) $I-V$ characteristic observed as a result of impact ionization.

in the saturation region. In p-channel MOS devices the effect occurs at substantially higher field strengths.

The impact ionization has several potentially damaging side effects. One serious negative consequence shown in Fig. 2.19*b* is degradation of the transistor output impedance, which results in reduced gain in amplifier stages that use transistors as active loads. One way to deal with this problem is through circuit techniques, where a shielding n-channel device is placed in series with the transistor, preventing it from having a V_{dS} greater than half the supply voltage [4]. The second damaging effect is the possibility of triggering latchup due to the ohmic drop induced by the ionization current that flows into the substrate. *Latchup* is a phenomenon caused by the parasitic lateral *pnp* and *npn* bipolar transistors created on the chip. The collectors of each transistor feed the other's base, and this creates an unstable device similar to a *pnpn* thyristor [5]. This causes a sustained dc current that may cause the chip to stop functioning and may even destroy it. Latchup may be prevented by proper substrate strapping and using guard rings to surround some critical transistors on the chip. Another strategy is to reduce the substrate resistance. In this method the p- and n-channel transistors are formed in a lightly doped epitaxial layer that is grown on a low-resistivity substrate. Finally, the third serious side effect of impact ionization is the threshold shift of the MOS device due to the continuous operation in the impact ionization mode. This phenomenon is due to the high electric field which creates high energy carries that can be trapped in the gate oxide, resulting in long-term threshold shift. Several process modifications have been proposed that are effective at raising the voltage at which impact ionization becomes a problem. One technique is to lower the impurity gradient in the drain junction using lightly doped drain (LDD) structures.

2.7. NOISE IN MOSFETS

There are three distinct sources of noise in solid-state devices: shot noise, thermal noise, and flicker noise.

Shot Noise

Since electric currents are carried by randomly propagating individual charge carriers (electrons or holes), superimposed on the nominal (average) current I, there is always a random variation i_{nS}. This is due to fluctuation in the number of carriers crossing a given surface in the conductor in any time interval. It can be shown that the mean square of i_{nS} is given by

$$\overline{i_{ns}^2} = 2qI\,\Delta f, \tag{2.24}$$

where $q = 1.6 \times 10^{-19}$ C is the magnitude of the electron charge and Δf is the bandwidth. This formula only holds, however, if the density of the charge carriers is so low and the external electric field is so high that the interaction between the

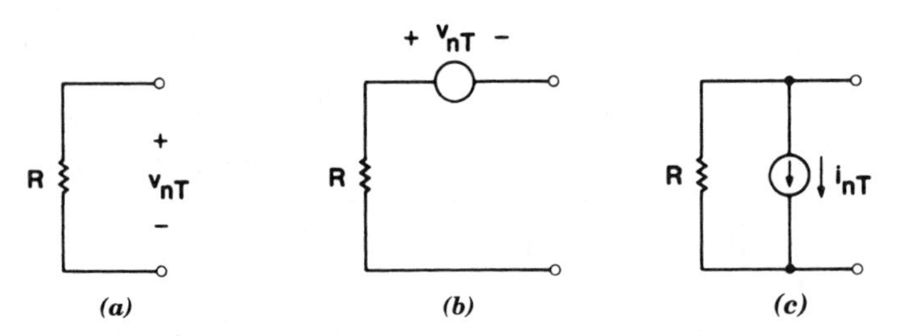

Figure 2.20. Thermal noise in a resistor: (*a*) noisy resistor; (*b*) and (*c*) equivalent circuits.

carriers is negligible. Otherwise, the randomness of their density and velocity is reduced due to the correlation introduced by the repulsion of their charges. The noise current is then much smaller than predicted by Eq. (2.24).

In a conducting MOSFET channel, the charge density is usually high and the electric field is low. Therefore, Eq. (2.24) does not hold. The noise current due to random carrier motions is hence better described as *thermal noise,* which is discussed next.

Thermal Noise

In a real resistor *R,* the electrons are in random thermal motion. As a result, a fluctuating voltage v_{nT} appears across the resistor even in the absence of a current from an external circuit (Fig. 2.20*a*). Thus the Thévenin model of the real (noisy) resistor is that shown in Fig. 2.20*b*. Clearly, the higher the absolute temperature *T* of the resistor, the larger v_{nT} will be. In fact, it can be shown that the *mean square* of v_{nT} is given by

$$\overline{v_{nT}^2} = 4kTR\,\Delta f. \tag{2.25}$$

Here *k* is the ubiquitous Boltzmann's constant, and Δf is the bandwidth in which the noise is measured, in hertz. (The value of $4kT$ at room temperature is about 1.66×10^{-20} V · C.)

If Eq. (2.25) was true for any bandwidth, the energy of the noise would be infinite. In fact, however, for very high frequencies ($\approx 10^{13}$ Hz) other physical phenomena enter, which cause $\overline{v_{nT}^2}$ to decrease with increasing frequency so that the overall noise energy is finite.

The average value (dc component) of the thermal noise is zero. Since its spectral density $\overline{v_{nT}^2}/\Delta f$ is independent of frequency (at least for lower frequencies), it is a *white noise.* Clearly, Fig. 2.20*b* may be redrawn in the form of a Norton equivalent that is as a (noiseless) resistor *R* in parallel with a noise current source i_{nT} (Fig. 2.20*c*). The value of the latter is given by

$$\overline{i_{nT}^2} = 4kTG\,\Delta f, \tag{2.26}$$

where $G = 1/R$.

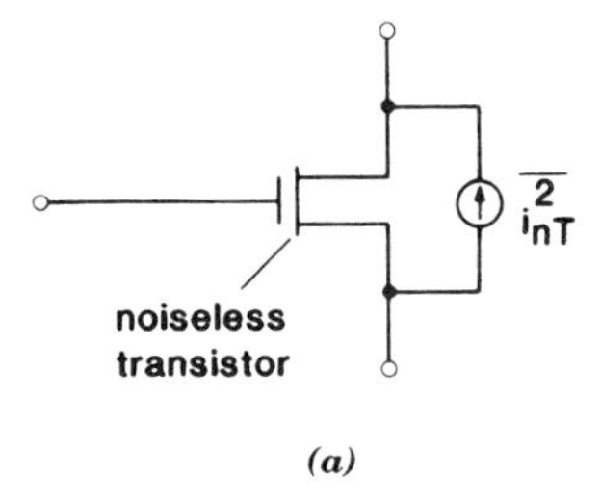

(a)

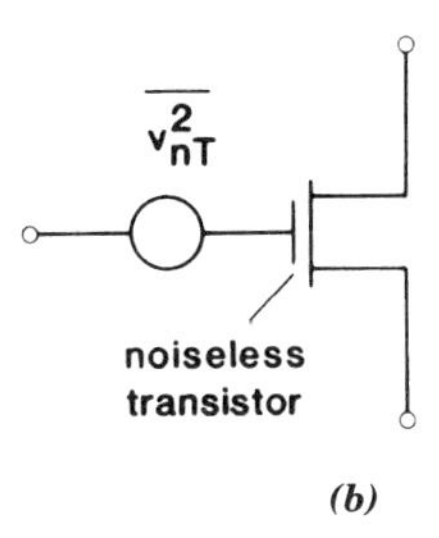

(b)

Figure 2.21. Equivalent models of the thermal noise in a MOSFET.

Since the channel of a MOSFET in conduction contains free carriers, it is subject to thermal noise. Therefore, Eqs. (2.25) and (2.26) will hold, with R given by the *incremental channel resistance.* The noise can then be modeled by a current source, as shown in Fig. 2.21*a.* If the device is in saturation, its channel tapers off (Fig. 2.8) and the approximation $R \approx 3/2g_m$ can be used in Eq. (2.26).

In most circuits it is convenient to model the effect of i_{nT} caused by a voltage source connected to the gate of an (otherwise noiseless) MOSFET (Fig. 2.21*b*). This "gate-referred" noise voltage source is then given by

$$\overline{v_{nT}^2} \approx \overline{\left(\frac{i_{\mathrm{nT}}}{g_m}\right)^2} = \frac{8}{3}\frac{kT}{g_m}\Delta f. \tag{2.27}$$

Both i_{nT} and v_{nT} depend thus on the dimensions, bias conditions, and temperature of the device. As an example of their orders of magnitude, for a transistor with $W = 200\ \mu\mathrm{m}$, $L = 10\ \mu\mathrm{m}$, and $C_{\mathrm{ox}} = 4.34 \times 10^{-8}\ \mathrm{F/cm^2}$ (corresponding to an oxide thickness of 800 Å) which is operated in saturation at a drain current $i_D = 200\ \mu\mathrm{A}$, the gate-referred noise voltage at room temperature is about $9\ \mathrm{nV}/\sqrt{\mathrm{Hz}}$.

If the device is switched off, R becomes very high, and the equivalent noise circuit will be a current source with a value given by Eq. (2.26). Clearly, $\overline{i_{nT}^2}$ is very small; hence for usual (low or moderate) external impedance levels, the MOSFET can be regarded as a noiseless open circuit if it is turned off.

Flicker (1/*f*) Noise

In an MOS transistor, extra electron energy states exist at the boundary between the Si and SiO_2. These can trap and release electrons from the channel, and hence introduce noise [6,7]. Since the process is relatively slow, most of the noise energy will be at low frequencies. As before, a possible model of this noise phenomenon is a current source in parallel with the channel resistance. The dc value of noise current is again zero. Its mean-square value increases with temperature and the density of the surface states; it decreases with the gate area $W \times L$ and the gate oxide capacitance per unit area C_{ox}. For devices fabricated with a "clean" process, the gate-referred noise voltage is nearly independent of the bias conditions and is given by the approximating formula

$$\overline{v_{nf}^2} = \frac{K}{C_{ox}WL}\frac{\Delta f}{f}. \tag{2.28}$$

Here K depends on the temperature and the fabrication process; a typical value [8, p. 31] is $3 \times 10^{-24}\ \mathrm{V^2 \cdot F}$. For the transistor described in the preceding example, the formula gives a noise voltage of 83 $\mathrm{nV/\sqrt{Hz}}$ at $f = 1$ kHz. As before, the equivalent channel current noise is related to $\overline{v_{nf}^2}$ by the formula $\overline{i_{nf}^2} = g_m^2\overline{v_{nf}^2}$.

The noise process described is usually called *flicker noise* or (in reference to the $1/f$ factor in $\overline{v_{nf}^2}$ and $\overline{i_{nf}^2}$) *1/f noise.* As the example given illustrates, at low frequencies (say, below 1 kHz) it is usually the dominant noise mechanism in a MOSFET.

In conclusion, the channel noise in a MOSFET can be modeled by an equivalent noise current generator, as in Fig. 2.21*a*. In the small-signal model this generator will be in parallel with the current sources $g_m v_{GS}$ and $g_{mb} v_{BS}$ (Fig. 2.16). Its value can be chosen as the root-mean-square (RMS) noise current, which from Eqs. (2.26)–(2.28) is

$$\begin{aligned} i_n &= \sqrt{\overline{i_{nT}^2} + \overline{i_{nf}^2}} \\ &= \sqrt{\left(4kTG + \frac{Kg_m^2}{C_{ox}WLf}\right)\Delta f}. \end{aligned} \tag{2.29}$$

Note that the mean squares of the noise currents are added, since the different noise mechanisms are statistically independent. Alternatively, the noise can be represented by its gate-referred voltage source (Fig. 2.21*b*), in series with the gate terminal. The value of the source is i_n/g_m, with i_n given by Eq. (2.29).

2.8. CMOS PROCESS

The CMOS process provides the most flexibility to the circuit designer, due to the availability of complementary MOS devices on the same chip. The original motivation for developing the CMOS technology was the need for low-power and high-speed logic gates for digital circuits. The required isolation between the two different device types is accomplished by the use of "wells," that is, large, low-doping-level deep diffusions, which serve as the substrates for one of the two device types. As an example, Fig. 2.22 shows part of an *n*-well CMOS chip, where high-resistivity *p*-type substrate is used for the *n*-channel devices, and diffused *n* wells for the *p*-channel devices.

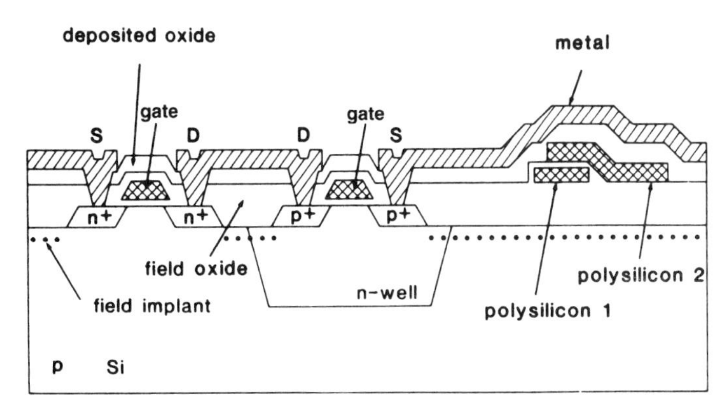

Figure 2.22. Device structure fabricated in a high-performance *n*-well CMOS process.

As will be shown in later chapters, a CMOS circuit can be operated with a single power supply, and it can be used to realize high-speed, high-gain, low-power analog amplifier stages. An additional advantage is that for the devices in the well (in Fig. 2.22, the PMOS transistor), the source can be connected to the well, thereby eliminating the body effect, and if the device is used in an amplifier, increasing the gain of the circuit. This, however, results in a large stray capacitance between the source and the substrate, due to the large size of the well-to-body interface. Another important advantage of the CMOS process is the availability of transmission gates made of a parallel connection of complementary transistors that can be used as switches. When such transmission gates are used, the signal is no longer limited to a level, which is a threshold voltage below that of the high clock signal, as is the case when single-channel switches are used. In addition, in CMOS chips a bipolar transistor can be fashioned from a source diffusion, the well, and the substrate. This can be used in an emitter-follower buffer stage (described later), in a bandgap voltage reference circuit, and so on.

In addition to transistors, analog MOS circuits usually require on-chip capacitors, and sometimes also on-chip resistors. In a silicon-gate "double-poly" process, a second layer of low-resistivity polysilicon is available for use as an interconnect or for the formation of a floating gate for memory applications. These two layers of polysilicon can also be used as the top and bottom electrodes of a monolithic capacitor. Figure 2.22 shows the construction of a capacitor with two polysilicon electrodes. Resistors can be created on an MOS chip using a diffused or implanted layer on the surface of the substrate. Since the sheet resistance of these resistive layers is relatively low (typically 25 to 70 Ω for a square layer), the size of the resistors obtainable on a reasonably small area is limited to about 100 kΩ. The higher resistivity well diffusion is also available as a resistor. This resistor, however, has much higher voltage and temperature coefficients compared to diffused or implanted ones.

PROBLEMS

2.1. A *pn* junction diode is connected to an external voltage v in the forward direction (Fig. 2.3). Reversing the polarity of the voltage reduces the current by a factor

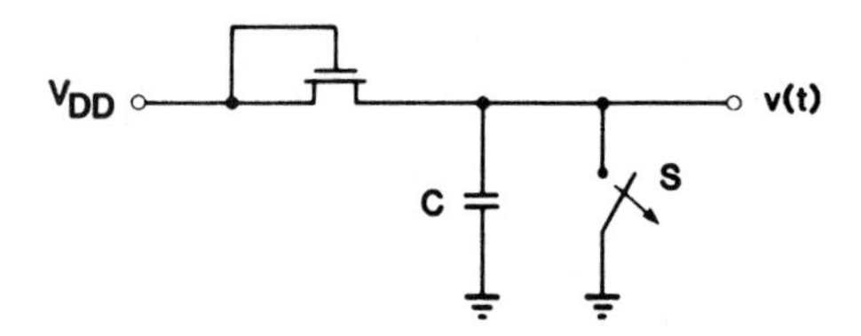

Figure 2.23. Circuit for Problem 2.8.

10^6. Assume that the diode satisfies Eq. (2.1) and is at room temperature. What is v?

2.2. For a *pn* junction (Fig. 2.3), $N_A = N_D = 10^{16}$ ions/cm^3, $|v| = 5$ V, $A = 0.34$ mm^2, and the measured value of C is 27 pF. How much is x_d, the width of the depletion layer? How much is ϕ_i?

2.3. Using the definition $R = 1/(\partial i_D/\partial V_{DS})$, calculate the channel resistance of an NMOS transistor from **(a)** Eq. (2.6), **(b)** Eq. (2.7), and **(c)** Eq. (2.9).

2.4. For an NMOS transistor, $\mu_n = 10^3$ cm^2/V·s, the thickness of the gate oxide is 10^3 Å (1 Å $= 10^{-8}$ cm), $W = 25$ μm, and $L = 5$ μm. The threshold voltage is 4 V. Calculate i_D for $v_S = v_B = 0$ V and $v_G = 6$ V, and **(a)** $v_D = 0.1$ V, **(b)** $v_D = 2$ V, and **(c)** $v_D = 4$ V.

2.5. Repeat the calculations of Problem 2.4 if $v_S = 0$, $v_B = -3$ V, and $\phi_p = 0.3$ V. What conclusions can you draw from your results regarding the body effect?

2.6. For an NMOS transistor, $k' = 2$ μA/V^2, $W = 30$ μm, $L = 10$ μm, $\phi_p = 0.3$ V, $\gamma = 1.5$ V$^{1/2}$, and $\lambda = 0.03$ V^{-1}. Find the incremental conductances g_m, g_d, and g_{mb} for $v_{SB} = 0$ V, $v^0_{DS} = 5$ V, and $i^0_D = 10$ μA. Repeat your calculations for $v_{SB} = 2$ V!

2.7. An NMOS switch transistor has a gate-to-source voltage $v_{GS} > V_T$. Its drain is open-circuited. How much is v_{DS}? Why?

2.8. In the circuit of Fig. 2.23, the switch S is opened at $t = 0$. **(a)** Is the transistor operating in its linear or saturation region? **(b)** Neglecting body effect and channel-length modulation, find $v(t)$ by solving the appropriate differential equation for the circuit.

2.9. In the circuit of Fig. 2.24 a noise voltage v_n is generated due to thermal and

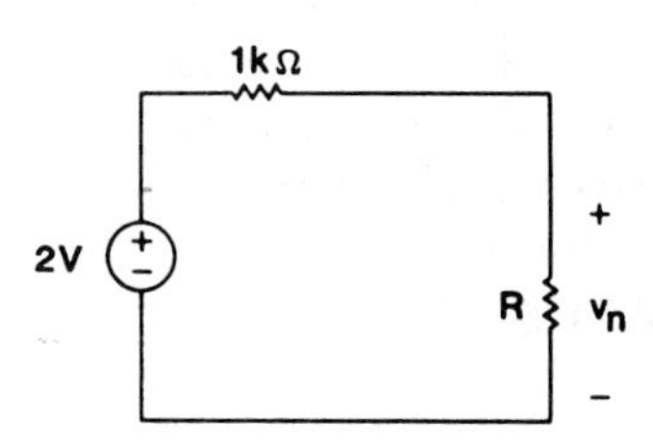

Figure 2.24. Circuit for Problem 2.9.

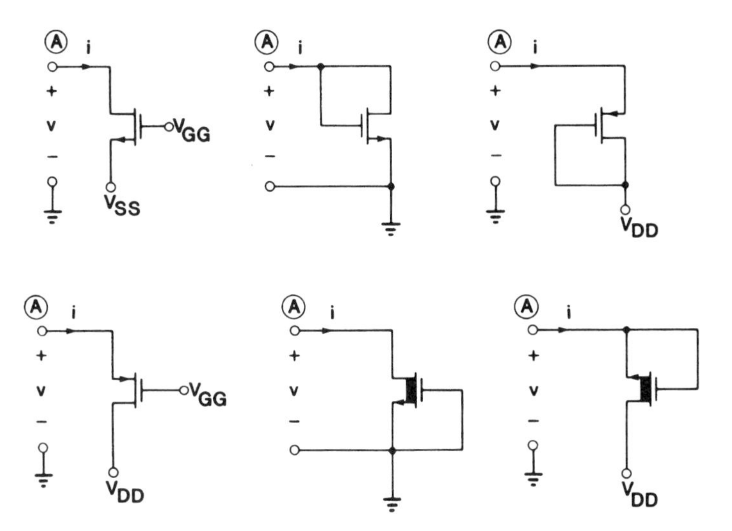

Figure 2.25. Circuit for Problem 2.10.

shot-noise effects. For what value of R will the two noise voltages v_{nT} and v_{nS} be equal?

2.10. Calculate the incremental impedance $\partial v/\partial i$ seen at node A of the circuits shown in Fig. 2.25.

2.11. Show that the transconductance g_m in the saturation region is equal to the drain conductance in the triode region for a given device and a fixed V_G.

REFERENCES

1. R. S. Muller and T. I. Kamins, *Device Electronics for Integrated Circuits,* Wiley, New York, 1977.
2. A. S. Grove, *Physics and Technology of Semiconductor Devices,* Wiley, New York, 1967.
3. Y. P. Tsividis and R. W. Ulmer, A CMOS voltage reference, *IEEE J. Solid-State Circuits, SC-13*(6), 774–778 (1978).
4. C. A. Laber, C. F. Rahim, S. F. Dryer, G. T. Uehara, P. T. Kwoh, and P. R. Gray, *IEEE J. Solid-State Circuits, SC-22*(2), 181–189 (1987).
5. S. M. Sze (Ed.), *VLSI Technology,* McGraw-Hill, New York, 1983.
6. M. B. Das and J. M. Moore, IEEE *Trans. Electron. Devices, ED-21*(2), 247–257 (1974).
7. P. R. Gray and R. G. Meyer, *Analysis and Design of Analog Integrated Circuits,* Wiley, New York, 1977.
8. P. R. Gray and D. A. Hodges, and R. W. Brodersen (Eds.), *Analog MOS Integrated Circuits,* IEEE Press, New York, 1980.

CHAPTER 3

BASIC ANALOG CMOS SUBCIRCUITS

In this chapter some of the basic subcircuits commonly utilized in analog MOS integrated circuits are examined. These blocks include a variety of bias circuits, current mirrors, single-stage amplifiers, source followers, and differential stages. These subcircuits are typically combined to synthesize a more complex circuit function. The operational amplifier and comparator, covered in later chapters, are examples of how simple subcircuits are combined to form more complex functions.

The first part of this chapter covers the subject of the bias circuits in CMOS technology and the current mirrors. Next, the CMOS gain stage is introduced, with particular emphasis on the use of *active devices* as *active loads.* The current mirror subcircuit covered as a biasing element is utilized as a dynamic load to obtain very high voltage gains from a single-stage amplifier. The differential amplifier, which represents a broad class of circuits, is discussed next. The differential amplifier is one of the most widely used gain stages, whose basic function is to amplify the difference between two input signals. Finally, the last part of the chapter deals with the small-signal analysis and frequency response of CMOS amplifier stages. A good understanding of the topics presented in this chapter is essential for the analog CMOS designer, as most designs start at the subcircuit level and progress upward to realize a more complex function.

3.1. BIAS CIRCUITS IN MOS TECHNOLOGY

Op-amp and amplifier stages, described in detail later, need various dc bias voltages and currents for their operation. An ideal voltage or current bias is independent of the dc power supply voltages ($V_{DD} > 0$ and $V_{SS} \leq 0$) and of temperature.

To obtain the dc bias voltages $V_{o1}, V_{o2}, \ldots, V_{on}$, where $V_{SS} < V_{o1} < V_{o2} < \cdots$

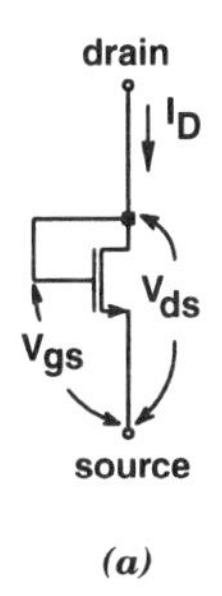

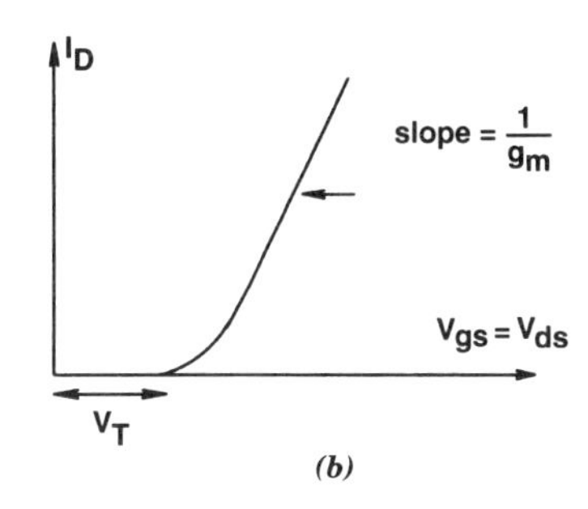

Figure 3.1. (*a*) Diode-connected NMOS transistor; (*b*) current–voltage characteristic of diode-connected transistor.

$V_{on} < V_{DD}$, voltage division can be used. Pure resistive dividers are seldom used in MOS technology because the resulting voltages cannot be used directly to establish bias currents in MOS transistors. Instead, combinations of MOSFETs and resistors are often used. A MOS transistor with its gate connected to the drain forms a two-terminal device, as shown in Fig. 3.1*a*. Its current–voltage characteristics are shown in Fig. 3.1*b*. Since $V_{DS} = V_{GS}$, the dynamic resistance r_{ds} is characterized by

$$r_{ds} = \frac{\partial v_{ds}}{\partial i_{ds}} = \frac{\partial v_{gs}}{\partial i_{ds}} = \frac{1}{g_m}. \tag{3.1}$$

Therefore, a useful approximation valid for low frequencies, small signals, and negligible substrate effects is that the device behaves like a resistor of value $1/g_m$. Any number of *n*- and *p*-channel devices and resistors can be combined to form a voltage divider, as shown in Fig. 3.2, where both *n*- and *p*-channel transistors are used

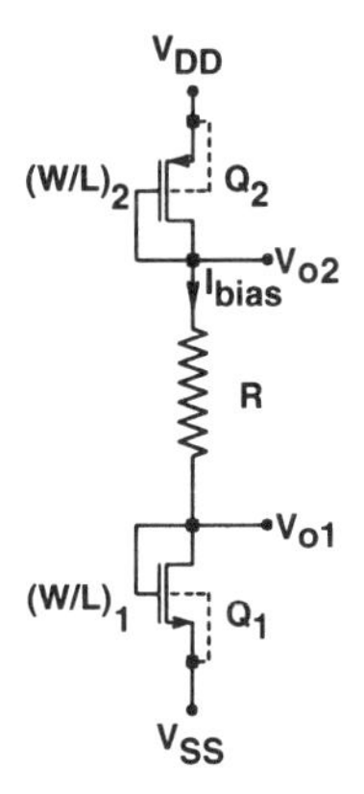

Figure 3.2. Voltage-divider-based bias circuit.

and $V_{SS} = 0$ is chosen. Since $V_{GS} = V_{DS}$ here for both devices, the condition for saturation,

$$V_{DS} > V_{GS} - V_T, \tag{3.2}$$

is satisfied. Hence the common value of the drain currents is given approximately by

$$\begin{aligned} I_{\text{bias}} &= k_n' \left(\frac{W}{L}\right)_1 (V_{o1} - V_{Tn})^2 \\ &= k_p' \left(\frac{W}{L}\right)_2 [(V_{o2} - V_{DD}) - V_{Tp}]^2 \\ &= \frac{V_{o2} - V_{o1}}{R} \end{aligned} \tag{3.3}$$

Here I_{bias} is usually specified; then V_{o1} and V_{o2} can be selected and Eq. (3.3) used to find the W/L ratios of the devices and the value R of the resistor.

An undesirable feature of this configuration is that the bias voltages and current depend on the supply voltages V_{DD} and V_{SS}. In fact, the bias current increases rapidly with increasing power supply voltage. Since such bias strings are used to provide bias for other devices in the circuit, the dc power consumption of the overall circuit becomes heavily dependent on the supply voltages.

A CMOS circuit with (theoretically) perfect supply independence is shown in Fig. 3.3. If Q_3 and Q_4 are matched transistors so that $(W/L)_3 = (W/L)_4$, they ideally carry equal currents. Choosing $(W/L)_1 = (W/L)_2$ will then result in $V_{GS1} = V_{GS2}$,

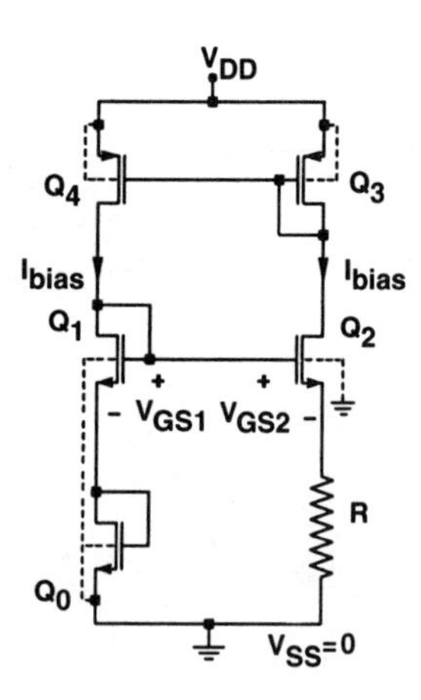

Figure 3.3. V_T-referenced supply-independent CMOS bias source.

and the voltage across the resistor, $I_{\text{bias}}R$, will be equal to V_{GS0}. This equilibrium condition leads to the equation

$$RI_{\text{bias}} = V_{GS0} = \sqrt{\frac{I_{\text{bias}}}{k'(W/L)}} + V_T. \tag{3.4}$$

Equation (3.4), which is independent of V_{DD}, can be solved to obtain I_{bias}. Note that in this analysis we neglected the effects of channel-length modulation (i.e., we assumed that I_D is independent of V_{DS}).

An alternative version of the bias circuit shown in Fig. 3.3 that uses the base–emitter voltage (V_{BE}) of a bipolar transistor as the reference voltage is shown in Fig. 3.4*a*. In a CMOS process, the substrate, well, and the source–drain junction inside the well can be used to form a vertical bipolar transistor. For example, Fig. 3.4*b*

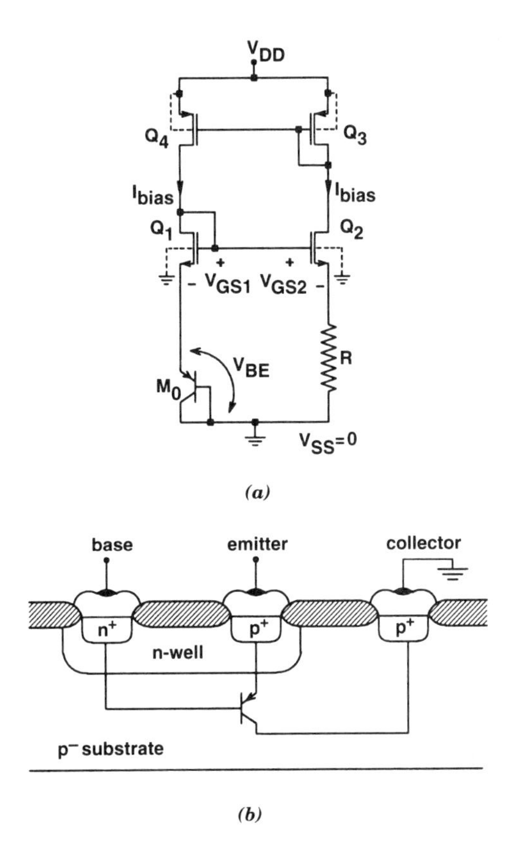

Figure 3.4. (*a*) V_{BE}-based supply-independent bias circuit; (*b*) vertical *pnp* bipolar transistor in an *n*-well CMOS process.

shows a vertical *pnp* device that is formed in an *n*-well process. The collector of this *pnp* device is the p^- substrate that is permanently connected to the most negative voltage on the chip. For the bipolar transistor, the collector current is given by

$$I_c = I_s e^{V_{BE}/V_T}, \tag{3.5}$$

where V_{BE} is the base–emitter voltage, $V_T = kT/q$, and I_s is a constant current, which is proportional to the cross-sectional area of the emitter, which is used to describe the transfer characteristic of the transistor in the forward active region.

In Fig. 3.4*a*, as in Fig. 3.3, if $(W/L)_1 = (W/L)_2$ and $(W/L)_3 = (W/L)_4$, equal currents are forced through the two branches of the bias circuit and the voltage drop across resistor R equals V_{BE}. Thus the bias current is given by

$$I_{\text{bias}} = \frac{V_{BE}}{R}. \tag{3.6}$$

Combining Eqs. (3.5) and (3.6), we have

$$I_{\text{bias}} = \frac{V_T}{R} \ln \frac{I_{\text{bias}}}{I_s}. \tag{3.7}$$

This equation can be solved iteratively for I_{bias}.

Both supply-independent bias circuits have a second trivial steady-state condition, cutoff when $I_{\text{bias}} = 0$. To prevent the bias circuit from settling to the wrong steady-state condition, a startup circuit is necessary in all practical applications. The circuit to the right of the dashed line in Fig. 3.5 functions as a startup circuit. If $I_{\text{bias}} = 0$, Q_5 is off and the voltage at node A is high, causing Q_6 to turn on and draw a current through Q_3, forcing the circuit to move to its other equilibrium state. Once the circuit settles in the desired state, Q_5 turns on and node A goes low, turning off Q_6. At this state the startup circuit is essentially out of the picture.

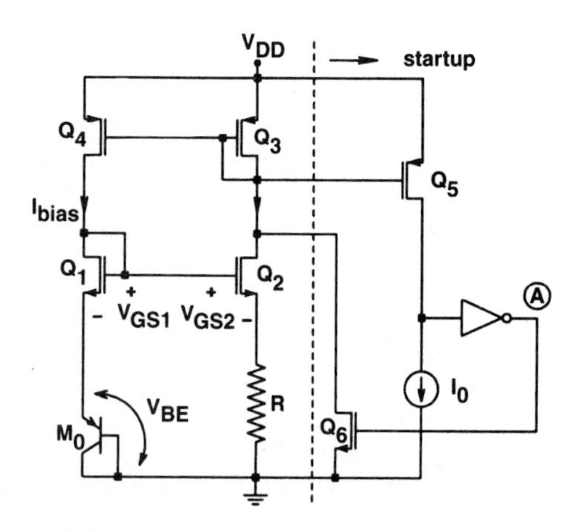

Figure 3.5. Supply-independent bias circuit with startup.

Another important performance aspect of the bias circuits is their temperature dependence. Unfortunately, supply-independent bias circuits are not necessarily temperature independent, because the base–emitter voltage (V_{BE}), and gate-to-source voltage (V_{GS}) are both temperature dependent. If the temperature coefficient T_{CF} is defined as the relative change of the bias current per degree Celsius temperature variation, we have [1]

$$T_{CF} = \frac{1}{I_{\text{bias}}} \frac{\partial I_{\text{bias}}}{\partial T} . \tag{3.8}$$

Using the definition above and Eq. (3.6), the relative temperature coefficient of the V_{BE}-based bias generator can be calculated:

$$T_{\text{CF}} = \left(\frac{\partial V_{BE}}{\partial T} \frac{1}{R} - \frac{V_{BE}}{R^2} \frac{\partial R}{\partial T} \right) \frac{1}{I_{\text{bias}}} \tag{3.9}$$

$$T_{CF} = \frac{1}{V_{BE}} \frac{\partial V_{BE}}{\partial T} - \frac{1}{R} \frac{\partial R}{\partial T} . \tag{3.10}$$

Since the temperature coefficient of the base–emitter junction voltage is negative (-2 mV/°C) while resistors typically have a positive temperature coefficient, the two terms in Eq. (3.9) add, resulting in a net T_{CF} that is quite high. The temperature behavior of the threshold-based bias generator of Fig. 3.3 is similar to the V_{BE}-based circuit.

An alternative supply-independent bias generator is the ΔV_{BE}-based circuit shown in Fig. 3.6*a*. The operation of this circuit is based on the difference between the base–emitter voltages of two transistors operated at different current densities. In Fig. 3.6*a*, as in Figs. 3.3 and 3.4, $(W/L)_1 = (W/L)_2$ and $(W/L)_3 = (W/L)_4$. Therefore, equal currents flow through the two branches of the circuit and $V_{GS1} = V_{GS2}$. Also, the *pnp* transistor M_1, has an emitter area that is m times the emitter area of, M_0. The voltage across the resistor R is $\Delta V_{BE} = V_{BE0} - V_{BE1}$. From Eq. (3.5),

$$I_{\text{bias}} = I_s e^{V_{BE0}/V_T},$$

$$I_{\text{bias}} = m I_s e^{V_{BE1}/V_T}, \tag{3.11}$$

and hence

$$\Delta V_{BE} = V_{BE0} - V_{BE1} = V_T \ln (m). \tag{3.12}$$

ΔV_{BE} appears across R and produces a current of value

$$I_{\text{bias}} = \frac{\Delta V_{BE}}{R} = \frac{V_T \ln m}{R} . \tag{3.13}$$

Obviously, the resulting bias current is independent of the power supply V_{DD}. This circuit also has two operating states: one at the desired operating current given by Eq. (3.13) and the other at zero. To prevent the circuit from operating in the cutoff state, a startup circuit similar to the one shown in Fig. 3.5 is required.

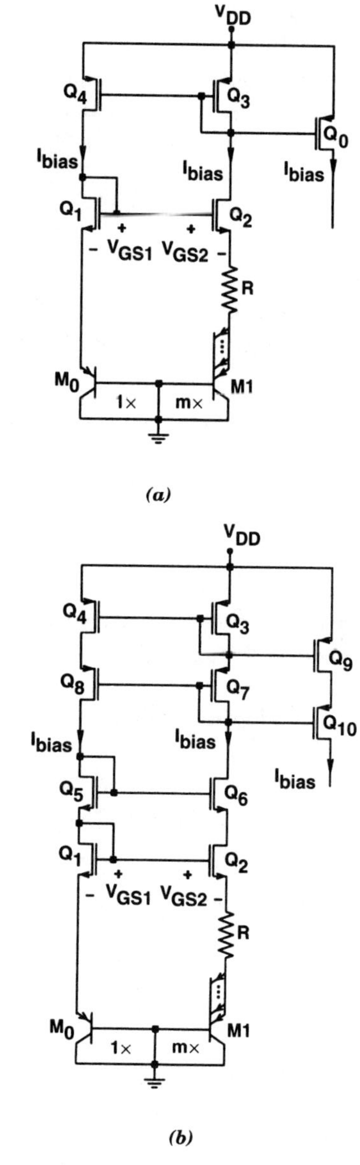

Figure 3.6. (*a*) ΔV_{BE}-based supply independent bias generator; (*b*) high-performance ΔV_{BE}-based supply-independent bias generator.

The temperature coefficient of the bias current can be calculated from Eq. (3.13):

$$T_{CF} = \left(\frac{\partial V_T}{\partial T} \frac{\ln m}{R} - \frac{V_T \ln m}{R^2} \frac{\partial R}{\partial T} \right) \frac{1}{I_{\text{bias}}} \tag{3.14}$$

$$T_{CF} = \frac{1}{V_T} \frac{\partial V_T}{\partial T} - \frac{1}{R} \frac{\partial R}{\partial T} \, . \tag{3.15}$$

Since $\partial V_T/\partial T$ and $\partial R/\partial T$ are both positive, the two terms in the temperature coefficients tend to cancel each other. Thus compared to V_{BE}-or threshold-based bias circuits, the ΔV_{BE}-based bias circuit can produce a much smaller temperature coefficient.

One drawback of the ΔV_{BE}-based bias generator is the strong dependence of I_{bias} on the mismatches between Q_3–Q_4 and Q_1–Q_2 device pairs. The mismatch between Q_3–Q_4 will result in different currents to flow in the two branches of the circuit. If $1 + \epsilon$ represents the ratio of the two currents, ΔV_{BE} will become

$$\Delta V_{BE} = V_T \ln\,[m(1 + \epsilon)], \tag{3.16}$$

which is equivalent to modifying m, the ratio of the emitter areas, by $1 + \epsilon$. The mismatch between Q_1 and Q_2 and the current difference due to the mismatch of Q_3 and Q_4 will make the V_{GS} values of Q_1 and Q_2 different. This is equivalent to a dc offset voltage $V_{os} = \Delta V_{GS}$, which modifies Eq. (3.13) to

$$I_{\text{bias}} = \frac{V_T \ln\,[m(1 + \epsilon)] - \Delta V_{\text{GS}}}{R}. \tag{3.17}$$

Assuming that $m = 8$, $\epsilon = 0.01$, and $V_T = 26$ mV at room temperature, $\Delta V_{BE} = 26 \times \ln[8(1 + 0.01)] = 54.3$ mV results. For a $\Delta V_{GS} = 5$ mV offset voltage, from Eq. (3.17), I_{bias} will change by 10%. To reduce this variation special care should be taken in the layout of Q_1–Q_4. For better geometrical matching, these devices should use a common-centroid layout strategy [2].

The current-matching accuracy of the bias generator of Fig. 3.6*a* is further degraded due to the mismatch between the drain-to-source voltages of Q_3–Q_4 and Q_1–Q_2 transistor pairs. The circuit can be made symmetrical, and the drain-to-source voltage drops equalized, by adding transistors Q_5 to Q_8 to the two branches of the circuit (Fig. 3.6*b*). The improved configuration also uses the cascode current mirror principle, described in Section 3.2, to improve the power supply rejection. On the other hand, the minimum power supply voltage is increased compared to the circuit of Fig. 3.6*a*, due to the extra voltage drops required by the two cascode devices. This becomes a major shortcoming in advanced submicron process technologies, or in low-power/low-voltage applications where the power supply voltage is limited to 3.3 V.

3.2. MOS CURRENT MIRRORS AND CURRENT SOURCES

As will be seen in later sections, constant current sources and current mirrors are important components in MOS amplifiers. The MOS current sources are quite similar to the bipolar sources [1,3], where the current mirrors work on the principle that identical devices with equal gate-to-source and drain-to-source voltages carry equal

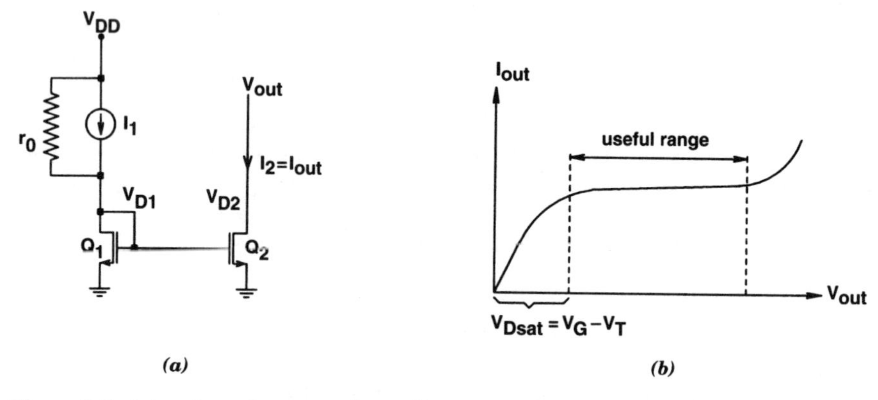

Figure 3.7. (*a*) *n*-channel current mirror; (*b*) output voltage–current relationship and useful output range of the current source.

drain currents. An NMOS realization of a current mirror is shown in Fig. 3.7*a*. In this circuit Q_1 is forced to carry a current I_1, since its input resistance at its shorted gate–drain terminals is low (much lower than r_0), and its gate potential V_{D1} adjusts accordingly. If Q_1 and Q_2 are in saturation, their drain currents I_1 and I_2 are determined to a large extent by their V_{GS} values. Since $V_{GS1} = V_{GS2} = V_G$, the condition

$$\frac{I_1}{I_2} \approx \frac{k_1}{k_2} \tag{3.18}$$

will therefore hold. More accurately, since drain saturation current is given by

$$i_p = k' \frac{W}{L}(V_G - V_T)^2(1 + \lambda V_D), \qquad V_G > V_T, \tag{3.19}$$

if the transistors have the same λ, k', and V_T, then

$$\frac{I_1}{I_2} = \frac{(W/L)_1}{(W/L)_2}\frac{1 + \lambda V_{D1}}{1 + \lambda V_{D2}}. \tag{3.20}$$

The current I_1 is thus ''mirrored'' in I_2.

Using Eq. (3.19) and ignoring the effect of λ, the gate voltage V_G is given by

$$V_G = V_T + \sqrt{\frac{i_D}{k'(W/L)}}. \tag{3.21}$$

For the transistors to operate in the saturation region, $V_D \geq V_G - V_T$ must hold. Using Eq. (3.21), therefore, the drain saturation voltage is

$$V_{\text{Dsat}} = \sqrt{\frac{i_D}{k'(W/L)}}, \tag{3.22}$$

where $V_{D\text{sat}}$ is the minimum drain voltage that keeps the transistors in saturation.

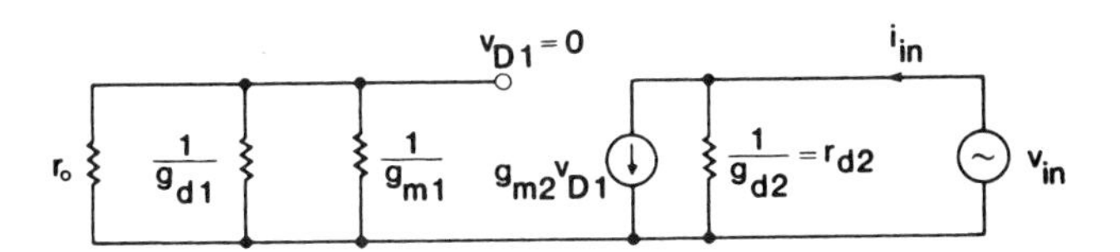

Figure 3.8. Small-signal equivalent circuit of the MOS current mirror.

For the MOS current source of Fig. 3.7*a*, the output voltage v_{out} has to be greater than $V_{D\text{sat}}$ to keep Q_2 in the saturation region. Figure 3.7*b* shows the output voltage–current relationship of the current source of Fig. 3.7*a*.

For small-signal analysis, the equivalent circuit of Fig. 2.14 can be used to model Q_1 and Q_2. The resulting circuit is shown in Fig. 3.8. Here r_o is the incremental resistance of the current source I_1 and v_{in} is the test voltage connected to the drain of Q_2 for measuring the output impedance of the circuit. The small-signal output impedance is simply

$$r_{\text{out}} = \frac{v_i}{i_{\text{in}}} \approx r_{d2} = \frac{1}{g_{d2}} = \frac{1 + \lambda V_{D2}^0}{\lambda i_{D2}^0}, \tag{3.23}$$

where Eq. (3.19) was used.

Clearly, the current source of Fig. 3.7*a* has an output impedance that is not better (i.e., higher) than the output impedance of a simple MOS transistor, also inaccurate current matching due to the variation of the drain voltage of Q_2 with the output voltage, and finally a reasonably wide output voltage range, which is limited at the lower end by $V_{D\text{sat}}$.

The output impedance r_{out} can be increased, and thus the circuit made to perform more like an ideal current source, by adding one more device and modifying the connections slightly. The resulting circuit (Fig. 3.9) is the MOS equivalent of *Wilson's current source*[1,3]. In this circuit, if I_2 increases, Q_2 causes v_1 to become larger. This results in a drop of v_3 which then counteracts the increase of I_2. Thus a negative feedback loop exists, which tries to hold I_2 constant. The small-signal

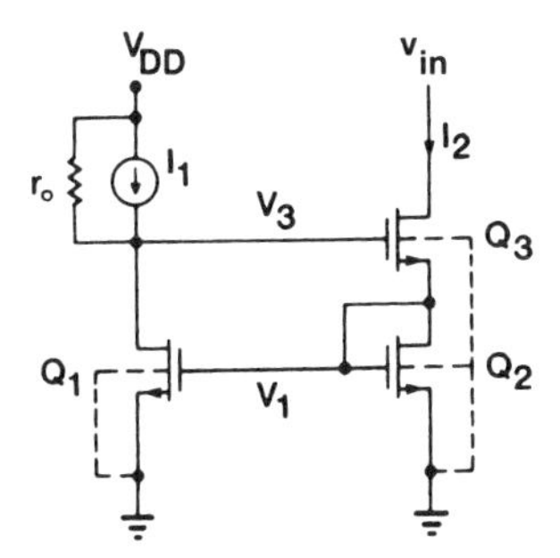

Figure 3.9. MOS version of Wilson's current source.

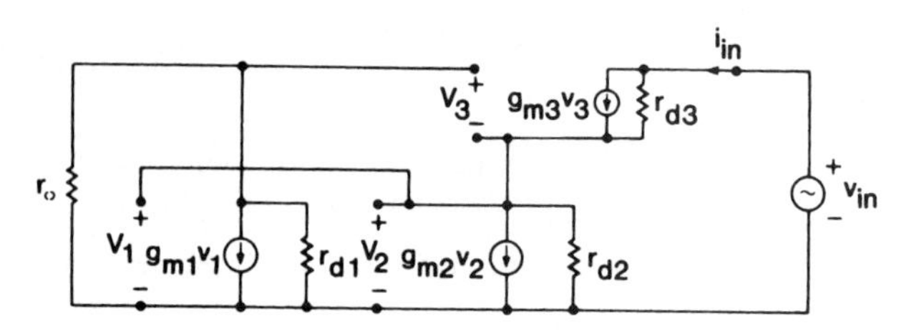

Figure 3.10. Small-signal equivalent circuit of Wilson's current source.

equivalent circuit is shown in Fig. 3.10; a simplified circuit is shown in Fig. 3.11. The latter was obtained by combining r_o and r_{d1} into $r_1 = (r_o^{-1} + r_{d1}^{-1})^{-1}$, by replacing the self-controlled current source $g_{m2}v_2$ by a resistor $1/g_{m2}$, and by neglecting r_{d2}, which is now parallel with the (usually much smaller) resistor $1/g_{m2}$. Solving for i_{in} in Fig. 3.11 gives

$$i_{in} = \frac{v_{in}}{r_{d3} + [1 + g_{m3}r_{d3}(1 + g_{m1}r_1)]/g_{m2}}. \tag{3.24}$$

Typical values for g_m are around 1 mA/V, while r_d is on the order of hundreds of kiloohms. Hence $g_{m3}r_{dB} \gg 1$ and for reasonably large r_o, $g_{m1}r_1 \gg 1$. Then, from Eq. (3.24),

$$r_{out} = \frac{v_{in}}{i_{in}} \approx \frac{g_{m1}r_1g_{m3}r_{d3}}{g_{m2}} = g_{m1}r_1 \frac{g_{m3}}{g_{m2}} r_{d3}. \tag{3.25}$$

Here, on the right-hand side the value of the first factor is around 100, while that of the second is around 1. Hence the output impedance r_{out} is two orders of magnitude larger than r_{d3}. The output impedance of the current source drops as soon as transistor Q_3 enters the linear (triode) region. The minimum level of the output voltage swing of the Wilson's current source is thus limited to $V_{min} = V_{GS2} + V_{Dsat3}$. In summary, the characteristics of Wilson's current source are the following: high output impedance, restricted output voltage swing, and poor current-matching accuracy, due to the mismatch between the drain-to-source voltages of transistors Q_1 and Q_2.

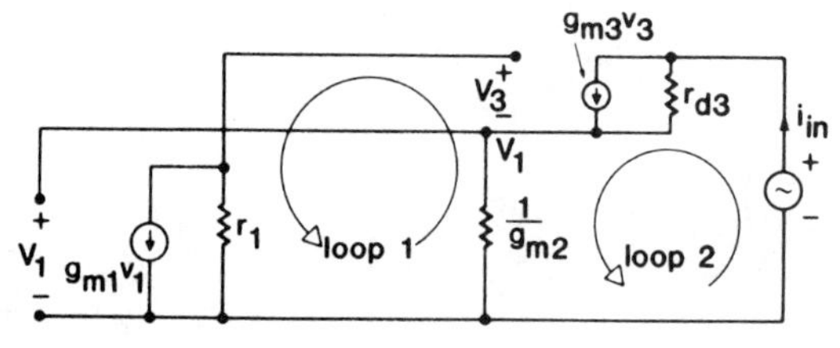

Figure 3.11. Simplified small-signal equivalent circuit of Wilson's current source.

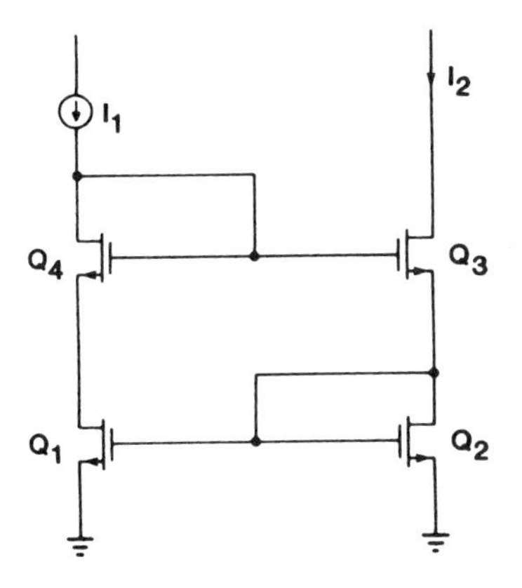

Figure 3.12. Improved MOS Wilson's current source.

The circuit can be made symmetrical, and the drain–source voltage drops of Q_1 and Q_2 equalized, by adding another transistor, Q_4 (Fig. 3.12). It can easily be shown that the output impedance of the resulting *improved* MOS *Wilson's current source* is again given by Eq. (3.25). The detailed analysis of the circuit is left to the reader (Problem 3.6a).

A slightly better version of the circuit of Fig. 3.12 (often called *cascode current source*) is shown in Fig. 3.13*a*. Its small-signal equivalent circuit is shown in Fig. 3.13*b* and in a simplified form in Fig. 3.13*c*. This circuit also uses feedback to maintain I_2 constant, and it also equalizes the drain potentials of Q_1 and Q_2, thus improving the current-matching properties of the current mirror. It can easily be shown (Problem 3.6b) that now

$$r_{\text{out}} = \frac{v_{\text{in}}}{i_{\text{in}}} = r_{d2} + r_{d3} + g_{m3}r_{d3}r_{d2} \approx (g_{m3}r_{d2})r_{d3}. \tag{3.26}$$

Hence there is again a 100-fold increase over the single-MOSFET output resistance. In addition, now the internal impedance r_o of the current source I_1 is in parallel with $1/g_{m1} + 1/g_{m4}$, a low input impedance, rather than with r_{d1}. Hence its loading effect is much reduced.

The cascode current source is similar to the improved Wilson's current source: It is characterized by high output impedance and accurate current mirroring capability. However, a common disadvantage of both circuits is that the minimum level of the output voltage swing is higher than that of the simple current mirror of Fig. 3.7. This reduces the available voltage swing of the stage(s) driven by the mirror. For the circuit of Fig. 3.13, the minimum voltage swing before Q_3 makes the transition from the saturation region to the linear region is

$$V_{\min} = V_{GS2} + V_{D\text{sat}3} = V_T + 2V_{D\text{sat}}. \tag{3.27}$$

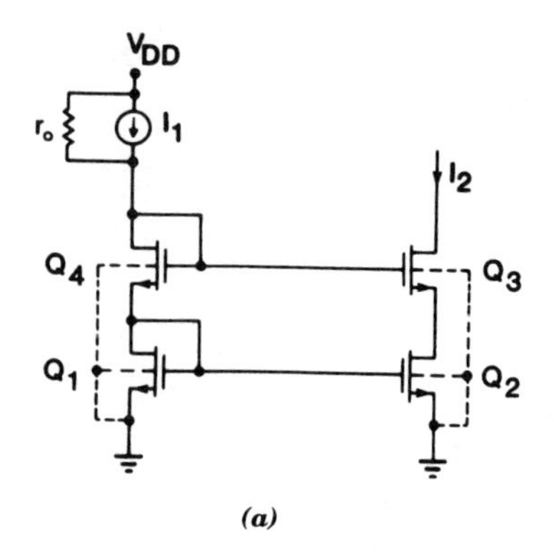

(a)

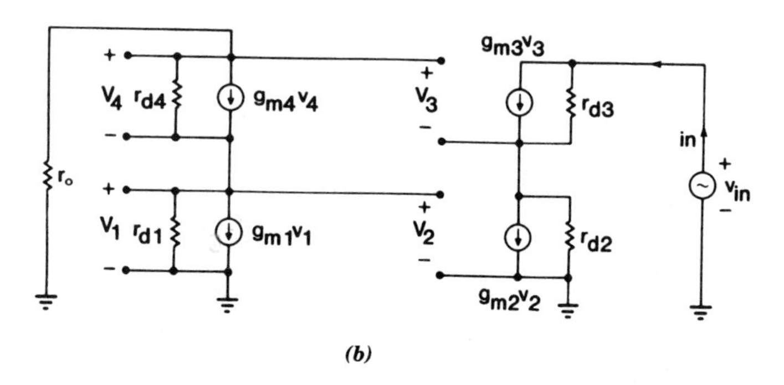

(b)

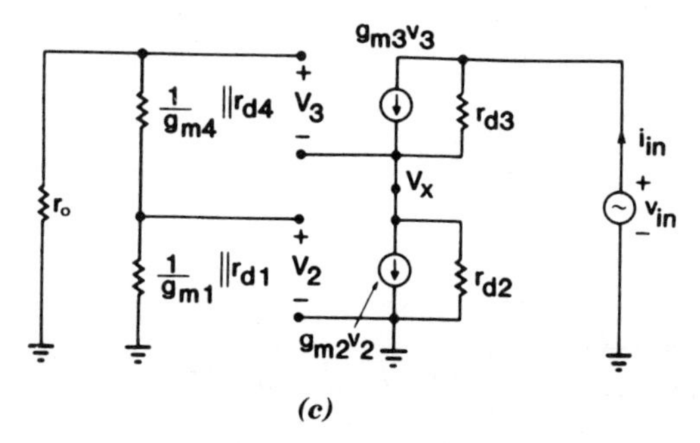

(c)

Figure 3.13. (*a*) Cascode current source; (*b*) equivalent small-signal circuit of the cascode current source; (*c*) simplified small-signal circuit of the cascode current source.

The output voltage–current plot for the cascode current source is shown in Fig. 3.14. The plot shows three operating regions. In the region where both Q_2 and Q_3 are in saturation ($V_D \geq V_T + 2V_{D\text{sat}}$) the output impedance has the highest value. In the region where Q_2 is in saturation and Q_3 is the linear region ($2V_{D\text{sat}} \leq V_D < V_T + 2V_{D\text{sat}}$), the output impedance is substantially lower. When $V_D < 2V_{D\text{sat}}$, both transistors are in their linear regions, and the output impedance is very low.

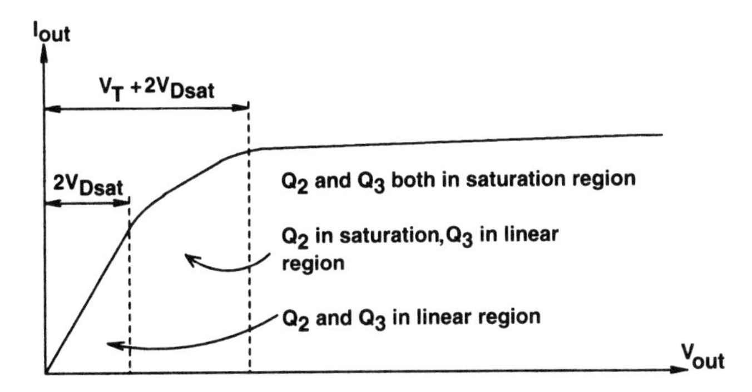

Figure 3.14. Output voltage–current relationship for cascode current source.

As mentioned earlier, a major drawback of the cascode current source is its limited output voltage swing. As Eq. (3.27) shows, the minimum voltage at the output of the current source that keeps both Q_2 and Q_3 in saturation is now $V_T + 2V_{Dsat}$. This loss of voltage swing is especially important when the current source is used as the load of a gain stage. To reduce the loss and increase the output swing, we can bias the drain of the lower transistor Q_2 at the edge of the saturation region. The resulting high-swing cascode current source is shown in Fig. 3.15 [4].

In this circuit a source follower (Q_5 and Q_6) has been inserted between the gates of Q_3 and Q_4 in Fig. 3.13*a*, and the *W/L* ratio of Q_3 has been reduced by a factor of 4. Using the simplified MOS *I–V* equation $I_D = k' \, W/L(V_{GS} - V_T)^2$ for the saturation region, and the *W/L* ratios given in Fig. 3.15, we have

$$V_{GS3} = V_T + 2V_{Dsat} \tag{3.28}$$

$$V_{GS1} = V_{GS4} = V_{GS6} = V_T + V_{Dsat}$$

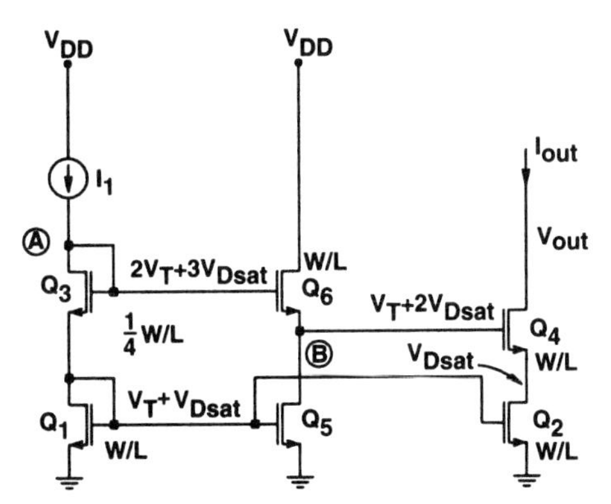

Figure 3.15. High-swing cascode current source.

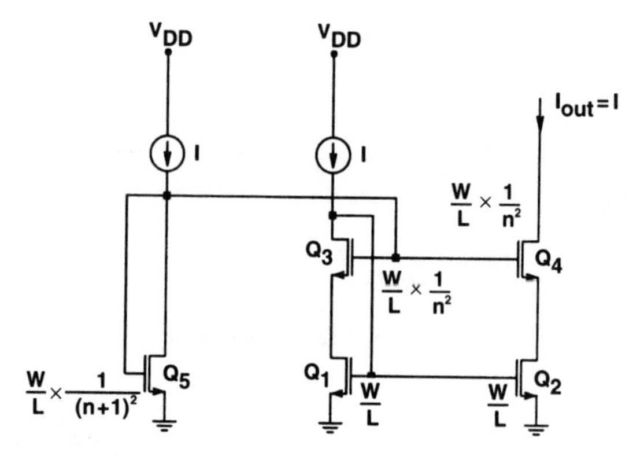

Figure 3.16. High-swing improved current source.

where $V_{Dsat} = \sqrt{I/(k'W/L)}$. From Fig. 3.15, the node voltages V_A, V_B, and V_{DS2} can be found:

$$\begin{aligned} V_A &= V_{GS1} + V_{GS3} = 2V_T + 3V_{Dsat}, \\ V_B &= V_A - V_{GS6} = V_T + 2V_{Dsat}, \\ V_{DS2} &= V_B - V_{GS4} = V_{Dsat}. \end{aligned} \tag{3.29}$$

Q_2 is therefore biased at the edge of saturation. The lowest level of the output voltage swing is now limited to $2V_{Dsat}$, which is a major improvement compared to that of the circuit of Fig. 3.13*a*. The high-swing cascode current source exhibits a high output impedance, similar to that of the cascode current source, and an improved output voltage swing. The current matching, however, suffers due to the mismatch in the drain-to-source voltages of the mirror transistors Q_1 and Q_2. For Q_1, $V_{DS1} = V_T + V_{Dsat}$, while for Q_2, $V_{DS2} = V_{Dsat}$.

An improved high-swing cascode current source, along with its biasing circuit, is shown in Fig. 3.16. Here n is a positive-integer number [5, p. 560]. Once again, using the simplified MOS I–V equation results in

$$V_{GS1} = V_{GS3} = V_T + \sqrt{\frac{I}{k'(W/L)}} = V_T + (V_{Dsat})_{W/L}, \tag{3.30}$$

where $(V_{Dsat})_{W/L}$ is the minimum drain-to-source voltage required to keep devices Q_1 and Q_3 with a current I and aspect ratio W/L in saturation. For devices Q_3 and Q_5 we have

$$\begin{aligned} V_{GS3} &= V_{GS4} = V_T + n\sqrt{\frac{I}{k'(W/L)}} = V_T + n(V_{Dsat})_{W/L}, \\ V_{GS5} &= V_T + (n+1)\sqrt{\frac{I}{k'(W/L)}} = V_T + (n+1)(V_{Dsat})_{W/L}. \end{aligned} \tag{3.31}$$

Using Eq. (3.31) yields

$$V_{DS1} = V_{DS2} = V_{GS5} - V_{GS3} \sqrt{\frac{I}{k'(W/L)}} = (V_{D\text{sat}})_{W/L}, \tag{3.32}$$

$$V_{DS3} = V_{GS1} - V_{DS1} = V_T.$$

Clearly, since $V_{DS1} = V_{DS2} = (V_{D\text{sat}})_{W/L}$, both Q_1 and Q_2 are biased at the edge of their saturation regions. Also, assuming that $V_{DS3} = V_T \geq V_{D\text{sat3}} = n(V_{D\text{sat}})_{W/L}$, Q_3 will be biased in the saturation region. As a result, the circuit will have a high output impedance. Notice that the output node v_{out} has high-swing capability. Actually, as long as v_{out} is greater than $(n + 1)(V_{D\text{sat}})_{W/L}$, the output will maintain its large output impedance. For the improved current mirror the devices Q_1 and Q_2 have equal V_{GS} as well as equal V_{DS} values and therefore good current-matching capability.

The formulas and numerical estimates given for the current sources are somewhat optimistic since they neglect the body effect of the floating devices (transistors Q_3 and Q_4 in Figs. 3.13 to 3.16). Also, real MOS transistors do not display an abrupt transition from the saturation to linear region. Therefore, it is necessary to bias the drain voltages of the mirror devices Q_1 and Q_2 slightly above the ideal saturation voltage produced by $V_{D\text{sat}} = \sqrt{I/(k'W/L)}$ to achieve the high output impedance derived earlier.

3.3. MOS GAIN STAGES [6–8]

A simple NMOS gain stage with resistive load is shown in Fig. 3.17. Q_1 is biased so that it operates in its saturation region. The low-frequency small-signal equivalent circuit is shown in Fig. 3.18. The voltage gain of the stage is clearly

$$A_v = \frac{v_{\text{out}}}{v_{\text{in}}} = -g_{m1} \frac{R_L\, r_{d1}}{R_L + r_{d1}}. \tag{3.33}$$

In integrated-circuit realization, the resistor R_L is undesirable since it occupies a large area and introduces a large voltage drop and hence is usually replaced by a

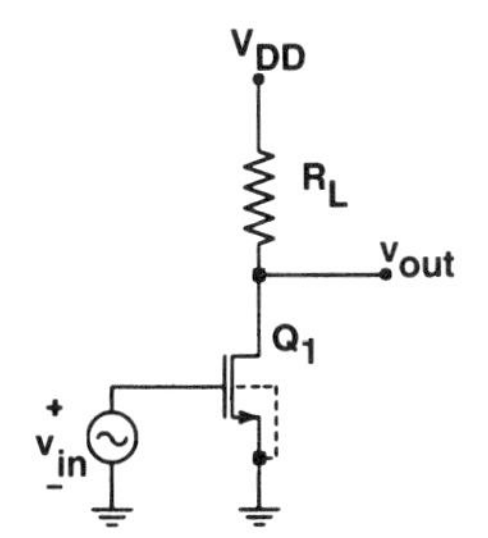

Figure 3.17. Resistive-load MOS gain stage.

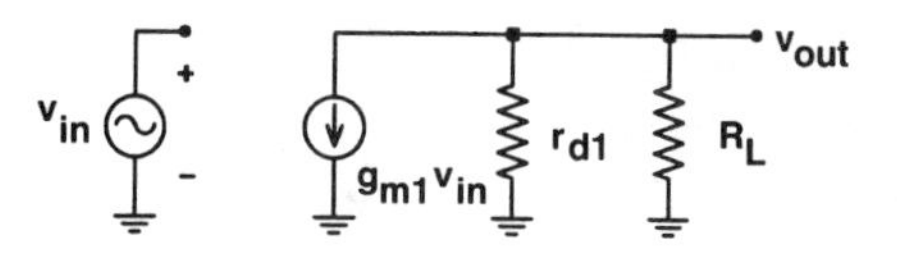

Figure 3.18. Small-signal low-frequency equivalent circuit of the resistive-load gain stage.

second MOSFET. If an NMOS enhancement-mode device is used as a load, the circuit of Fig. 3.19 results. The drain and gate of Q_2 are shorted to ensure that $v_{ds} > v_{gs} - V_T$, and hence the device is in saturation. The small-signal equivalent circuit of the load device Q_2 alone is shown in Fig. 3.20. Here the voltage-controlled current source $g_m v_{ds}$ is across the voltage v_{ds}; hence it behaves simply as a resistor of value $1/g_m$. Similarly, since $v_{sb} = v_{\text{out}}$, the source $g_{mb}v_{sb}$ corresponds to a resistor $1/|g_{mb}|$ [recall that by Eq. (2.19), $g_{mb} < 0$!]. In conclusion, Q_2 behaves like a resistor of value $1/(g_{m2} + |g_{mb2}| + g_{d2})$. Replacing R_L in Fig. 3.17 by this resistor and neglecting g_{d1} and g_{d2} in comparison with $g_{m2} + |g_{mb2}|$ gives

$$A_v \simeq \frac{-g_{m1}}{g_{m2} + |g_{mb2}|}. \tag{3.34}$$

If the body effect is small, so that $|g_{mb2}| \ll g_{m2}$, then using Eq. (2.18),

$$\begin{aligned} A_v &\simeq -\frac{g_{m1}}{g_{m2}} = -\frac{2\sqrt{k_1(1 + \lambda_1 v_{DS1}^0)i_{D1}^0}}{2\sqrt{k_2(1 + \lambda_2 v_{DS2}^0)i_{D2}^0}} \\ &\simeq -\sqrt{\frac{k_1}{k_2}} = -\sqrt{\frac{(W/L)_1}{(W/L)_2}} \end{aligned} \tag{3.35}$$

results. Here the channel-modulation terms $\lambda_i v_{DS}^0$ were also neglected, and the relations $i_{D1} = i_{D2}$, and $k_1' = k_2'$ utilized.

The sad message conveyed by Eq. (3.35) is that a large gain can be obtained only if the aspect ratio W/L of Q_1 is many times that of Q_2. If, for example, a gain of 10 is required, $(W/L)_1 = 100(W/L)_2$ must hold. This is possible only if a large silicon area is used. In addition, the body effect also reduces the gain significantly.

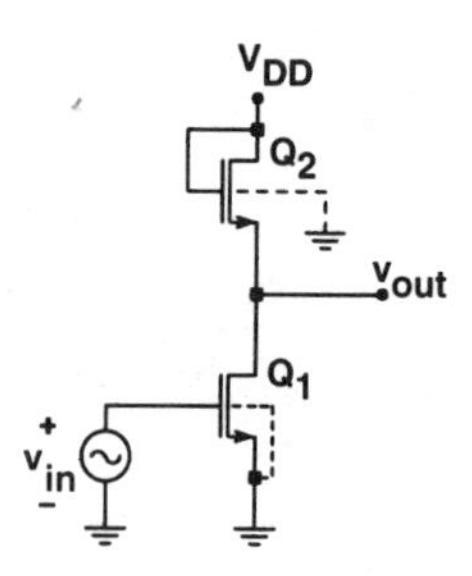

Figure 3.19. Enhancement-load NMOS gain stage.

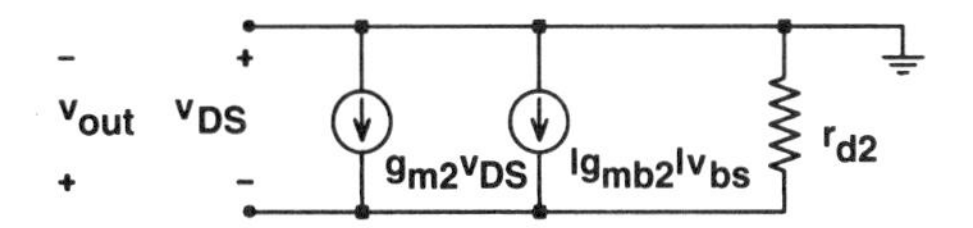

Figure 3.20. Small-signal equivalent circuit of the enhancement-load device Q_2.

Including body effect (but still neglecting channel-length modulation), using Eqs. (2.12), (2.19), and (3.34),

$$A_v \simeq \frac{-g_{m1}/g_{m2}}{1 + \gamma/(2\sqrt{2|\phi_p| + |v_{out}^0|})} \tag{3.36}$$

results. For $|\phi_p| = 0.3$ V, v_{out}^o, $= 5$ V, and $\gamma = 1$, the denominator is 1.21; hence the gain is reduced from 10 to 8.26.

In conclusion, the NMOS enhancement-load gain stage provides a low gain. This stage is nonetheless often useful in wideband amplifiers, where a low but predictable gain can be tolerated and the gain stage exhibits a wide bandwidth due to the low resistance of the load. For high-gain applications, however, the stage needs a large silicon area and since the load device has a high resistance (small W/L ratio), it has a large dc voltage drop across it which reduces the signal-handling capability and hence the dynamic range of the stage.

To improve the performance and increase the gain of the MOS amplifiers, a current–source load can be used. Any of the current sources described in the earlier sections can serve as a load. A common-source MOS gain stage that uses an NMOS input device and the p-channel version of the simple current source of Fig. 3.7*a* is shown in Fig. 3.21. The operation of this circuit is similar to the resistive-load gain stage of Fig. 3.17, but the resistive load is replaced by the small-signal output impedance r_{d2} of the PMOS current source. Using Eq. (3.33), the gain of the CMOS amplifies stage is

$$A_v = -g_{m1}\frac{r_{d1}r_{d2}}{r_{d1} + r_{d2}}. \tag{3.37}$$

For $g_{m1} = 0.2$ mA/V and $r_{d1} = r_{d2} = 1$ MΩ, the small-signal low-frequency gain is $A_v = -100$. Obviously, the gain is proportional to the transconductance of the

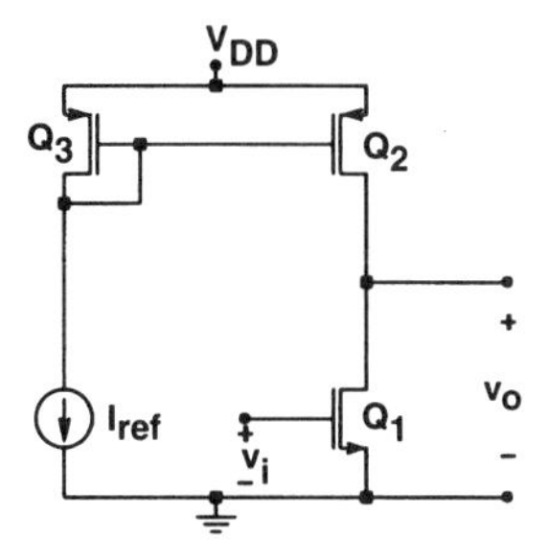

Figure 3.21. Common-source gain stage with NMOS input and p-channel current source as active load.

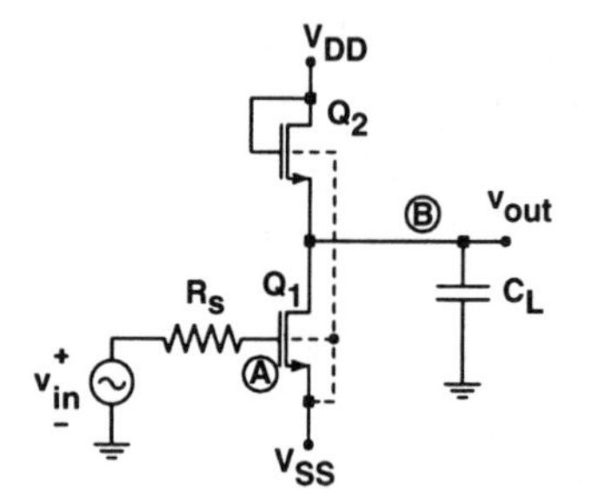

Figure 3.22. Enhancement-load gain stage with capacitive load.

input device and the small-signal output resistance of the stage $r_O = r_{d1} \parallel r_{d2}$. Since $r_d \gg 1/g_m$ for a given size of the load device, large values of gain can be achieved in a moderately small silicon area. Using a cascode current source with significantly higher output impedance will not increase the gain of the amplifier directly, because the output resistance of the stage is limited by the output impedance of the input device.

For the gain stage of Fig. 3.21 to operate properly, both input and load devices should operate in their saturation regions. The output signal swing is thus limited to $V_{DD} - |V_{D\text{sat}2}|$ and $V_{D\text{sat}1}$ on the positive and negative side, respectively. The output signal must therefore remain in the range

$$V_{D\text{sat}1} \leq v_{\text{out}} \leq V_{DD} - |V_{D\text{sat}2}|, \tag{3.38}$$

and the total output swing is

$$V_{\text{o,range}} = V_{DD} - |V_{D\text{sat}2}| - V_{D\text{sat}1}. \tag{3.39}$$

For high-frequency applications, all gain stages discussed so far have a common shortcoming. Consider the circuit of Fig. 3.22, which includes the source resistance R_S and the capacitive load C_L of the gain stage. Including the parasitic capacitances in the small-signal equivalent circuit, the diagram shown in Fig. 3.23*a* is obtained. Combining parallel-connected elements, the circuit of Fig. 3.23*b* results, where

$$\begin{aligned} G_{L\text{eq}} &= g_{d1} + g_{d2} + g_{m2} + |g_{mb2}|, \\ C_{L\text{eq}} &= C_{db1} + C_{gs2} + C_{sb2} + C_L. \end{aligned} \tag{3.40}$$

The node equations for nodes A and B are

$$\begin{aligned} (V_1 - V_{\text{in}})G_s + V_1 sC_{gs1} + (V_1 - V_{\text{out}})sC_{gd1} &= 0, \\ (V_{\text{out}} - V_1)sC_{gd1} + g_{m1}V_1 + V_{\text{out}}(G_{L\text{eq}} + sC_{L\text{eq}}) &= 0, \end{aligned} \tag{3.41}$$

where all voltages are Laplace-transformed functions.

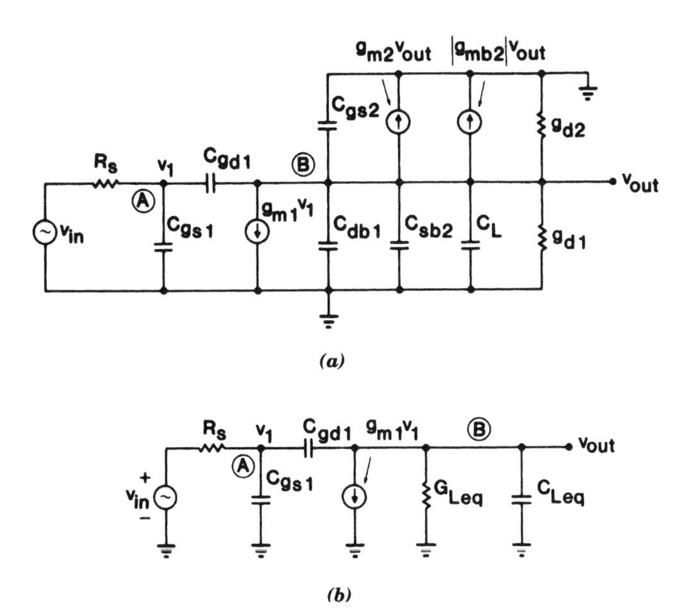

Figure 3.23. (*a*) Equivalent circuit of the MOS gain stage; (*b*) simplified equivalent circuit of the MOS gain stage.

Solving Eq. (3.41) gives

$$A_v(s) = \frac{V_{\text{out}}(s)}{V_{\text{in}}(s)}$$

$$= \frac{G_s(sC_{gd1} - g_{m1})}{[s(C_{gs1} + C_{gd1}) + G_s][G_{Leq} + s(C_{gd1} + C_{Leq})] - sC_{gd1}(sC_{gd1} - g_{m1})} . \tag{3.42}$$

To obtain the frequency response, s must be replaced by $j\omega$. For moderate frequencies, $g_{m1} \gg \omega C_{gd1}$, $G_{Leq} \gg \omega(C_{gd1} + C_{leq})$ hold. Then a good approximation is

$$A_v(j\omega) \approx \frac{-g_{m1}G_s}{G_sG_{Leq} + j\omega[G_{Leq}(C_{gs1} + C_{gd1}) + g_{m1}C_{gd1}]}$$

$$= \frac{-g_{m1}/G_{Leq}}{1 + j\omega R_s[C_{gs1} + C_{gd1}(1 + g_{m1}/G_{Leq})]} \tag{3.43}$$

$$= \frac{A_v^0}{1 + j\omega R_s C_{\text{in}}} .$$

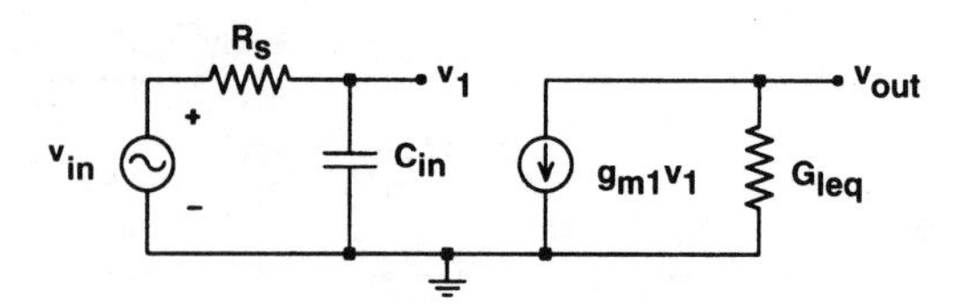

Figure 3.24. Approximate equivalent circuit of the MOS gain stage.

Here $A^0{}_v = -g_{m1}/G_{Leq}$ is the dc value of $A_v(j\omega)$, and

$$C_{in} = C_{gs1} + C_{gd1}\left(1 + \frac{g_{m1}}{G_{Leq}}\right) = C_{gs1} + C_{gd1}(1 + |A_v^0|). \tag{3.44}$$

$A_v(j\omega)$ in Eq. (3.43) can be recognized as the transfer function of the circuit shown in Fig. 3.24. Thus the capacitor C_{gd1} which is connected between the input and output terminals of the gain stage (Fig. 3.23*a*) behaves like a capacitance $(1 + |A_v^0|)$ times its real size, loading the input terminal. This is the well-known *Miller effect* [2]. For $|A_v^0|) \gg 1$, the high-frequency gain will be seriously affected and the bandwidth considerably reduced by this phenomenon.

To prevent the Miller effect, the *cascode gain stage* of Fig. 3.25 can be used. Here Q_2 is used to isolate the input and output nodes. It provides a low input resistance $1/g_{m2}$ at its source and a high one at its drain to drive Q_3. Ignoring the body effect, the low-frequency small-signal equivalent circuit is in the form shown in Fig. 3.26. Neglecting the small g_d admittances, clearly

$$g_{m1}\, v_{in} = -g_{m2}\, v_1 = -g_{m3}\, v_{out}. \tag{3.45}$$

Hence, for low frequencies,

$$\begin{aligned} v_1 &\simeq -\frac{g_{m1}}{g_{m2}} v_{in}, \\ v_{out} &\simeq \frac{g_{m2}}{g_{m3}} v_1 \simeq -\frac{g_{m1}}{g_{m3}} v_{in}. \end{aligned} \tag{3.46}$$

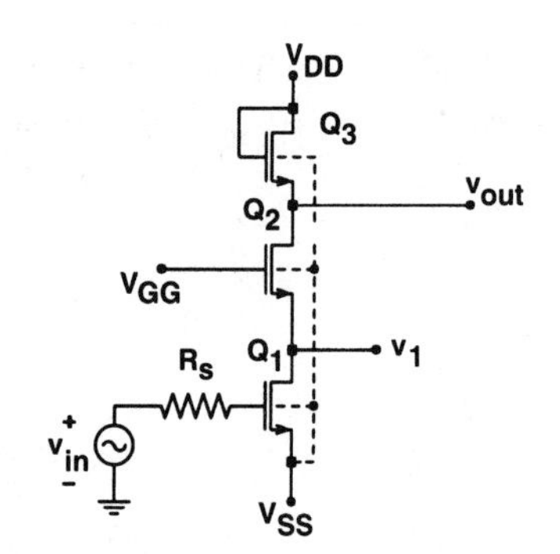

Figure 3.25. Cascode gain stage with enhancement load.

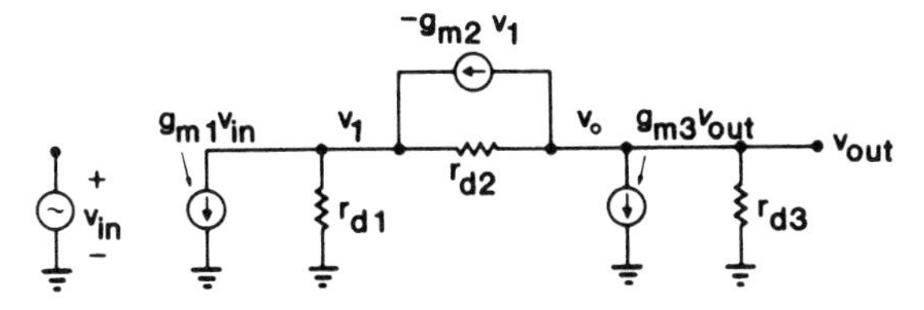

Figure 3.26. Low-frequency small-signal circuit of cascode gain stage.

The gate-to-drain gain of Q_1 is $-g_{m1}/g_{m2}$, and therefore the C_{gd1} of the driver transistor Q_1 is now multiplied by $(1 + g_{m1}/g_{m2})$. Choosing $g_{m1} = g_{m2}$, this factor is only 2. The overall voltage gain $-g_{m1}/g_{m3}$, however, can still be large, without introducing significant Miller effect, since there is no appreciable capacitance between the input and output terminals.

As before, the gain of the cascode gain stage can be increased by using a current mirror as an active load device. A cascode amplifier with a p-channel current source as active load is shown in Fig. 3.27. It can readily be shown that the low-frequency voltage gain of this circuit is given by

$$A_v = -g_{m1}(r_{\text{out1}} \parallel r_{\text{out2}}), \tag{3.47}$$

where, as derived earlier [cf. Eqs. (3.23) and (3.26)],

$$\begin{aligned} r_{\text{out1}} &= r_{o3}, \\ r_{\text{out2}} &= r_{o1} + r_{o2}(1 + g_{m2}r_{o1}) \approx r_{o2}g_{m2}r_{o1}. \end{aligned} \tag{3.48}$$

The value of r_{out2} is much larger than r_{out1} because of the local feedback. The effective output impedance is therefore given by

$$r_0 = r_{\text{out1}} \parallel r_{\text{out2}} \sim r_{\text{out1}} = r_{o3} \tag{3.49}$$

and the gain of the stage is

$$A_v \approx -g_{m1}r_{o3}. \tag{3.50}$$

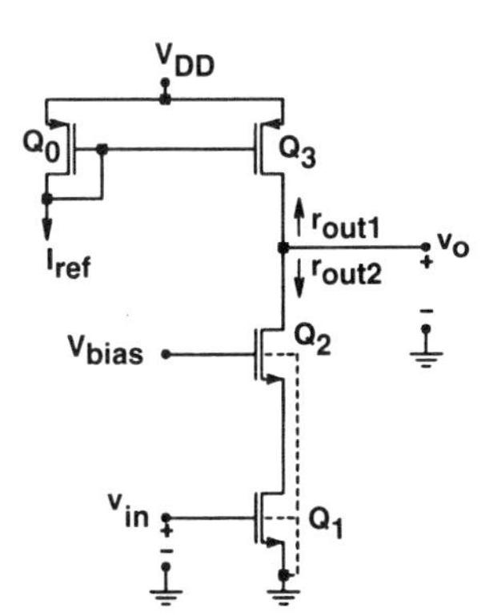

Figure 3.27. Cascode stage with p-channel current source as active load.

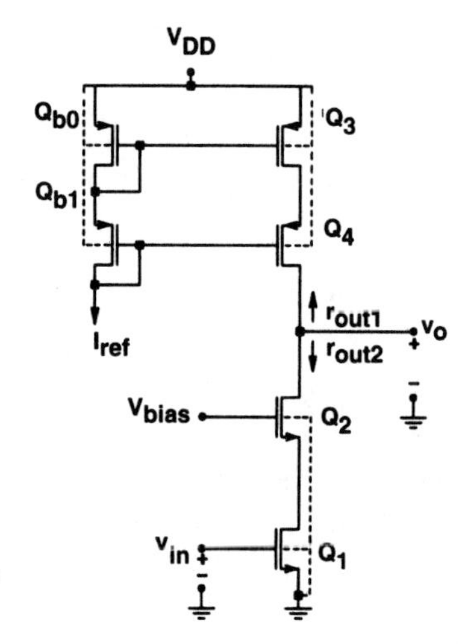

Figure 3.28. Cascode stage with *p*-channel cascode current source as active load.

To exploit fully the availability of the high output impedance r_{out2} for high gain, a cascode current source must be used for the active load. A cascode gain stage with a *p*-channel cascode current source as active load is shown in Fig. 3.28. The value of the load resistance r_{out1} is given by

$$r_{\text{out1}} = r_{o3} + r_{o4}\,(1 + g_{m4}\,r_{o3}) \approx r_{o4}\,r_{o3}\,g_{m4}. \tag{3.51}$$

Assuming that $r_{\text{out1}} = r_{\text{out2}}$, we have

$$r_o = r_{\text{out1}} \parallel r_{\text{out2}} = \frac{1}{2}\,r_{\text{out1}} = \frac{1}{2}\,r_{o4}\,r_{o3}\,g_{m4}, \tag{3.52}$$

and the gain is given by

$$A_v = -\frac{1}{2}\,g_{m1}\,r_{o4}\,g_{m4}\,r_{o3}. \tag{3.53}$$

Since r_o is much larger than r_{o3}, the gain A_v can be much larger than the gain of a stage with a simple current source.

To improve the output signal swing, a biasing strategy similar to the high-swing cascode current source should be used, so that both Q_1 and Q_3 are biased only slightly above the saturation region. The maximum output signal swing is given by

$$V_{D\text{sat1}} + V_{D\text{sat2}} \leq V_{\text{out}} \leq V_{DD} - \left|V_{D\text{sat3}} + V_{D\text{sat4}}\right|. \tag{3.54}$$

The use of a cascode current source as an active load in the gain stage of Fig. 3.28 provides a high voltage gain that is sufficient for the majority of the applications.

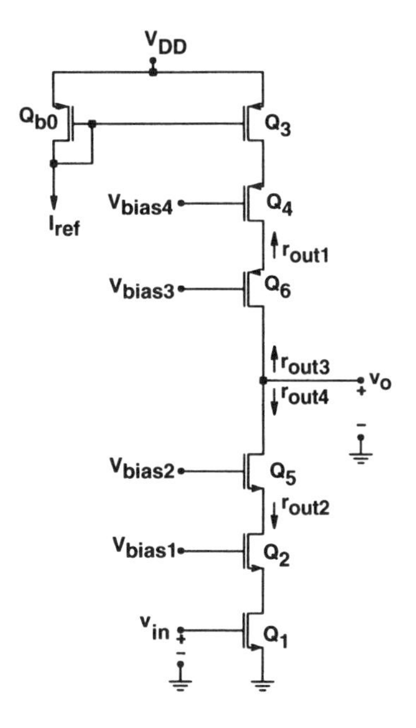

Figure 3.29. Gain stage with double-cascode input and double-cascode load.

However, in cases where an extremely high gain is required, the cascoding concept for the current sources can be extended by stacking more devices in the cascode configuration, which increases the output impedance and hence the voltage gain. A double cascode gain stage with double cascode active load is shown in Fig. 3.29. Compared to Fig. 3.28, two additional devices in a common-gate configuration have been added to the cascode input and cascode load. Using the results expressed in Eq. (3.51), the output impedance r_{out3} can be calculated as

$$r_{out3} \approx r_{out1}\, g_{m6}\, r_{o6}. \tag{3.55}$$

Combining Eqs. (3.51) and (3.55), we have

$$r_{out3} \approx r_{o3}\, r_{o4}\, g_{m4}\, g_{m6}\, r_{o6}. \tag{3.56}$$

Similarly, r_{out4} can be calculated as

$$r_{out4} \approx r_{o1}\, r_{o2}\, g_{m2}\, g_{m5}\, r_{o5}. \tag{3.57}$$

Once again if we assume that $r_{out3} \approx r_{out4}$, the voltage gain becomes

$$A_v = -\frac{1}{2}\, g_{m1}\, r_{o1}\, g_{m2}\, r_{o2}\, g_{m5}\, r_{o5}. \tag{3.58}$$

The resulting gain is quite high and is proportional to $g_m r_o$ raised to the third power. The high gain is achieved at the cost of reduced output swing. To maximize the output voltage swing, once again Q_1, Q_2, Q_3, and Q_4 should be biased slightly above the saturation region. The bias voltages are therefore given by

$$V_{\text{bias1}} = V_{GS2} + V_{D\text{sat1}}, \tag{3.59}$$

$$V_{\text{bias2}} = V_{GS5} + V_{D\text{sat2}} + V_{D\text{sat1}}. \tag{3.60}$$

Similar equations can be derived for V_{bias3} and V_{bias4}. Using these results, the maximum output signal swing is limited to the following range:

$$V_{D\text{sat1}} + V_{D\text{sat2}} + V_{D\text{sat5}} \leq v_{\text{out}} \leq V_{DD} - |V_{D\text{sat3}} + V_{D\text{sat4}} + V_{D\text{sat6}}|. \tag{3.61}$$

Another technique to improve the output impedance of the cascode current source of Fig. 3.28 is to place Q_2 in a feedback loop that senses the drain voltage of Q_1 and adjust the gate of Q_2 to minimize the variation of the drain current. A simplified form of this circuit is shown in Fig. 3.30, where the gain stage reduces the feedback from the drain of Q_2 to the drain of Q_1 [9]. Thus the output impedance of the circuit is increased by the gain of the additional gain stage, A:

$$r_{\text{out2}} \approx g_{m2} r_{o2} A r_{o1}. \tag{3.62}$$

Similarly, the output impedance r_{out1} can be calculated as

$$r_{\text{out1}} \approx g_{m4} r_{o4} A r_{o3}. \tag{3.63}$$

Assuming that $r_{\text{out1}} \approx r_{\text{out2}}$, the voltage gain of the stage is given by

$$A_v \approx -\frac{1}{2} g_{m1}\, g_{m2}\, r_{o2}\, r_{o1}\, A, \tag{3.64}$$

which has been increased by several orders of magnitude.

A simple implementation of the circuit of Fig. 3.30 is shown in Fig. 3.31, where the feedback amplifier is realized as a common-source amplifier consisting of transistor Q_5 and the corresponding current source I_{B1} [10]. The operation principle of the circuit of Fig. 3.31 is described briefly as follows. The transistor Q_1 converts the input voltage v_i into a drain current that flows through Q_1 and Q_2 to the output terminal. For high output impedance the drain voltage of Q_1 must be kept stable. This is accomplished by the feedback loop consisting of the amplifier (Q_5 and I_{B1}) and Q_2 as a source follower. In this way the drain–source voltage of Q_1 is regulated to a fixed value. The disadvantage of the circuit of Fig. 3.31 is that it limits the output swing, because the drain voltage of Q_1 is V_{GS5}, where as it can be as low as $V_{D\text{sat1}} = V_{GS1} - V_T$. Additionally, the gate-to-drain capacitance of Q_5 multiplied

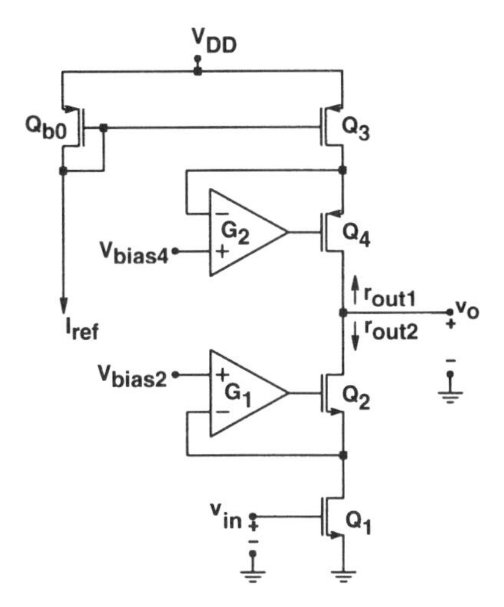

Figure 3.30. Cascode gain stage with enhanced output impedance.

by the gain of the feedback amplifier (Q_5 and I_{B1}) forms a low-frequency pole with $1/g_{m2}$, which degrades the high-frequency performance of the main amplifier stage.

Folded-cascode op amps, presented in Chapter 4, can be used for the feedback amplifier. One example of this type of implementation is shown in Fig. 3.32, where PMOS and NMOS input folded-cascode op-amps have been used to enhance the output impedance of the NMOS and PMOS cascode stages [9]. To maximize the output voltage swing, $V_{\text{bias1}} = V_{GS1} - V_{Tn}$ and $V_{\text{bias3}} = V_{DD} + (V_{GS4} - V_{Tp})$ should hold. Folded-cascode op-amps are described in detail in Chapter 4.

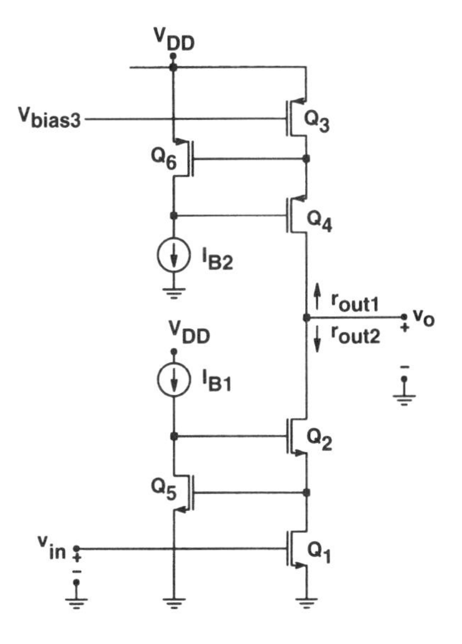

Figure 3.31. Simple implementation of Fig. 3.30.

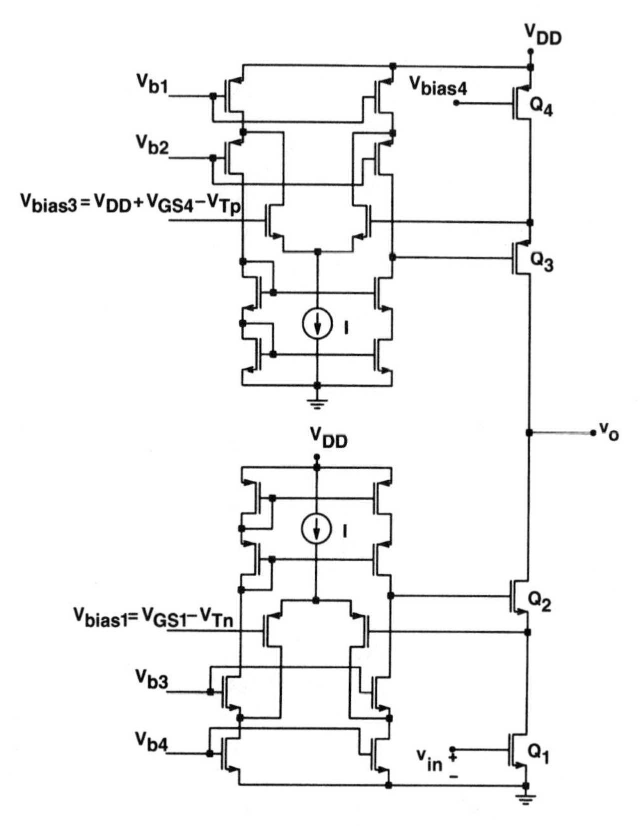

Figure 3.32. Complete circuit diagram of a gain-enhanced cascode amplifier employing folded-cascode op-amps as feedback amplifiers.

3.4. MOS SOURCE FOLLOWERS [5–8]

MOS source followers are similar to bipolar emitter followers. They can be used as buffers or as dc level shifters. The basic source follower, with an NMOS input device and an NMOS current source as an active load, is shown in Fig. 3.33 and its small-signal low-frequency equivalent circuit in Fig. 3.34. The node current equation for the output node is

$$(g_{d1} + g_{d2})v_{\text{out}} + |g_{mb1}|v_{\text{out}} - g_{m2}\, v_{gs1} = 0. \tag{3.65}$$

Substituting $v_{gs1} = v_{\text{in}} - v_{\text{out}}$ and solving yields

$$A_v = \frac{v_{\text{out}}}{v_{\text{in}}} = \frac{g_{m1}/(g_{d1} + g_{d2}|g_{mb1}|)}{g_{m1}/(g_{d1} + g_{d2} + |g_{mb1}|) + 1}. \tag{3.66}$$

Hence $A_v \approx 1$ if $g_{m1} \gg g_{d1} + g_{d2} + |g_{mb1}|$.

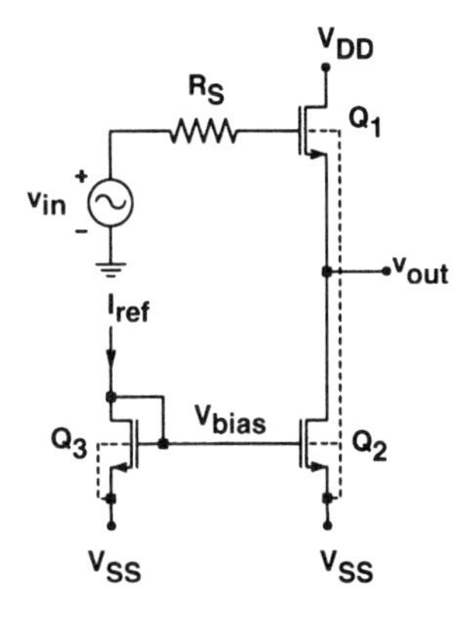

Figure 3.33. Basic structure of MOS source follower.

The output impedance of the source follower can be calculated by applying a test source v_x at its output (Fig. 3.35). The current law gives

$$i_x = (g_{d1} + g_{d2})v_x + |g_{mb1}|\, v_x - g_{m1}v_{gs1}. \tag{3.67}$$

Here $v_{gs1} = -v_x$, and hence Eq. (3.67) gives

$$R_{\text{out}} = \frac{v_x}{i_x} = \frac{1}{g_{d1} + g_{d2} + g_{m1} + |g_{mb1}|} \approx \frac{1}{g_{m1}} \tag{3.68}$$

since usually $g_{m1} \gg g_{d1}$, g_{d2}, and $|g_{mb1}|$. Thus R_{out} has a relatively low value, on the order of 1 kΩ.

The dc bias current of the stage is determined by the current source Q_2, which drives Q_1 at its low-impedance source terminal. Thus the dc drop V_{GS1} between the input and output terminals is determined by V_{bias} and the dimensions of Q_1 and Q_2; these parameters can be used to control the level shift provided by the stage.

The gate of the load device Q_2 may be connected to its drain to eliminate the

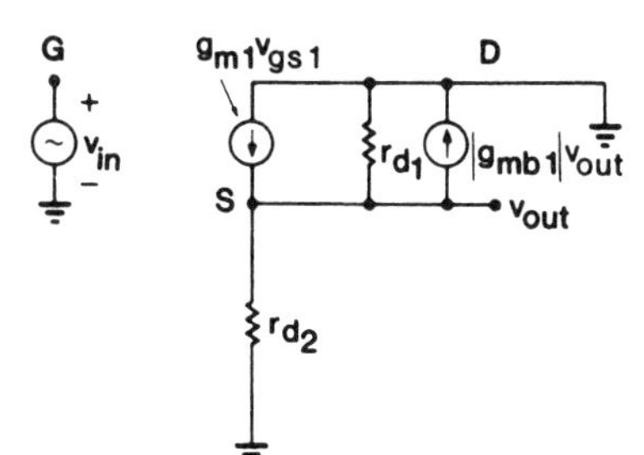

Figure 3.34. Small-signal low-frequency equivalent circuit of the source follower.

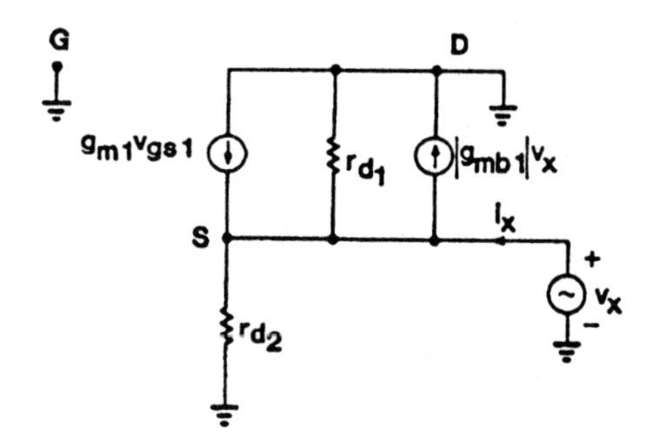

Figure 3.35. Equivalent circuit to calculatc the output impedance of the source follower.

gate bias voltage (Fig. 3.36). Analysis shows that for $g_{m2} \gg g_{d1}, |g_{mb1}|$, the voltage gain is then $(1 + g_{m2}/g_{m1})^{-1}$. This can be close to 1 only if $g_{m1}, \gg g_{m2}$, which as discussed in connection with Fig. 3.35, requires a large area for the stage. Hence this stage is rarely used.

The large-signal operation of the circuit can be analyzed simply if the load device is regarded as a current source. Figure 3.37 shows the redrawn circuit; r_o is the average large-signal output resistance of the current source and R_L is the load resistor. From Eq. (2.9), ignoring the body effect,

$$i_{D1} = I_o + v_{\text{out}}(g_o + G_L) = k_1(v_{\text{in}} - v_{\text{out}} - V_{T1})^2. \tag{3.69}$$

If $(g_o + G_L)/k_1 \ll 2|v_{\text{in}} - V_{T1}|$, then

$$v_{\text{out}} \simeq v_{\text{in}} - V_{T1} - \sqrt{\frac{I_o}{k_1}}, \tag{3.70}$$

so that the circuit operates as a linear buffer with a constant offset. To achieve this, $(W/L)_1$ must be sufficiently large.

A major disadvantage of this stage is the following. If $v_{\text{out}} < 0$, the load *supplies*

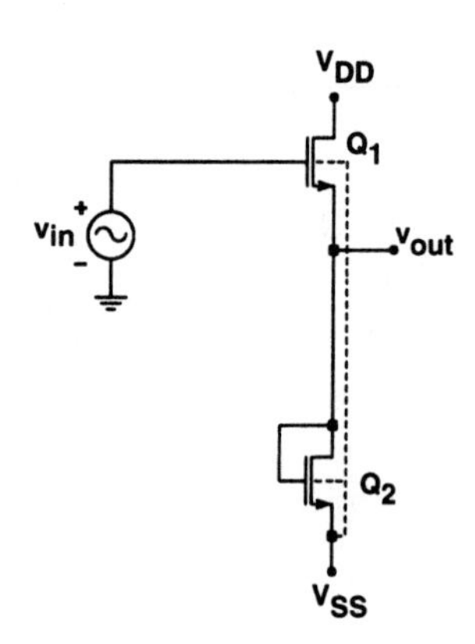

Figure 3.36. Enhancement-load source follower.

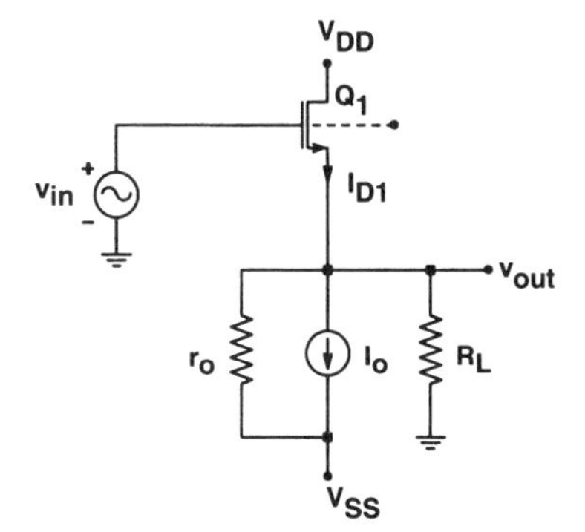

Figure 3.37. Source follower with a current source as load.

current to the output stage. However, the latter can sink (absorb) an output current only if it is less than I_o. This represents a serious limitation. Also, for $v_{out} > 0$, Q_1 must supply the output current *plus* I_o. In addition, there is a voltage drop greater then V_{T1} between the input and output terminals. Thus if v_{in} comes from a gain stage such as that shown in Fig. 3.19, where the output voltage must be less than $V_{DD} - V_{T2}$, the maximum positive output voltage swing is $V_{DD} - 2V_T$. The negative swing is limited by the requirement that the device(s) in the current source must remain in saturation for the smallest output voltage.

3.5. MOS DIFFERENTIAL AMPLIFIERS

The input stage of an operational amplifier must provide a high input impedance, large common-mode rejection ratio (CMRR) and power supply rejection ratio (PSRR), low dc offset voltage and noise, and most (or all) of the op-amp's voltage gain (accurate definitions of these terms are given in Chapter 4). The output signal of the input stage is much larger than the input one and so is no longer as sensitive to noise and offset voltage generated in the following stages. (Note that a large common-mode rejection is desirable even if the noninverting terminal is grounded in normal operation, to suppress noise in the ground line.)

The requirements above can often be met by using the source-coupled stage shown in Fig. 3.38. Since this circuit operates in a differential mode, it can provide high differential gain along with a low common-mode gain and hence ensure a large CMRR. The differential configuration also helps in achieving a large PSRR, since variations of V_{DD} are, to a large extent, canceled in the differential output voltage $v_{o1} - v_{o2}$.

An approximate analysis of the amplifier can readily be performed. We assume that the current source I is ideal, that is, that its internal conductance g is zero. We also assume ideal symmetry between Q_1 and Q_2 and Q_3 and Q_4, and all devices operate in saturation. Then the incremental drain currents satisfy $i_{d1} \approx g_{mi}(v_{in1} - v)$, $i_{d2} \approx g_{mi}(v_{in2} - v)$, and $i_{d1} + i_{d2} \approx 0$. This gives $v \approx (v_{in1} + v_{in2})/2$ for the

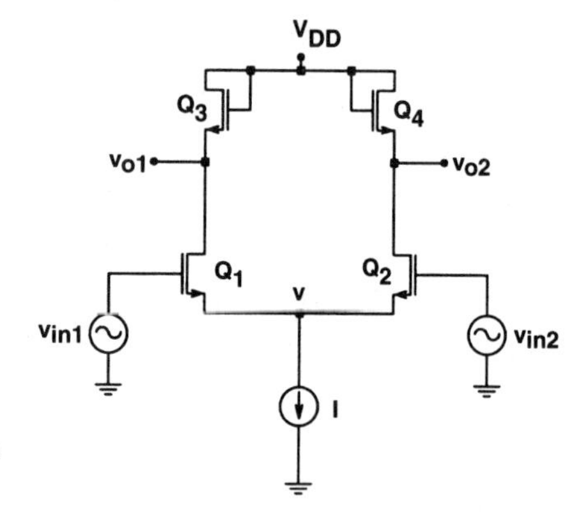

Figure 3.38. Source-coupled differential stage with diode-connected NMOS load devices.

source voltages of Q_1 and Q_2, and $i_{d1} \approx -i_{d2} \approx g_{mi}(v_{\text{in1}} - v_{\text{in2}})/2$ for their drain currents. Hence the output voltages are $v_{o1} \approx -v_{o2} = -i_{d1}/g_l = g_{mi}(v_{\text{in1}} - v_{\text{in2}})/2g_l$, where g_l is the load conductance. Defining the differential gain by $A_{dm} \triangleq (v_{o1} - v_{o2})/(v_{\text{in1}} - v_{\text{in2}})$, we obtain the simple result $A_{dm} \approx -g_{mi}/g_l$. Thus the differential gain is the same as for a simple inverter; however, the stage also provides a rejection of common-mode signals and of noise in the power supplies V_{DD} and V_{SS}, all of which are canceled (or, for actual circuits, reduced) by the differential operation of the stage. A more detailed analysis follows.

The low-frequency small-signal equivalent circuit of the source-coupled stage is shown in Fig. 3.39. In the circuit, the body-effect transconductances of the input devices Q_1 and Q_2 are ignored to simplify the discussions. It will also be assumed that the circuit is perfectly symmetrical, so that the parameters of Q_1 and Q_2 are identical, as are those of Q_3 and Q_4. The load conductance g_l of Q_3 and Q_4 can be found as was done in connection with Fig. 3.20: the result is

$$g_l = g_{m3} + g_{d3} + |g_{mb3}| = g_{m4} + g_{d4} + |g_{mb4}|. \tag{3.71}$$

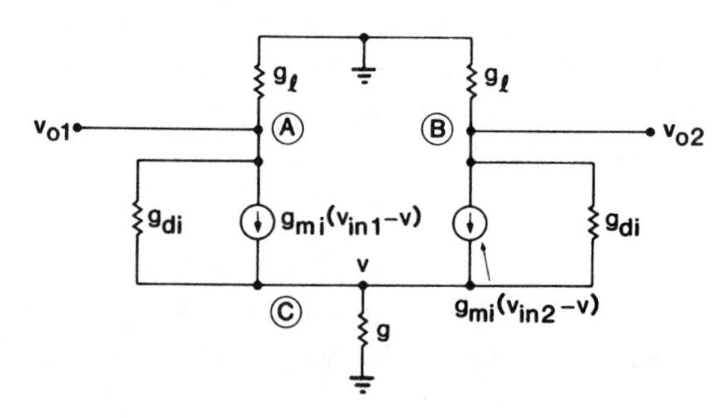

Figure 3.39. Small-signal equivalent circuit of the source-coupled pair.

Applying the current law at nodes A and B,

$$\begin{aligned} v_{o1}g_l + (v_{o1} - v)g_{di} + g_{mi}(v_{\text{in}1} - v) &= 0, \\ v_{o2}g_l + (v_{o2} - v)g_{di} + g_{mi}(v_{\text{in}2} - v) &= 0 \end{aligned} \tag{3.72}$$

result. (Here the subscripts i and l refer to the input and load devices, respectively.) The current law at node C gives

$$(v - v_{o1})g_{di} - g_{mi}(v_{\text{in}1} - v) + v_g + (v - v_{o2})g_{di} - (v_{\text{in}2} - v)g_{mi} = 0. \tag{3.73}$$

Equations (3.72) and (3.73) represent three equations in the three unknowns v_{o1}, v_{o2}, and v. Solving them for v_{o1} and v_{o2}, we get

$$\begin{aligned} v_{o1} &= -\frac{g_l g_{mi}(g_{di} + g_{mi})(v_{\text{in}1} - v_{\text{in}2}) + g g_{mi}(g_l + g_{di})v_{\text{in}1}}{(g_l + g_{di})[2g_l(g_{di} + g_{mi}) + g(g_l + g_{di})]}, \\ v_{o2} &= -\frac{g_l g_{mi}(g_{di} + g_{mi})(v_{\text{in}2} - v_{\text{in}1}) + g g_{mi}(g_l + g_{di})v_{\text{in}2}}{(g_l + g_{di})[2g_l(g_{di} + g_{mi}) + g(g_l + g_{di})]}. \end{aligned} \tag{3.74}$$

The differential and common-mode input voltages are

$$\begin{aligned} v_{\text{in},d} &= v_{\text{in}1} - v_{\text{in}2}, \\ v_{\text{in},c} &= \frac{v_{\text{in}1} + v_{\text{in}2}}{2}. \end{aligned} \tag{3.75}$$

The differential and common-mode output voltages can be defined similarly:

$$\begin{aligned} v_{o,d} &= v_{o1} - v_{o2}, \\ v_{o,c} &= \frac{v_{o1} + v_{o2}}{2}. \end{aligned} \tag{3.76}$$

Then the differential-mode gain can be obtained from Eq. (3.74):

$$\begin{aligned} A_{dm} &= \frac{v_{o,d}}{v_{\text{in},d}} = \frac{v_{o1} - v_{o2}}{v_{\text{in}1} - v_{\text{in}2}} \\ &= -\frac{2g_l g_{mi}(g_{di} + g_{mi}) + g g_{mi}(g_l + g_{di})}{(g_l + g_{di})[2g_l(g_{di} + g_{mi}) + g(g_l + g_{di})]}. \end{aligned} \tag{3.77}$$

For $g = 0$ and $g_{di} \ll g_l$, $A_{dm} \approx -g_{mi}/g_l$, as predicted earlier. Similarly, the common-mode gain can be found:

$$\begin{aligned} A_{cm} &= \frac{v_{o,c}}{v_{\text{in},c}} = \frac{(v_{o1} + v_{o2})/2}{(v_{\text{in}1} + v_{\text{in}2})/2} \\ &= -\frac{g g_{mi}}{2g_l(g_{di} + g_{mi}) + g(g_l + g_{di})}. \end{aligned} \tag{3.78}$$

Hence the common-mode rejection ratio is

$$\text{CMRR} = \left|\frac{A_{dm}}{A_{cm}}\right| = 1 + 2\frac{g_l}{g}\frac{g_{di} + g_{mi}}{g_{di} + g_l}. \tag{3.79}$$

Normally, g, $g_{di} \ll g_l$, g_{mi}, and approximations

$$\begin{aligned} A_{dm} &\approx -\frac{g_{mi}}{g_l}, \\ A_{cm} &\approx -\frac{g}{2g_l}, \\ \text{CMRR} &\approx -\frac{2g_{mi}}{g} \end{aligned} \tag{3.80}$$

can be used. Clearly, to obtain a large CMRR, g must be small; that is, the current source should have a large output impedance. The circuits described earlier and shown in Figs. 3.7 to 3.16 are suitable to achieve this. All, however, require a dc voltage drop for operation which limits the achievable output voltage swing.

As Eq. (3.80) indicates, with the described approximation (including the assumed absence of the body effect), the differential gain can be obtained from

$$A_{dm} \approx -\frac{g_{mi}}{g_{ml}} = -\sqrt{\frac{(W/L)_i}{(W/L)_l}}. \tag{3.81}$$

Obviously, a large gain can be achieved only if the aspect ratio $(W/L)_i$ of the input devices is many times that of the load devices $(W/L)_l$. A somewhat improved version of the differential stage of Fig. 3.38 can be obtained by using PMOS devices Q_3 and Q_4 as loads (Fig. 3.40). For this circuit Eqs. (3.77) to (3.80) remain valid; however, the differential gain A_{dm} is larger and is given by

$$A_{dm} \approx -\frac{g_{mi}}{g_{ml}} = -\sqrt{\frac{\mu_n(W/L)_i}{\mu_p(W/L)_l}}. \tag{3.82}$$

where μ_n and μ_p are the mobilities of the NMOS and PMOS devices, respectively.

The gain of the differential stage of Fig. 3.40 can be increased by using a con-

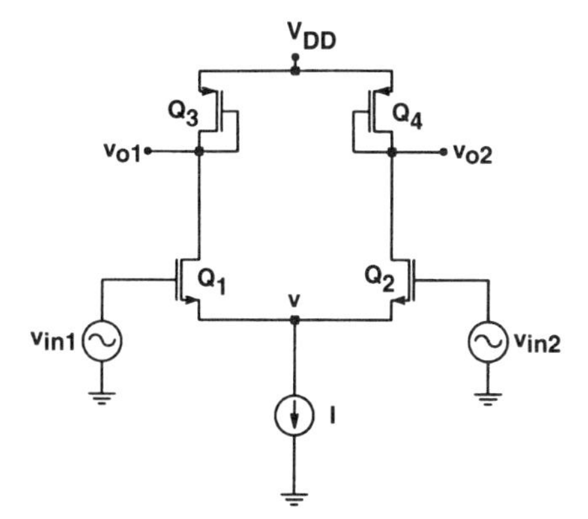

Figure 3.40. Source-coupled CMOS differential stage with diode-connected PMOS load devices.

trolled amount of positive feedback to increase the transconductance of the input device [11]. The resulting circuit is shown in Fig. 3.41 and the differential gain can be derived as

$$A_{dm} \approx -\frac{g_{mi}}{g_{ml}}(1 - \alpha)^{-1}, \tag{3.83}$$

where $\alpha = (W/L)_5/(W/L)_3$. As an example, if $\alpha = \frac{3}{4}$, the differential gain will be increased by a factor of 4.

All the differential stages described thus far have low gain and a differential output voltage. For high gain the circuit of Fig. 3.42*a* can be used. This circuit has differential input but single-ended output. Hence it performs as a combination of a differential gain stage and a differential-to-single-ended converter. In Fig. 3.42*a* transistors Q_1–Q_2 and Q_3–Q_4 form matched transistor pairs. They have equal *W/L* ratios. All current levels are determined by the current source I_o, half of which flows through Q_1–Q_3 and the other half flows through Q_2–Q_4. All transistors have their substrates connected to their sources to eliminate body effect and improve matching.

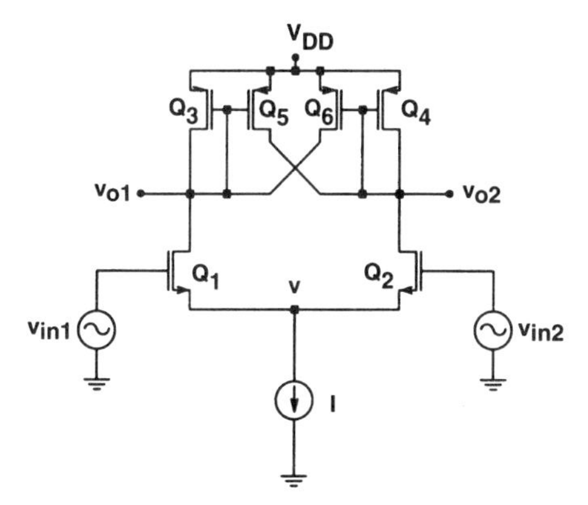

Figure 3.41. Source-coupled CMOS differential stage with positive feedback to increase gain.

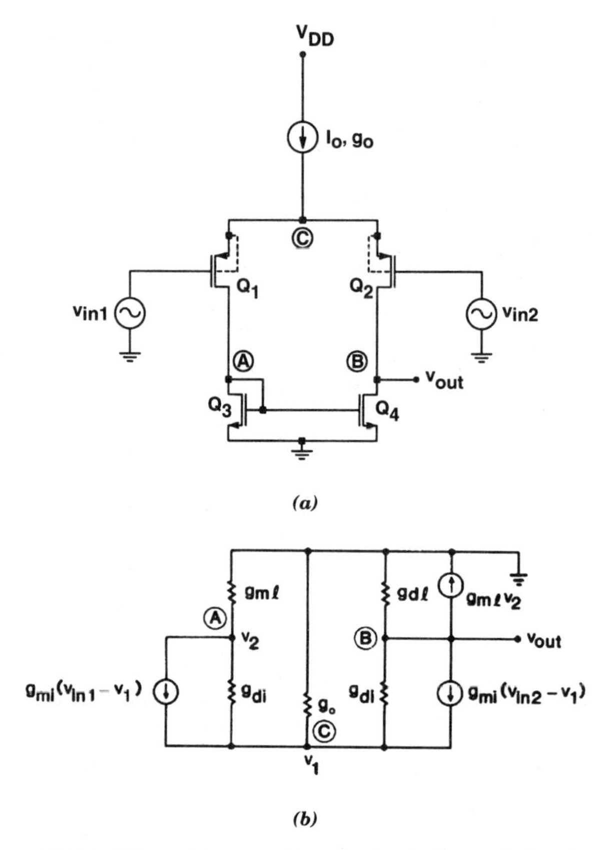

Figure 3.42. (*a*) CMOS differential stage with active load; (*b*) small-signal equivalent circuit for CMOS differential stage.

An approximate analysis of the circuit can readily be performed as follows. Assuming that the current source I_0 is ideal, the incremental drain currents of Q_1 and Q_2 must satisfy $i_{d1} + i_{d2} = 0$. Also, if both Q_1 and Q_2 are in saturation, then $i_{d1} \approx g_{mi}(v_{\text{in}1} - v_1)$ and $i_{d2} \approx g_{mi}(v_{\text{in}2} - v_1)$. Combining these equations, $v_1 \approx (v_{\text{in}1} + v_{\text{in}2})/2$ results. Hence $i_{d1} = -i_{d2} \approx g_{mi}(v_{\text{in}1} - v_{\text{in}1})/2$. The current i_{d1} is easily imposed on Q_3 by Q_1, since the impedance at the common terminal of the gate and drain of Q_3 is only $1/g_{m3}$.

Transistors Q_3 and Q_4 form a current mirror similar to that shown in Fig. 3.7*a*, and hence the current through Q_4 satisfies $i_{d4} = i_{d3} = i_{d1}$. Thus both Q_2 and Q_4 send a current $i_{d1} = g_{mi}(v_{\text{in}1} - v_{\text{in}2})/2$ into the output terminal. Since the output is loaded by the drain resistances of Q_2 and Q_4, the differential gain is thus $A_{dm} \triangleq v_{\text{out}}/(v_{\text{in}1} - v_{\text{in}2}) \approx g_{mi}/(g_{di} + g_{dl})$. A more exact analysis follows next.

The small-signal equivalent circuit of the stage is shown in Fig. 3.42*b*. It was drawn under the assumption that both input devices Q_1 and Q_2 have the same conductances g_{mi} and g_{di}, and that both load devices have the parameters g_{ml} and g_{dl}; also, that separate "wells" are provided for the PMOS devices. Then $v_{BS} = 0$ for all devices, and hence no body effect occurs. The output conductance of the current source is denoted by g_o.

Writing and solving the current law equations for nodes A, B, and C (Problem 3.11), we obtain

$$v_{\text{out}} = \frac{g_{mi}g_{ml}}{D}\left\{2(g_{di} + g_{mi})(v_{\text{in1}} - v_{\text{in2}}) + g_o\left[v_{\text{in1}} - \left(\frac{g_{di}}{g_{ml}} + 1\right)v_{\text{in2}}\right]\right\}, \tag{3.84}$$

where

$$D = (g_{di} + g_{mi})[g_{dl}g_{di} + 2g_{ml}(g_{dl} + g_{di})] + g_o(g_{di} + g_{ml})(g_{dl} + g_{di}).$$

We define, as before, the differential and common-mode input signals by Eq. (3.75). Then the differential gain A_{dm} and the common-mode gain A_{cm} can be defined by

$$v_{\text{out}} = A_{dm}v_{\text{in},d} + A_{cm}v_{\text{in},c}. \tag{3.85}$$

From Eqs. (3.84) and (3.85),

$$A_{dm} = \frac{g_{mi}g_{ml}}{D}\left[2(g_{di} + g_{mi}) + g_o\left(1 + \frac{g_{di}}{2g_{ml}}\right)\right], \tag{3.86}$$

$$A_{cm} = -\frac{g_{mi}g_{di}g_o}{D},$$

$$\text{CMRR} = \left|\frac{A_{dm}}{A_{cm}}\right| = \frac{g_{ml}}{g_{dl}g_o}[2(g_{di} + g_{mi}) + g_o] + \frac{1}{2}. \tag{3.87}$$

For g_{mi}, $g_{ml} \gg g_o$, g_{di}, and g_{dl}, the approximations

$$A_{dm} \approx \frac{g_{mi}}{g_{dl} + g_{di}},$$

$$A_{cm} \approx \frac{-g_o g_{di}}{2g_{ml}(g_{dl} + g_{di})}, \tag{3.88}$$

$$\text{CMRR} \approx 2\frac{g_{mi}g_{ml}}{g_o g_{di}}$$

can be used. Note that A_{dm} is the same as that obtainable from a CMOS differential-input/differential-output stage (Problem 3.12). Thus the single-ended output signal

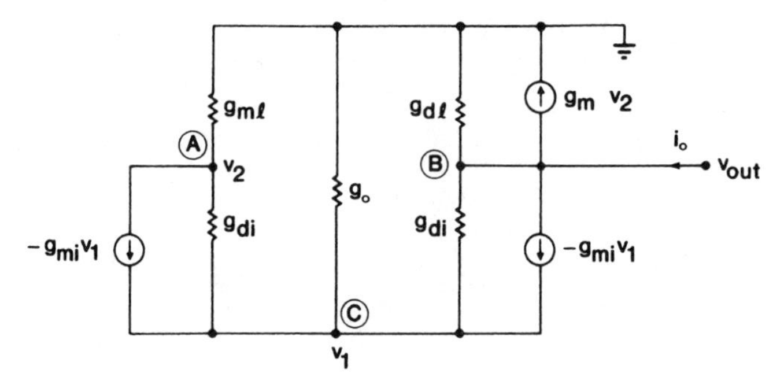

Figure 3.43. Equivalent circuit of the output impedance of the CMOS differential stage.

does not result in lower gain for the stage. By contrast, the CMRR is higher by a factor of g_{ml}/g_{di}, which (for usual values) is much greater than 1.

To calculate the small-signal output impedance of the CMOS stage, a test source i_o can be applied at the output of the small-signal equivalent circuit and the input voltages v_{in1} and v_{in2} set to zero (Fig. 3.43). Analysis shows (Problem 3.13) that

$$r_{\text{out}} = \frac{v_{\text{out}}}{i_o} = \frac{1}{D}\{(g_{di} + g_{ml})[2(g_{di} + g_{mi}) + g_o] - g_{di}(g_{mi} + g_{di})\} \tag{3.89}$$

where D is defined in Eq. (3.84).

For $g_{mi}, g_{ml} \gg g_o, g_{di}, g_{dl}$, the approximation

$$r_{\text{out}} \approx \frac{1}{g_{dl} + g_{di}} \tag{3.90}$$

can be used.

3.6 FREQUENCY RESPONSE OF MOS AMPLIFIER STAGES

In previous sections the linearized (small-signal) performance of MOS amplifier stages was analyzed at low frequencies. Thus the parasitic capacitances illustrated in the equivalent circuit of Fig. 2.16 were ignored. For high-frequency signals, however, the admittances of these branches are no longer negligible, and hence neither are the currents which they conduct. Then the gains and the input impedances of the various circuits all become functions of the signal frequency ω. These effects are analyzed next.

Consider again the all-NMOS single-ended amplifier (Fig. 3.22), discussed in Section 3.3. Using the equivalent circuit of Fig. 2.16, the high-frequency small-signal equivalent circuit of Fig. 3.23a resulted. In the circuit, R_s is the output impedance of

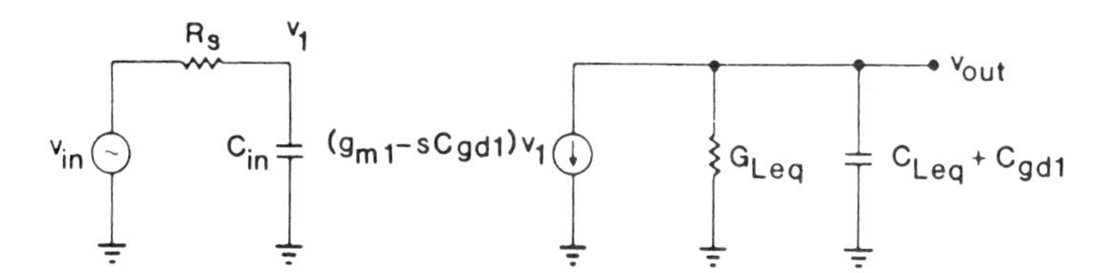

Figure 3.44. Simplified equivalent circuit of the MOS gain stage using Miller's theorem.

the signal source and C_L is the load capacitance. This circuit was then simplified to that of Fig. 3.23*b*, which (in the Laplace transform domain) was shown to have the frequency response given in Eq. (3.42). With the approximations $g_{m1} \gg \omega\, C_{gd1}$, $G_{Leq} \gg \omega\,(C_{gd1} + C_{Leq})$, the frequency response of Eq. (3.49) resulted. It corresponded to the simplified equivalent circuit of Fig. 3.24, which was then used to introduce the Miller effect.

The accuracy of the simplified circuit can be improved by restoring the two capacitances C_{Leq} and C_{gd1} which load the output node in the exact circuit of Fig. 3.23*b*. Also, in the numerator of Eq. (3.42), the term sC_{gd1} was neglected in comparison with g_{m1}. At higher frequencies, this is no longer justified. To restore the sC_{gd1} term, the gain of the controlled source g_{m1} can be changed to $g_{m1} - sC_{gd1}$ in the equivalent circuit. The resulting circuit is shown in Fig. 3.44. The corresponding transfer function is

$$A_v(s) \simeq \frac{G_s(sC_{gd1} - g_{m1})}{(sC_{\text{in}} + G_s)[s(C_{Leq} + C_{gd1}) + G_{Leq}]}, \tag{3.91}$$

where C_{in} is given by Eq. (3.44). This function has a right-half-plane (positive) real zero at

$$s_z = \frac{g_{m1}}{C_{gd1}} \tag{3.92}$$

and two left-half-plane (negative) poles at

$$\begin{aligned} s_{p1} &= -\frac{G_s}{C_{\text{in}}}, \\ s_{p2} &= -\frac{G_{Leq}}{(C_{leq} + C_{gd1})}. \end{aligned} \tag{3.93}$$

Normally, C_{gd1} is small. Hence $s_z \gg |s_{p1}|$, and if C_{Leq} is also small, then $|s_{p2}| \gg |s_{p1}|$. Then s_{p1} is closest to the $j\omega$ axis and is therefore the dominant pole of the circuit.

The frequency response $A_v(j\omega)$ can be obtained simply by replacing s by $j\omega$ in (3.91). It can be arranged in the form

$$A_v(j\omega) = \frac{G_s C_{gd1}}{C_{\text{in}}(C_{Leq} + C_{gd1})} \frac{j\omega - s_z}{(j\omega - s_{p1})(j\omega - s_{p2})} . \tag{3.94}$$

If $|s_{p1}| \ll |s_{p2}|$ and s_z, the 3-dB frequency (i.e., the frequency where $|A_v(j\omega)|$ is $1/\sqrt{2}$ times its dc value) is

$$\omega_{3\text{dB}} \simeq |s_{p1}| = \frac{G_s}{C_{\text{in}}} . \tag{3.95}$$

For high gain, an active (current-source) load can be used. A CMOS gain stage with a p-channel current source as active load and capacitive loading is shown in Fig. 3.45*a*; the corresponding high-frequency linear equivalent circuit in Fig. 3.45*b*. Defining

$$\begin{aligned} G_{Leq} &= g_{d1} + g_{d2}, \\ C_{Leq} &= C_{db1} + C_{db2} + C_{gd2} + C_L, \end{aligned} \tag{3.96}$$

the simplified circuit of Fig. 3.45*c* results.

Using the approximation yielding Fig. 3.44 for the NMOS gain stage, the circuit of Fig. 3.45*d* can be obtained. Obviously, the circuit in Fig. 3.45*d* is identical to that in Fig. 3.44*d*, and hence the transfer function of Eq. (3.91) is also valid for the circuit of Fig. 3.45*d*. The poles and the zero are also given by Eqs. (3.92) and (3.93). The dominant pole is normally $s_{p1} = -G_s/C_{\text{in}}$. Notice, however, that G_{Leq} in Fig. 3.45 is $g_{d1} + g_{d2}$, while in Fig. 3.44 it is $g_{d1} + g_{d2} + g_{m2} + g_{mb2}$. Since the former is much smaller, s_{p2} for the circuit of Fig. 3.45 is at a much lower frequency than that of Fig. 3.44.

Figure 3.46*a* shows a CMOS cascode gain stage with an active load. C_L represents the capacitive load of the stage. Figure 3.46*b* and *c* show the detailed and simplified high-frequency linearized equivalent circuits, respectively. Using Miller's theorem, the circuit of Fig. 3.46*d* results, where

$$\begin{aligned} g_2 &= g_{m2} + \frac{1}{r_{d1}}, \\ C_1 &= C_{gs1} + \left(1 + \frac{g_{m1}}{g_{m2}}\right) C_{gd1}, \\ C_2 &= C_{gd1} + C_{db1} + C_{gs2} + C_{sb2}, \\ C_{Leq} &= C_L + C_{gd2} + C_{db2} + C_{db3} + C_{gd3}, \\ g'_{m1} &= g_{m1} - sC_{gd1} . \end{aligned} \tag{3.97}$$

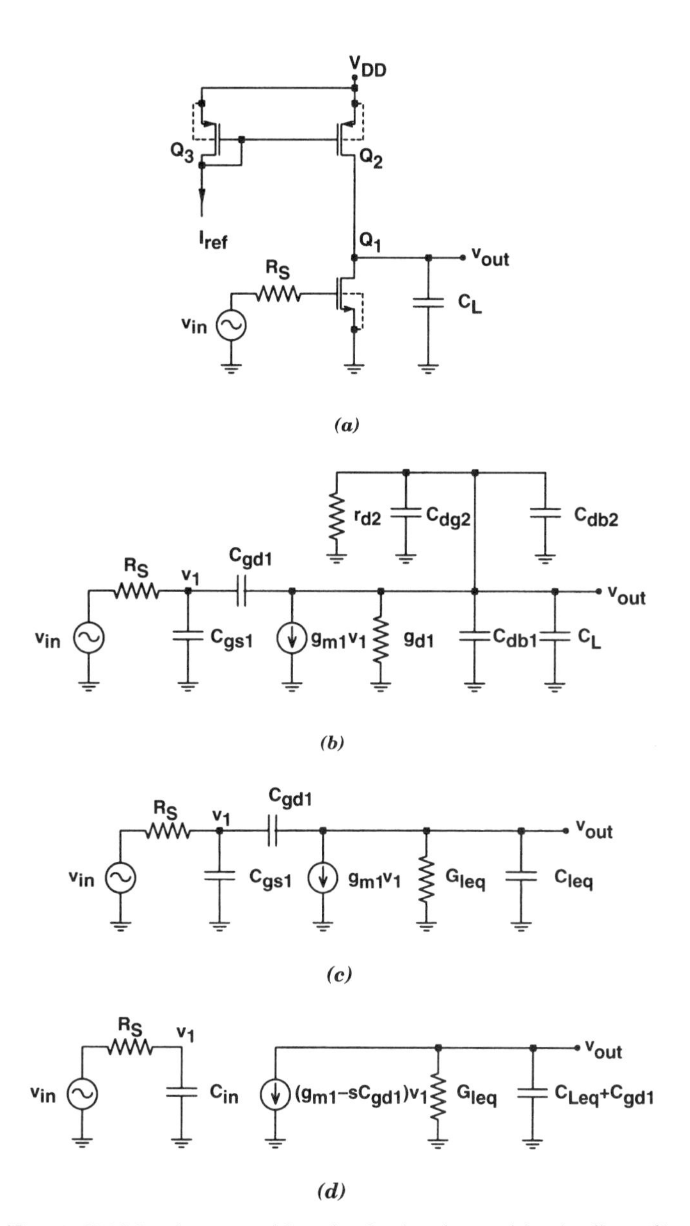

Figure 3.45. (*a*) CMOS gain stage with active load and capacitive loading; (*b*) equivalent circuit of the CMOS gain stage; (*c*) simplified equivalent circuit of CMOS gain stage; (*d*) simplified equivalent circuit of CMOS gain stage using Miller's theorem.

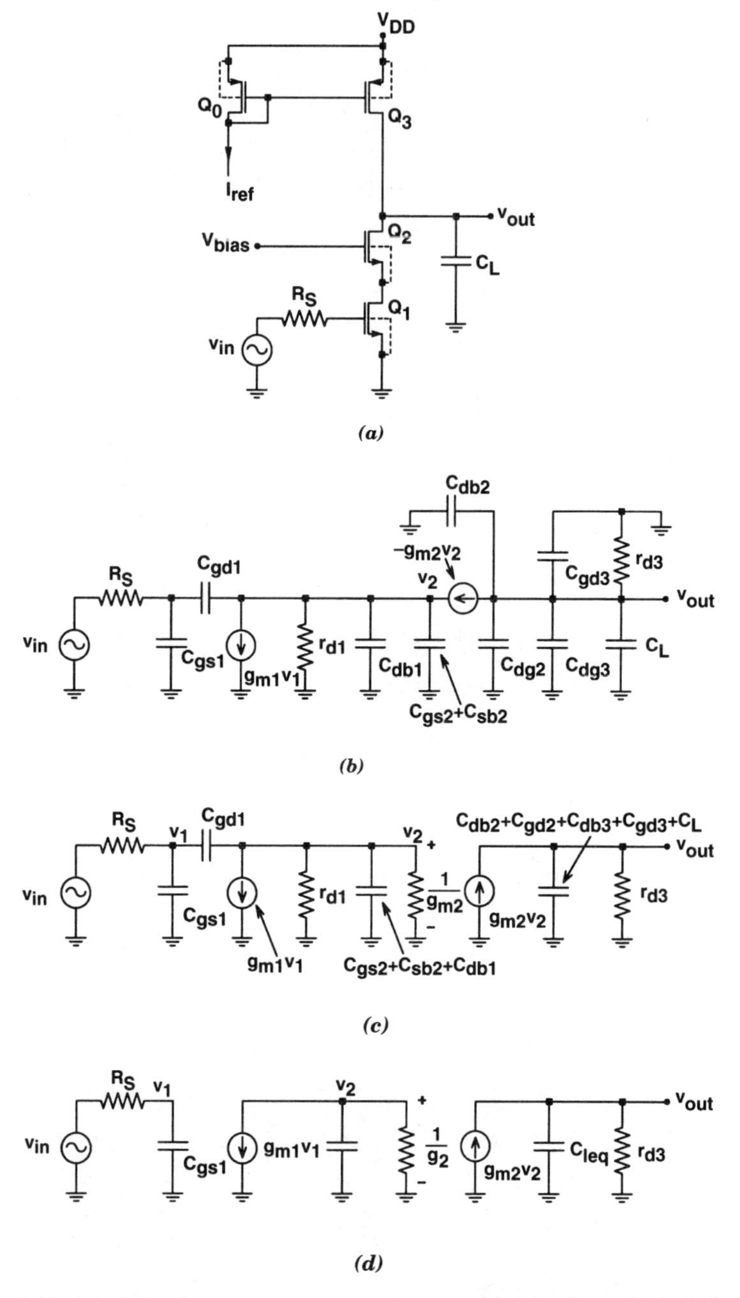

Figure 3.46. (*a*) Active-load cascode stage with capacitive loading; (*b*) high-frequency equivalent circuit of the cascode stage; (*c*) simplified high-frequency equivalent circuit of the cascode stage; (*d*) simplified equivalent circuit of the cascode stage obtained using Miller's theorem.

The circuit of Fig. 3.46*d* can easily be analyzed (thanks to the Miller approximation, which neatly partitioned it into buffered sections). The result is

$$A_v(s) = \frac{G_s g_{m2}(sC_{gd1} - g_{m1})}{(sC_1 + G_s)(sC_2 + g_2)(sC_{Leq} + g_{d3})} \tag{3.98}$$

from which the zero and the poles can be recognized directly:

$$\begin{aligned} s_z &= \frac{g_{m1}}{C_{gd1}}, \\ s_{p1} &= -\frac{G_s}{C_1}, \\ s_{p2} &= -\frac{g_2}{C_2}, \\ s_{p3} &= -\frac{g_{d3}}{C_{Leq}}. \end{aligned} \tag{3.99}$$

For practical values, usually $|s_{p1}| \ll s_z, |s_{p2}|, |s_{p3}|$. Then s_{p1} is the dominant pole, and the 3-dB frequency is given by

$$f_{3\text{dB}} \approx \frac{|s_{p1}|}{2\pi} = \frac{G_s}{2\pi C_1}. \tag{3.100}$$

Typically, $g_{m1} = g_{m2}$; then $C_1 = C_{gs1} + 2C_{gd1}$ and $f_{3\text{dB}} \approx G_s/[2\pi(C_{gs1} + 2C_{gd1})]$. By contrast, for the simple inverter stage of Fig. 3.45*a* the corresponding value is $G_s/\{2\pi[C_{gs1} + (1 + g_{m1}/G_{Leq})C_{gd1}]\}$, as Eqs. (3.91) and (3.52) show. Since g_{m1}/G_{Leq} is the magnitude of the dc gain of the stage, it is usually large. Hence the dominant pole (and thus the 3-dB frequency) is much smaller for the simple gain stage than for the cascode circuit. This confirms the effectiveness of the latter for high-frequency amplification.

The analysis of the NMOS source follower is straightforward. Figure 3.47*a* shows the actual circuit; Fig. 3.47*b* and *c* show the high-frequency small-signal equivalent circuits. The transfer function can be readily derived (Problem 3.15); the result is

$$A_v(s) = \frac{sC_{gs1} + g_{m1}}{s(C_{gs1} + C_{Leq}) + (g_{m1} + G_{Leq})}, \tag{3.101}$$

where

$$\begin{aligned} C_{Leq} &= C_L + C_{sb1} + C_{db2} + C_{gd2}, \\ G_{Leq} &= g_{d1} + g_{d2}. \end{aligned} \tag{3.102}$$

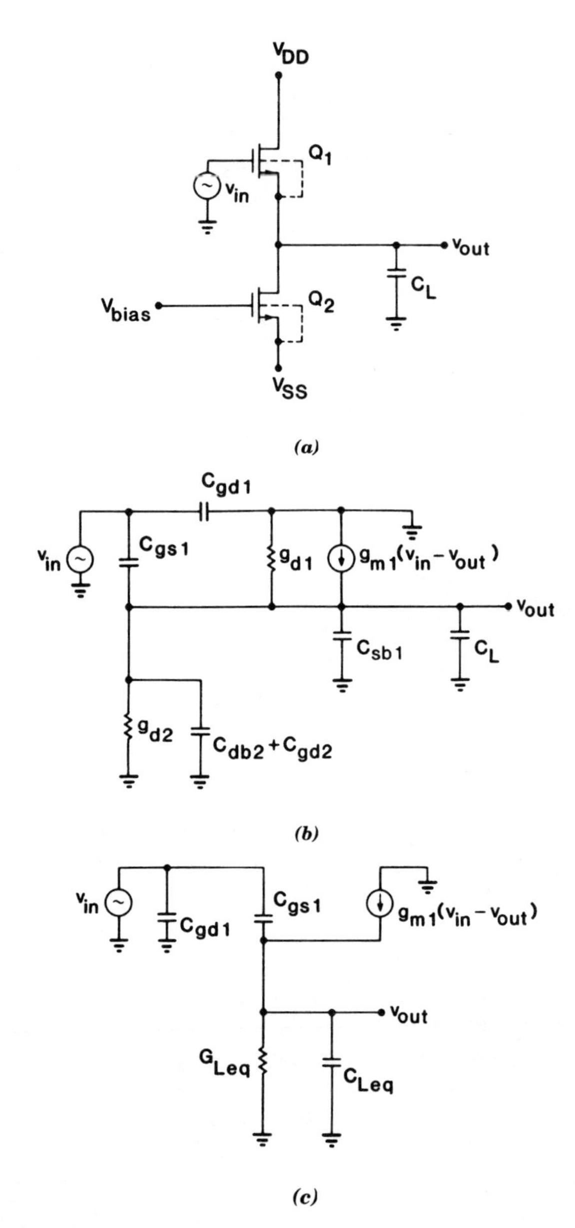

Figure 3.47. (*a*) NMOS source follower with capacitive loading; (*b*) equivalent circuit of the NMOS source follower; (*c*) simplified equivalent circuit of the NMOS source follower.

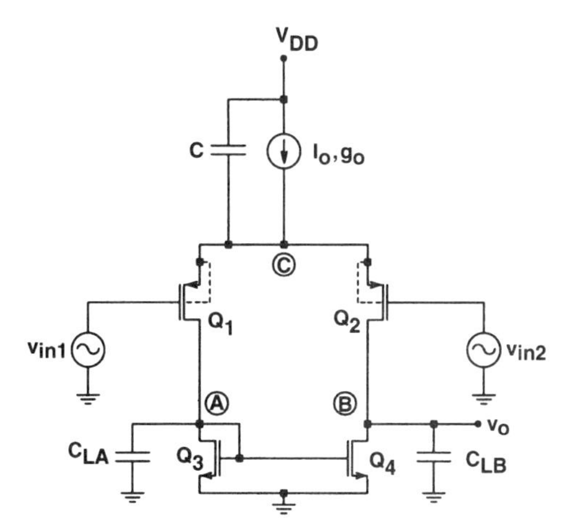

Figure 3.48. CMOS differential stage with capacitive loading. Circuit for Problem 3.2.

The zero and the pole are hence

$$s_z = \frac{-g_{m1}}{C_{gs1}}. \tag{3.103}$$

and

$$s_p = -\frac{g_{m1} + G_{Leq}}{C_{gs1} + C_{Leq}}$$

Choosing $G_{Leq}/g_{m1} \simeq C_{Leq}/C_{gs1}$, $s_z \simeq s_p$ can be achieved. Then $A_v(s) \simeq C_{gs1}/(C_{gs1} + C_{Leq})$, and hence the gain is constant up to very high frequencies, where higher-order effects cause it to drop. In the actual implementation, in order to meet the condition on C_{Leq}/C_{gs1}, it may be necessary to connect a capacitor C in parallel with C_{gs1}, that is, between the input and output terminals. Then C_{gs1} should be replaced by $C_{gs1} + C$ in the relations above.

The small-signal analysis of the differential amplifier stage of Fig. 3.42*a* can, in principle, be performed similarly. Thus in the small-signal equivalent circuit of each transistor the stray capacitances can be included and a nodal analysis performed in the Laplace transform domain. The process becomes quite complicated, however, since the numbers of both nodes and branches are high.

Consider now the CMOS differential stage shown in Fig. 3.48. If only Q_1 has an input voltage, while the output voltage is used only at node B, the load capacitances usually satisfy $C_{LB} \gg C_{LA}$. Furthermore, since Q_3 is driven with its gate and drain shorted, it presents a large load conductance $g_{m3} + g_{d3} \simeq g_{m3}$. By contrast, the conductance connected to node B is $g_{d2} + g_{d4}$, a small value. Hence the time constant of the admittance connected to node B, $\tau_B = C_{LB}/(g_{d2} + g_{d4})$, is likely to be several orders of magnitude larger than that at node A, $\tau_A \simeq C_{LA}/g_{m3}$. The time constant at node C is also small, since Q_1 and Q_2 load this node with the large conductance $g_{m1} + g_{m2}$.

Clearly, in a situation like that represented by the circuit of Fig. 3.48, the general nodal analysis is very complicated. Thus either a computer program (such as SPICE) that can perform the frequency analysis of linearized MOS circuits should be used or some simplifying assumptions made in the theoretical analysis. For the circuit of Fig. 3.48, it was verified above that the dominant pole is that corresponding to the largest time constant τ_B: its value is $|s_{p1}| \simeq (g_{d2} + g_{d4})/C_{LB}$. Therefore, for example, the differential-mode voltage gain can be approximated by

$$A_{dm}(s) \simeq \frac{A_{dm}(0)}{1 + s/|s_{p1}|} \simeq \frac{g_{mi}}{sC_{LB} + (g_{di} + g_{dl})}. \tag{3.104}$$

Here it was assumed that Q_1 and Q_2 as well as Q_3 and Q_4 are matched devices, and Eq. (3.88) was used. The 3-dB frequency can also be (approximately) predicted from Eq. (3.104) as $(g_{di} + g_{dl})/2\pi C_{LB}$.

The same approximation can be used to find the common-mode voltage gain:

$$A_{cm}(s) \approx \frac{A_{dm}(0)}{1 + s/|s_{p1}|} = -\frac{g_{di}}{2g_{ml}} \frac{g}{sC_{LB} + (g_{di} + g_{dl})}. \tag{3.105}$$

Here g is the output conductance of the current source. In a CMOS op-amp, the common source of the p-channel devices is tied to an n well. There is a large stray capacitance C between the n well and the V_{DD} lead, which reduces the impedance between node C and ground at high frequencies. The effect of C can be incorporated in (3.105) simply by replacing g by $g + sC$. Then

$$A_{cm}(s) \approx -\frac{g_{di}}{2g_{ml}} \frac{g + sC}{sC_{LB} + (g_{di} + g_{dl})} \tag{3.106}$$

results. The zero at $s = -g/C$ will cause $|A_{Cm}/A_{dm}|$ to increase by 20 dB/decade at high frequencies, thus causing a reduced CMRR.

PROBLEMS

3.1. For the circuit of Fig. 3.49, $\gamma = 2\ \text{V}^{1/2}$, $|\phi_p| = 0.3$ V, $V_T = 2$ V, $V_{DD} = 10$ V, and $V_{oi} = 2.5i$, $i = 1,2,3$. Find the W/L values for Q_1, Q_2, Q_3, and Q_4 if the currents drawn at the nodes V_{o1}, V_{o2}, V_{o3}, V_{o4} are negligible.

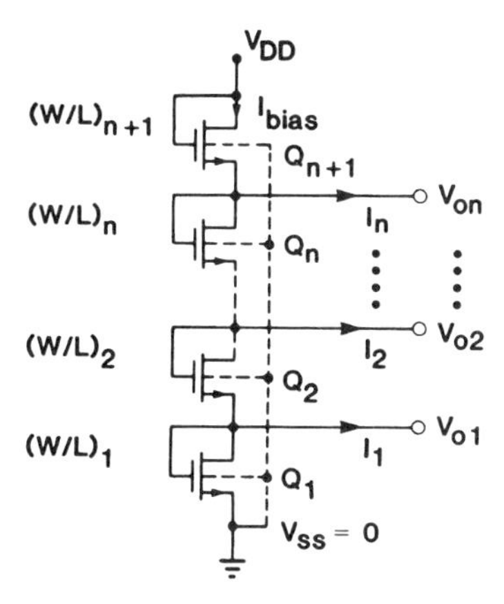

Figure 3.49. NMOS voltage divider.

3.2. Derive the formulas for the *W/L* ratios of Fig. 3.49 if the currents drawn at the nodes V_{o1}, V_{o2}, and V_{o3} are not negligible.

3.3. Prove Eq. (3.7) for the circuit of Fig. 3.3.

3.4. Prove Eq. (3.25) for the circuit of Fig. 3.11.

3.5. In Figs. 3.10 and 3.11, assume that $r_o \gg r_{d1}$ and $g_{m2} = g_{m3}$. Show that the output resistance is increased by the open-circuit voltage gain of Q_1.

3.6. **(a)** Prove that Eq. (3.25) holds for the circuit of Fig. 3.12. **(b)** Analyze the circuit of Fig. 3.13. How much is r_{out}? Show that the output resistance is that of Q_2, magnified by the voltage gain of Q_3.

3.7. Calculate the gain of the circuit of Fig. 3.23 without neglecting the r_{di}.

3.8. Prove Eqs. (3.56) and (3.57) for the circuit of Fig. 3.29.

3.9. Prove Eq. (3.64) for the circuit of Fig. 3.30.

3.10. **(a)** Derive the relations for $v_{o,d}$ and $v_{o,c}$ of the source-coupled stage (Fig. 3.38) if the circuit is not exactly symmetrical. **(b)** Rewrite your relations in the form

$$v_{o,d} = A_{dd}v_{\text{in},d} + A_{dc}V_{\text{in},c},$$

$$v_{o,c} = A_{cd}v_{\text{in},d} + A_{cc}V_{\text{in},c}.$$

What are A_{dd}, A_{dc}, and A_{cd}, and A_{cc}? **(c)** Let the maximum difference between symmetrically located elements in the small-signal equivalent circuit be 1%. How much are the maximum values of $|A_{cd}|$ and $|A_{dc}|$?

3.11. Derive Eq. (3.84) for the circuit of Fig. 42*a*. (*Hint:* Write and solve the current equations for nodes A, B, and C.)

3.12. Modify the CMOS differential stage of Fig. 3.42 so that it has differential output signals. Compare the differential gain with that of the original circuit!

3.13. Prove that the small-signal output impedance of the circuit of Fig. 3.43 is given by Eq. (3.89). (*Hint:* Write and solve the current law for nodes A, B, and C.)

3.14. Analyze the CMOS gain stage of Fig. 3.45 in the Laplace domain. **(a)** Find the exact transfer function $A_v(s)$ from Fig. 3.45*c*. **(b)** Use Miller effect approximation to derive the simplified circuit of Fig. 3.45*d*; analyze the simplified circuit to verify Eq. (3.91).

3.15. Analyze the NMOS source follower of Fig. 3.47; verify Eq. (3.101).

3.16. Analyze the cascode gain stage of Fig. 3.46 in the Laplace transform domain. **(a)** Verify the equivalent circuits of Fig. 3.46*b* to *d*. **(b)** Show that Eq. (3.98) holds for the circuit of Fig. 3.46*d*.

REFERENCES

1. P. R. Gray and R. J. Meyer, *Analysis and Design of Analog Integrated Circuits,* Wiley, New York, 1993.
2. J. L. McCreary and P. R. Gray, *IEEE J. Solid-State Circuits, SC-10*(6), 371–379 (1975).
3. D. J. Hamilton and W. G. Howard, *Basic Integrated Circuit Engineering,* McGraw-Hill, New York, 1975.
4. T. C. Choi et al., *IEEE J. Solid-State Circuits, SC-18*(6), 652–664 (1983).
5. Y. Tsividis and P. Antognetti, *Design of MOS VLSI Circuits for Telecommunications,* Prentice Hall, Upper Saddle River, N. J., 1985.
6. P. R. Gray, Part II, IEEE Press, New York, 1980.
7. Y. Tsividis, *IEEE J. Solid-State Circuits, SC-13*(3), 383–391 (1978).
8. D. Senderowicz, D. A. Hodges, and P. R. Gray, *IEEE J. Solid-State Circuits, SC-13*(6), 760–766 (1978).
9. K. Bult and G. J. Geelen, *IEEE J. Solid-State Circuits, SC-25*(6), 1379–1384 (1990).
10. E. Sackinger and W. Guggenbuhl, *IEEE J. Solid-State Circuits SC-25*(1), 289–298 (1990).
11. D. Allstot, *IEEE J. Solid-State Circuits, SC-17*(6), 1080–1087 (1984).

CHAPTER 4

CMOS OPERATIONAL AMPLIFIERS

The CMOS operational amplifier is the most intricate, and in many ways the most important, building block of linear CMOS and switched-capacitor circuits. Its performance usually limits the high-frequency application and the dynamic range of the overall circuit. It usually requires most of the dc power used up by the device. Without a thorough understanding of the operation and the basic limitations of these amplifiers, the circuit designer cannot determine or even predict the actual response of the overall system. Hence this chapter includes a fairly detailed explanation of the usual configurations and performance limitations of operational amplifiers.

The technology, and hence the design techniques used for MOS amplifiers, change rapidly. Therefore, the main purpose of the discussion is to illustrate the most important principles underlying the specific circuits and design procedures. Nevertheless, the treatment is detailed enough to enable the reader to design high-performance CMOS operational amplifiers suitable for most linear CMOS circuit applications.

4.1. OPERATIONAL AMPLIFIERS [1, Chap. 10; 2, Chap. 6]

In switched-capacitor circuits—in fact, in all linear CMOS circuits—the most commonly used active component is the operational amplifier, usually simply called the *op-amp*. Ideally, the op-amp is a voltage-controlled voltage source (Fig. 4.1) with infinite voltage gain and with zero input admittance as well as zero output impedance. It is free of frequency and temperature dependence, distortion, and noise. Needless to say, practical op-amps can only approximate such an ideal device. The main differences between the ideal op-amp and the real device are the following [2, Chap. 6]:

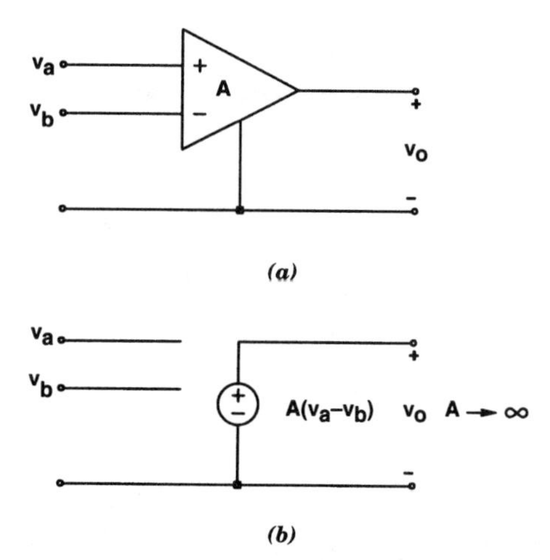

Figure 4.1. (*a*) Symbol for ideal op-amp; (*b*) equivalent circuit.

1. *Finite Gain.* For practical op-amps, the voltage gain is finite. Typical values for low frequencies and small signals are $A = 10^3$ to 10^5, corresponding to 60 to 100 dB gain.
2. *Finite Linear Range.* The linear relation $v_o = A(v_a - v_b)$ between the input and output voltages is valid only for a limited range of v_o. Normally, the maximum value of v_o for linear operation is somewhat smaller than the positive dc supply voltage; the minimum value of v_o is somewhat positive with respect to the negative supply.
3. *Offset Voltage.* For an ideal op-amp, if $v_a = v_b$ (which is easily obtained by short-circuiting the input terminals), $v_o = 0$. In real devices, this is not exactly true, and a voltage $v_{o,\text{off}} \neq 0$ will occur at the output for shorted inputs. Since $v_{o,\text{off}}$ is usually directly proportional to the gain, the effect can be more conveniently described in terms of the input offset voltage $v_{\text{in,off}}$, defined as the differential input voltage needed to restore $v_o = 0$ in the real device. For MOS op-amps, $v_{\text{in,off}}$ is typically ± 2 to 10 mV. This effect can be modeled by a voltage source of value $v_{\text{in,off}}$ in series with one of the input leads of the op-amp.
4. *Common-Mode Rejection Ratio (CMRR).* The common-mode input voltage is defined by

$$v_{\text{in},c} = \frac{v_a + v_b}{2}, \tag{4.1}$$

as contrasted with the differential-mode input voltage

$$v_{\text{in},d} = v_a - v_b. \tag{4.2}$$

Accordingly, we can define the differential gain A_D (which is the same as the gain A discussed earlier), and also the common-mode gain A_C, which

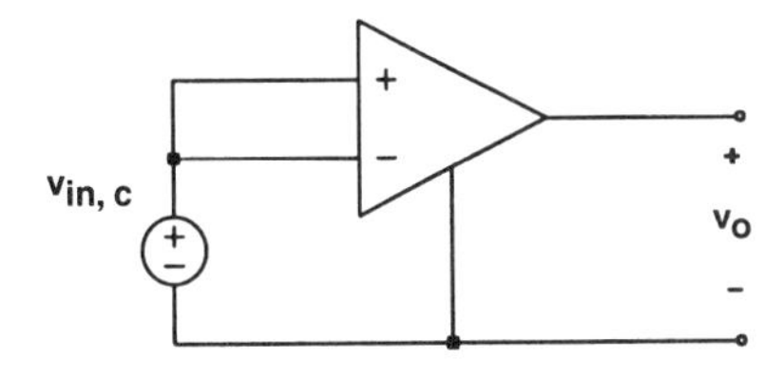

Figure 4.2. Op-amp with only common-mode input voltage.

can be measured as shown in Fig. 4.2, where $A_C = v_o/v_{\text{in},c}$. Here $v_{\text{in,off}} = 0$ is assumed; $|A_C|$ is usually around $1 \approx 10$.

The CMRR is now defined as A_D/A_C or (in logarithmic units) CMRR = $20 \log_{10} (A_D/A_C)$ in decibels. Typical CMRR values for CMOS amplifiers are in the range 60 to 80 dB. The CMRR measures how much the op-amp can suppress noise, and hence a large CMRR is an important requirement.

5. *Frequency Response.* Because of stray capacitances, finite carrier mobilities, and so on, the gain A decreases at high frequencies. It is usual to describe this effect in terms of the unity-gain bandwidth, that is, the frequency f_0 at which $|A(f_0)| = 1$. For CMOS op-amps, f_0 is usually in the range 1 to 100 MHz. It can be measured with the op-amp connected in a voltage-follower configuration (Problem 4.13).
6. *Slew Rate.* For a large input step voltage, some transistors in the op-amp may be driven out of their saturation regions or cut off completely. As a result, the output will follow the input at a slower finite rate. The maximum rate of change dv_o/dt is called the *slew rate.* It is not directly related to the frequency response. For typical CMOS op-amps, slew rates of 1 to 20 V/μs can be obtained.
7. *Nonzero Output Resistance.* For a real CMOS op-amp, the open-loop output impedance is nonzero. It is usually resistive and is on the order of 0.1 to 5 kΩ for op-amps with an output buffer; it can be much higher (≈ 1 MΩ) for op-amps with unbuffered output. This affects the speed with which the op-amp can charge a capacitor connected to its output, and hence the highest signal frequency.
8. *Noise.* As explained in Section 2.7, the MOS transistor generates noise, which can be described in terms of an equivalent current source in parallel with the channel of the device. The noisy transistors in an op-amp give rise to a noise voltage v_{on} at the output of the op-amp; this can again be modeled by an equivalent voltage source $v_n = v_{\text{on}}/A$ at the op-amp input. Unfortunately, the magnitude of this noise is relatively high, especially in the low-frequency band, where the flicker noise of the input devices is high; it is about 10 times the noise occurring in an op-amp fabricated in bipolar technology. In a wide band (say, in the range 10 Hz to 1 MHz), the equivalent input noise source is usually on the order of 10 to 50 μV rms, in contrast to the 3 to 5 μV achievable for low-noise bipolar op-amps.
9. *Dynamic Range.* Due to the limited linear range of the op-amp, there is a *maximum* input signal amplitude $v_{\text{in,max}}$ which the device can handle without

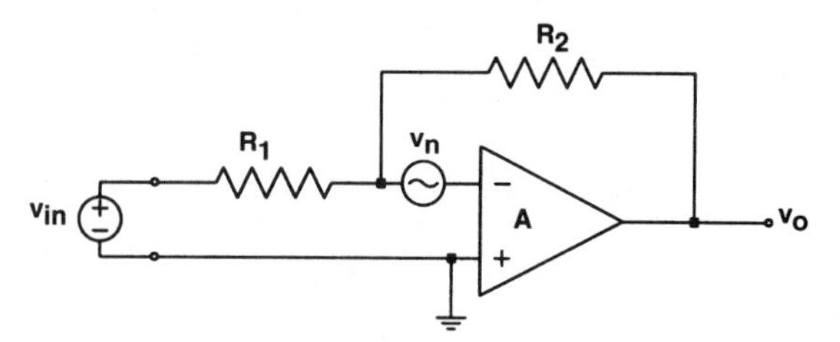

Figure 4.3. Noisy feedback amplifier.

generating an excessive amount of nonlinear distortion. If the power supply voltages of the op-amp are $\pm V_{CC}$, an optimistic estimate is $v_{\text{in,max}} \approx V_{CC}/A$, where A is the open-loop gain of the op-amp. Due to spurious signals (noise, clock feedthrough, low-level distortion such as crossover distortion, etc.) there is also a *minimum* input signal $v_{\text{in,min}}$ which still does not drown in noise and distortion. Usually, $v_{\text{in,min}}$ is on the same order of magnitude as the equivalent input noise v_n of the op-amp. The dynamic range of the op-amp is then defined as 20 $\log_{10}(v_{\text{in,max}}/v_{\text{in,min}})$ measured in decibels. When the op-amp is in open-loop condition, $v_{\text{in,max}} \approx V_{CC}/A$, which is on the order of a millivolt, while $v_{\text{in,min}} \approx \sqrt{\overline{v_n^2}}$, which is around 30 μV. Thus the open-loop dynamic range of the op-amp is only around 30 to 40 dB. However, the dynamic range of a circuit containing op-amps in negative feedback configuration can be much larger. As a simple illustration, consider the feedback amplifier shown in Fig. 4.3. It is easy to show (Problem 4.1) that the output due to the noise source v_n acting alone has the rms value

$$v_{\text{on}} = \frac{\sqrt{\overline{v_n^2}}}{1/A + R_1/(R_1 + R_2)} . \tag{4.3}$$

The voltage gain of the (noiseless) feedback circuit is

$$\frac{v_o}{v_{\text{in}}} = -\frac{1}{1/A + (R_1/R_2)(1 + 1/A)} . \tag{4.4}$$

The minimum input signal $v_{\text{in,min}}$ gives rise to an output voltage approximately equal to v_{on}. Hence

$$v_{\text{in,min}} \simeq \sqrt{\overline{v_n^2}}\,\frac{1/A + (R_1/R_2)(1 + 1/A)}{1/A + R_1/(R_1 + R_2)} , \tag{4.5}$$

and for $v_{o,\text{max}} \approx V_{CC}$,

$$v_{\text{in,max}} \simeq V_{CC}\left[\frac{1}{A} + \frac{R_1}{R_2}\left(1 + \frac{1}{A}\right)\right] . \tag{4.6}$$

Hence the dynamic range is given by

$$20 \log_{10}\left[\frac{V_{CC}}{\sqrt{\overline{v_n^2}}}\left(\frac{1}{A} + \frac{R_1}{R_1 + R_2}\right)\right] \approx 20 \log_{10}\left(\frac{V_{CC}}{\sqrt{\overline{v_n^2}}}\frac{R_1}{R_1 + R_2}\right), \tag{4.7}$$

where the indicated approximation in usually valid for $A \gg 1$. For typical values ($V_{CC} \sim 5$ V, $v_{\text{in,min}} \approx \sqrt{\overline{v_n^2}} \approx 30\ \mu$V, $R_2/R_1 \approx 5$), a dynamic range of about 90 dB results for the overall circuit.

In linear CMOS circuits, dynamic range values around 80 to 90 dB are readily achievable. Even higher values (up to 100 dB) are possible if the large low-frequency noise ($1/f$ noise) is canceled using a differential circuit configuration and chopper stabilization [3].

10. *Power Supply Rejection Ratio (PSRR).* If a power supply voltage contains an incremental component v due to noise, hum, and so on, a corresponding voltage $A_p v$ will appear at the op-amp output. The PSRR is defined as A_D/A_p, where $A_D = A$ is the differential gain. It is common to express the PSRR in decibels; then PSRR $= 20 \log_{10}(A_D/A_p)$. Usual PSRR values range from 60 to 80 dB for the op-amp alone; for a switched-capacitor circuit, 30 to 50 dB can be achieved.
11. *DC Power Dissipation.* Ideal op-amps require no dc power dissipated in the circuit; real ones do. Typical values for a CMOS op-amp range from 0.25 to 10 mW dc power drain.

To obtain near-ideal performance for a practical op-amp, the general structure of Fig. 4.4 is usually employed [1, Chap. 10]. The input differential amplifier (first block) is designed so that it provides a high input impedance, large CMRR and PSRR, low offset voltage, low noise, and high gain. Its output should preferably be single-ended, so that the rest of the op-amp need not contain symmetrical differential stages. Since the transistors in the input stage (and in subsequent stages) operate in their saturation regions, there is an appreciable dc voltage difference between the input and output signals of the input stage.

The second block in Fig. 4.4 may perform one or more of the following functions:

1. *Level Shifting.* This is needed to compensate for the dc voltage change occurring in the input stage, and thus to assure the appropriate dc bias for the following stages.

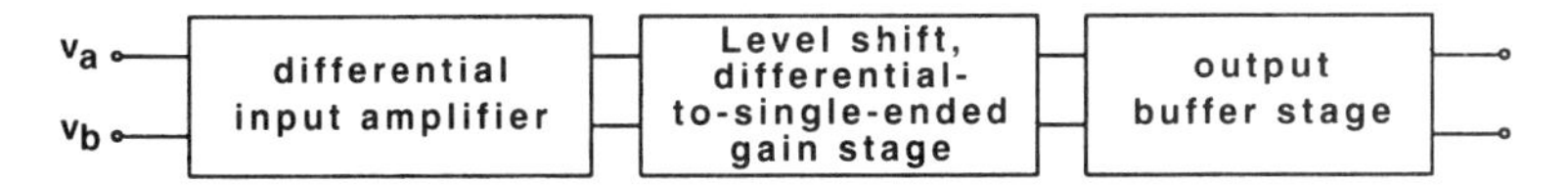

Figure 4.4. Block diagram for a practical op-amp.

2. *Added Gain.* In most cases the gain provided by the input stage is not sufficient, and additional amplification is required.
3. *Differential-to-Single-Ended Conversion.* In some circuits the input stage has a differential output, and the conversion to single-ended signals is performed in a subsequent stage.

The third block is the output buffer. It provides the low output impedance and larger output current needed to drive the load of the op-amp. It normally does not contribute to the voltage gain. If the op-amp is an internal component of a switched-capacitor circuit, the output load is a (usually small) capacitor, and the buffer need not provide a very large current or very low output impedance. However, if the op-amp is at the circuit output, it may have to drive a large capacitor and/or resistive load. This requires large current drive capability and very low output impedance, which can only be attained by using large output devices with appreciable dc bias currents. Thus the dc power drain will be much higher for such output op-amps than for interior ones.

As mentioned earlier, the ideal op-amp defined in Fig. 4.1 is a voltage-controlled voltage source, with zero output impedance. In fact, for practical op-amps, which do not have an output buffer, the output impedance may be very high, on the order of megohms. For such an amplifier, a better ideal representation can be found as a voltage-controlled current source, with a transconduction G_m value that is infinitely large. This ideal model is called an *operational transconductance amplifier* (OTA). If the op-amp has sufficiently high voltage gain and is in a stable feedback network, its output impedance is reduced to a very low value, and the difference between the performances of an op-amp and OTA can be neglected.

In a class of continuous-time filters, a finite-G_m transconductance is required. Here a low but accurately controlled value of G_m needs to be achieved. The corresponding active device is called a *transconductor;* it is not to be confused with an OTA. In the remainder of this chapter, the properties of typical CMOS op-amps and OTAs are described, and analysis and design techniques are given for them. Unless otherwise postulated, we assume that all devices are operated in the saturation region. Then i_D is to a good approximation independent of v_D and is given by $i_D \simeq k'(W/L)(v_{GS} - V_T)^2$. Here, due to body effect, V_T depends on the source-to-body voltage.

4.2. SINGLE-STAGE OPERATIONAL AMPLIFIERS

A practical block diagram of an MOS op-amp was shown in Fig. 4.4 and is reproduced in more detail in Fig. 4.5. The voltage gain required is obtained in the differential (G_1) and single-ended (G_2) gain stages. The output stage (G_3) is usually a wideband unity-gain low-output-impedance buffer, capable of driving large capacitive and/or resistive loads. If the op-amp is used in an internal (as opposed to output) stage of a switched-capacitor circuit, the load may be only a small capacitor, 2 pF or less. In such a situation, the output buffer (G_3) may be omitted, and the load may

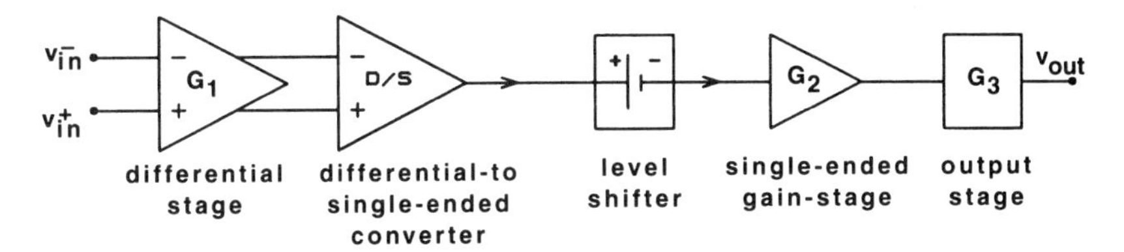

Figure 4.5. Basic building blocks of an operational amplifier.

then be connected directly to the output of G_2. It will then function as an operational transconductance amplifier (OTA). In Fig. 4.5, if the combination of the differential stage and differential-to-single-ended converter provides adequate gain and output-voltage swing, G_2 can also be omitted and the load may be driven directly by the differential stage. Again, the circuit will then realize an OTA.

The CMOS differential stage with an active load is shown again in Fig. 4.6. It was introduced in Chapter 3. This stage combines the functions of a differential amplifier and differential-to-single-ended converter. The role of the differential amplifier is to amplify the difference between the two input voltages, v_i^+ and v_i^-, regardless of the common-mode voltage. The differential stage is therefore characterized by its common-mode rejection ratio (CMRR), which is the ratio of the differential gain to the common-mode gain. The differential gain $A_{dm} \simeq v_{out}/(v_i^+ - v_i^-)$ was derived in Chapter 3 and is given by

$$A_{dm} \simeq g_{mi} r_o = g_{mi}(r_{d2} \| r_{d4}) = \frac{g_{mi}}{g_{d2} + g_{d4}}. \tag{4.8}$$

In this equation, $1/(g_{d2} + g_{d4})$ is the output impedance seen at the output of the differential stage, and g_{mi} is the transconductance of the input devices Q_1 and Q_2.

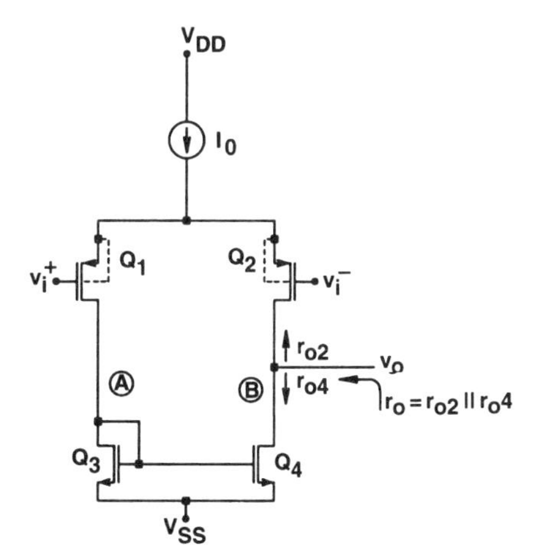

Figure 4.6. CMOS differential stage with active load.

The differential stage of Fig. 4.6 used as a single-stage op-amp has two major shortcomings. First, the total voltage gain is limited to the gain of a single-stage amplifier, which is typically about 50. Second, the output voltage swing is limited to the range

$$V_{D\text{sat}4} \leq v_o \leq v_{\text{inc}} + V_{TP}, \tag{4.9}$$

where v_{inc} is the common-mode input voltage defined as $v_{\text{inc}} = (v_i^+ + v_i^-)/2$, and V_{TP} is the threshold voltage of the p-channel devices. Obviously, in most cases the low voltage gain and the narrow output swing prevent the differential stage of Fig. 4.6 from being useful as a single-stage op-amp.

The gain of the differential stage can be increased in two ways, by increasing the transconductance of the input devices Q_1 and Q_2, or by increasing the output impedance seen at the output of the stage. As can be seen from the relation $g_{mi} = 2\sqrt{k'(W/L)I_0/2}$, transconductance can be increased by increasing the width of the input devices and/or by increasing the bias current. Notice that reducing L, the channel length of the input devices, can also increase the transconductance. This, however, also has the opposite effect of reducing the output impedance $1/g_{di}$ of the input devices (due to channel-length modulation effect), and hence by Eq. (4.8) reduces the gain. Increasing the width or the current of the stage will increase the size or the power dissipation of the circuit. Therefore, a more efficient way to increase the gain is to increase the output impedance r_o.

As is evident from Fig. 4.6, to increase the output impedance r_o, both r_{d2} and r_{d4} have to be increased. This can be achieved by using the cascode current source as load. Figure 4.7 illustrates a differential stage that uses cascode transistors to increase the voltage gain by increasing the output impedance. Here devices Q_1,Q_{1c} and Q_2,Q_{2c} form two source-couple cascode amplifiers, while Q_3, Q_{3c}, Q_4, and Q_{4c}

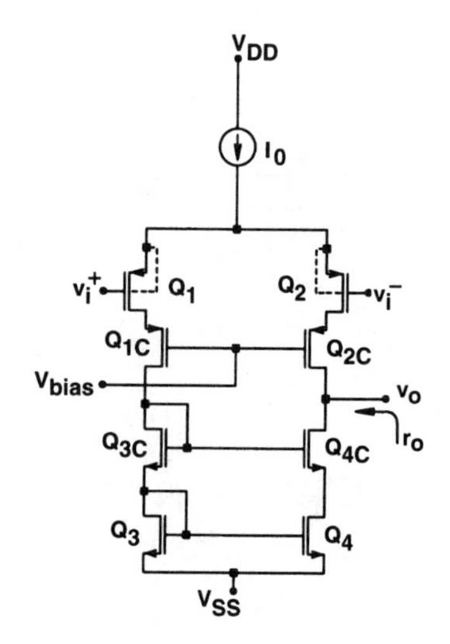

Figure 4.7. CMOS differential stage with cascode load.

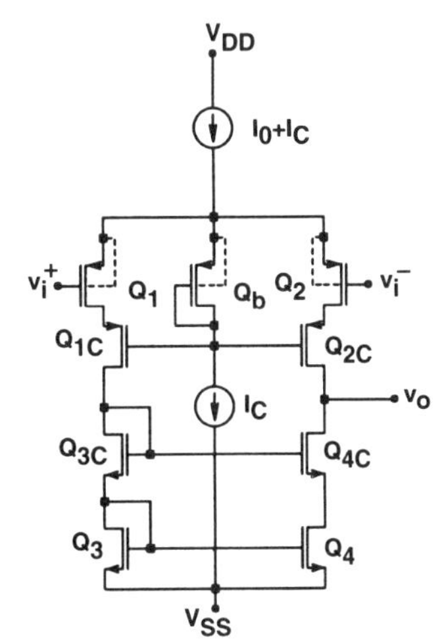

Figure 4.8. CMOS differential stage with cascode load and common-mode biasing scheme.

form a cascode current source that acts as an active load. For symmetrical dimensions $(W/L)_1 = (W/L)_2$, $(W/L)_3 = (W/L)_4$, $(W/L)_{1c} = (W/L)_{2c}$, and $(W/L)_{3c} = (W/L)_{4c}$ the output impedance of the stage is,

$$r_o = g_{m2c} r_{d2c} r_{d2} \parallel g_{m4c} r_{d4c} r_{d4}, \tag{4.10}$$

where $r_1 \parallel r_2$ denotes parallel-connected r_1 and r_2 (Problem 4.3). Since $g_{m2c} r_{d2c}$ and $g_{m4c} r_{d4c}$ are normally much greater than 1, $r_o \gg (r_{d2} \parallel r_{d4})$, which is the output impedance of the differential stage of Fig. 4.6. The differential voltage gain of the stage of Fig. 4.7 is given by

$$A_{dm} = g_{m1} r_o = g_{m1}(g_{m2c} r_{d2c} r_{d2} \parallel g_{m4c} r_{d4c} r_{d4}). \tag{4.11}$$

The use of cascoding increases the gain of the differential stage substantially. The disadvantage is, however, that the voltage drops across the additional transistors Q_{1c} and Q_{3c} result in a reduction in the allowable input common-mode range and output voltage swing. The swing performance can be improved by using high-swing biasing of the cascode, as discussed in Section 3.3. The input common-mode range can also be improved, by using a bias voltage for Q_{1c} and Q_{2c} that tracks the input common-mode voltage. One circuit that accomplishes this is shown in Fig. 4.8, where Q_b and I_c have been added to bias the gates of Q_{1c} and Q_{2c}. The W/L ratio of Q_b and the value of current I_c can be selected in such a way that Q_1 and Q_2 remain biased at the edge of the saturation region as the input common-mode voltage changes. Obviously, the bias voltage V_{bias} will be one V_{GS} drop (of Q_b) below the voltage V_c of the common source. Even though the performance of the circuit of Fig. 4.8 improved over that of Figs. 4.6 and 4.7, due to the very limited output voltage swing the stage is normally not useful as a single-stage op-amp.

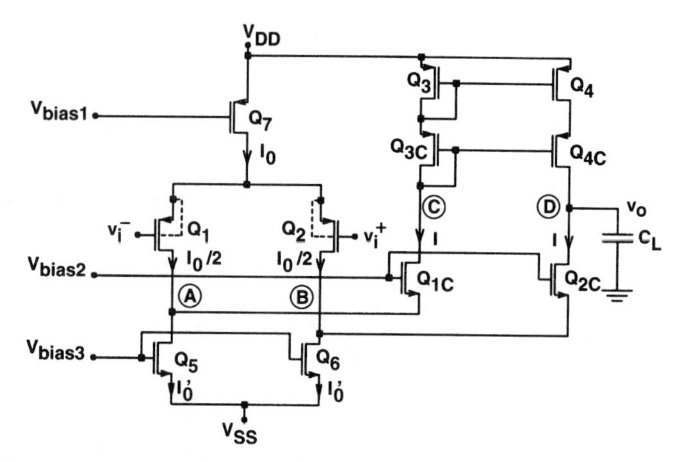

Figure 4.9. Folded-cascode op-amp consisting of a cascade of common-source and common-gate amplifiers.

Some of the problems described with the differential stage of Fig. 4.8 can be eliminated by using the *folded-cascode* configuration [4]. Consider the circuit of Fig. 4.8; let the bottom terminals (i.e., the sources of Q_3 and Q_4) of the composite load Q_3, Q_4 and Q_{1c} to Q_{4c} be disconnected from V_{SS}, folded up, and connected to V_{DD} instead. To assure proper dc bias currents, all NMOS devices must be replaced by PMOS types, and vice versa in the cascode loads; also, two additional current sources (Q_5 and Q_6) must be added between V_{SS} and the drains of Q_1 and Q_2 to supply bias currents to these input devices. The resulting single-stage op-amp is shown in Fig. 4.9. The basic operation of the circuit is as follows. The dc current I_0 of the current source Q_7 is shared equally by Q_1 and Q_2. Also, the matched sources Q_5 and Q_6 draw equal bias currents I'_o from nodes A and B. Hence Q_{1c} and Q_{2c} also carry equal bias currents $I'_o - I_o/2$. A differential input voltage $\Delta v_{in}^+ = \Delta v_{in}/2$ and $\Delta v_{in}^- = -\Delta v_{in}/2$ applied to the gates of Q_1 and Q_2 will offset their drain currents by $\pm\Delta I_o = \pm g_{mi}\,\Delta v_{in}/2$. Since the currents I'_o of Q_5 and Q_6 remain unchanged, the currents of Q_{1c} and Q_{2c} (which are driven at their low-impedance source terminals) will also change by $\pm\Delta I_o$. The current mirror Q_3, Q_4, Q_{3c}, and Q_{4c} transfers the current change in Q_3 and Q_{3c} to Q_4 and Q_{4c}. Hence the output voltage increment is $g_{mi}R_o\,\Delta v_{in}$, where R_o is the output impedance at node D. It can be shown (Problem 4.4) that

$$R_o \simeq (r_{d4}r_{d4c}g_{m4c}) \parallel [(r_{d2}\|r_{d6})\; r_{d2c}g_{m2c}]. \tag{4.12}$$

The incremental gain is then

$$A_{dm} = -g_{mi}R_o = -(r_{d4}r_{d4c}g_{m4c}) \parallel [(r_{d2} \parallel r_{d6})\; r_{d2c}g_{m2c}]\; g_{mi}. \tag{4.13}$$

A disadvantage of the folded-cascode op-amp is the reduced output voltage swing due to the many (four) cascoded devices in the output branches. The swing can be

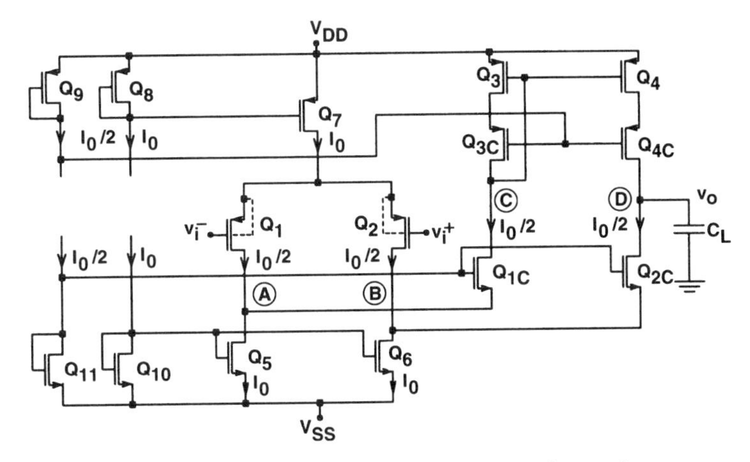

Figure 4.10. Folded cascode op-amp with improved biasing for maximum output voltage swing.

increased if one of the bias circuits of Fig. 3.15 or 3.16 is used to establish the gate voltages of Q_{1c} to Q_{4c} such that the drain-to-source voltages of Q_3 to Q_6 are only slightly larger than V_{Dsat}. In Fig. 4.10 the bias circuits of Fig. 3.16 has been added to the op-amp. It can be shown (Problem 4.5) that the necessary aspect ratios are given by

$$\left(\frac{W}{L}\right)_3 = \left(\frac{W}{L}\right)_4 = \left(\frac{W}{L}\right)_{3c} = \left(\frac{W}{L}\right)_{4c} = 4\left(\frac{W}{L}\right)_9, \tag{4.14}$$

$$\left(\frac{W}{L}\right)_5 = \left(\frac{W}{L}\right)_6 = \left(\frac{W}{L}\right)_{1c} = \left(\frac{W}{L}\right)_{2c} = (3 + 2\sqrt{2})\left(\frac{W}{L}\right)_{11}. \tag{4.15}$$

For this circuit it can also be shown that the maximum output voltage swing is within the range

$$V_{DD} - |V_{Dsat4}| - |V_{Dsat4c}| \leq v_{out} \leq V_{SS} + V_{Dsat6} + V_{Dsat2c}. \tag{4.16}$$

Thus the range lost at both the upper and lower limits is only $2|V_{Dsat}|$. The cascode op-amp shown in Fig. 4.10 has a large voltage gain and a reasonably large output voltage swing. Hence it can be used as a single-stage OTA.

Consider next the high-frequency behavior of the circuit of Fig. 4.10. The poles of the gain stage are contributed by the stray capacitances loading nodes A, B, C, and D. The dominant pole s_{p1} of the circuit is due to the load capacitance C_L in parallel with the output impedance R_o given by Eq. (4.12); hence its value is

$$s_{p1} = -\frac{1}{R_o C_L}. \tag{4.17}$$

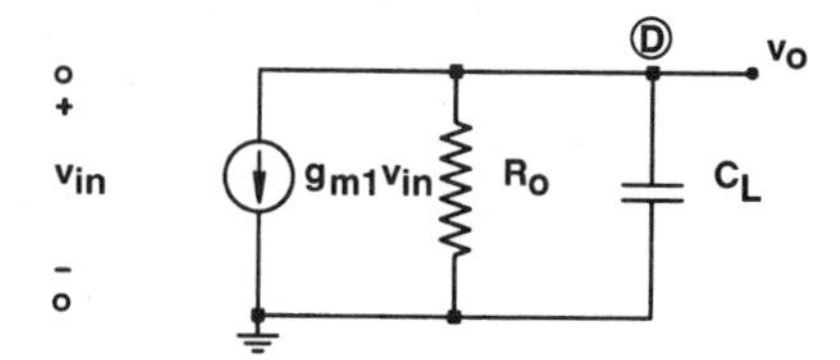

Figure 4.11. Small-signal low-frequency representation of the folded-cascode op-amp.

The resistance seen at node A is approximately $1/g_{m1c}$; at node B it is approximately $1/g_{m2c}$; and at node C it is approximately $1/g_{m3}$. Since $1/g_m$ is on the order of 1 kΩ and stray capacitances are much smaller then C_L, the corresponding poles s_{p2}, s_{p3}, and s_{p4} are usually at much higher frequencies then s_{p1}. The approximate low-frequency equivalent circuit is therefore that shown in Fig. 4.11. Here the input stage is represented by its simple Norton equivalent circuit, obtained using Eqs. (4.12) and (4.13). The overall transfer function is therefore

$$A_v(s) = \frac{V_{\text{out}}(s)}{V_{\text{in}}^+ - V_{\text{in}}^-} \approx \frac{-g_{mi}R_o}{1 - s/s_{p1}} = \frac{A_v(0)}{1 - (s/s_{p1})}. \tag{4.18}$$

The frequency response is obtained by replacing s by $j\omega$.

4.3. TWO-STAGE OPERATIONAL AMPLIFIERS

The single-stage operational amplifier was discussed in Section 4.2. Another widely used CMOS op-amp uses the two-stage configuration based on the system of Fig. 4.5. This implementation is derived directly from its bipolar-transistor counterpart [5]. A simple two-stage CMOS implementation of the scheme of Fig. 4.5 is shown in Fig. 4.12, where the second gain stage, G_2, drives a source follower. In this circuit

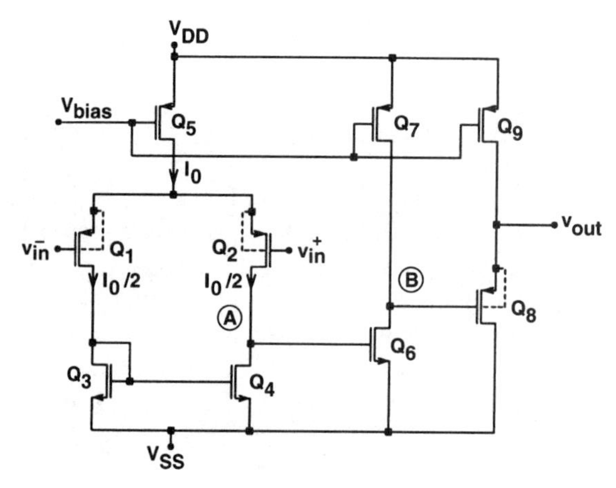

Figure 4.12. Uncompensated two-stage CMOS operational amplifier.

Q_5 acts as a simple current source, and devices Q_1 to Q_5 form a differential stage (cf. Fig. 4.6) with a single-ended output. Transistors Q_6 (acting as the driver device) and Q_7 (acting as the load) form the second gain stage, which also acts as a level shifter. Finally, the source follower consisting of Q_8 as driver and Q_9 as load realizes the output buffer. The low-frequency differential-mode gain of the input stage can be obtained from Eq. (4.8):

$$A_{v1} \approx \frac{g_{mi}}{g_{di} + g_{dl}}, \tag{4.19}$$

where the subscript i refers to input, and l to load device.

Here it is assumed that Q_1 is matched to Q_2, and Q_3 to Q_4. The low-frequency gain of the inverter formed by Q_6 and Q_7 is clearly

$$A_{v2} \approx \frac{-g_{m6}}{g_{d6} + g_{d7}}. \tag{4.20}$$

The overall voltage gain A_v is $A_{v1}A_{v2}$. For typical biasing conditions and device geometries, $A_v = 10{,}000$ to $20{,}000$ can be achieved. The output terminals A and B of both stages are high-impedance nodes; the low-frequency output impedance of the input stage driving node A is

$$R_{o1} \approx \frac{1}{g_{dl} + g_{di}}; \tag{4.21}$$

that of the second stage (Q_6, Q_7) is

$$R_{o2} \approx \frac{1}{g_{d6} + g_{d7}}. \tag{4.22}$$

An equivalent circuit showing these impedances and also the parasitic capacitances C_A and C_B loading nodes A and B, respectively, is shown in Fig. 4.13. It is

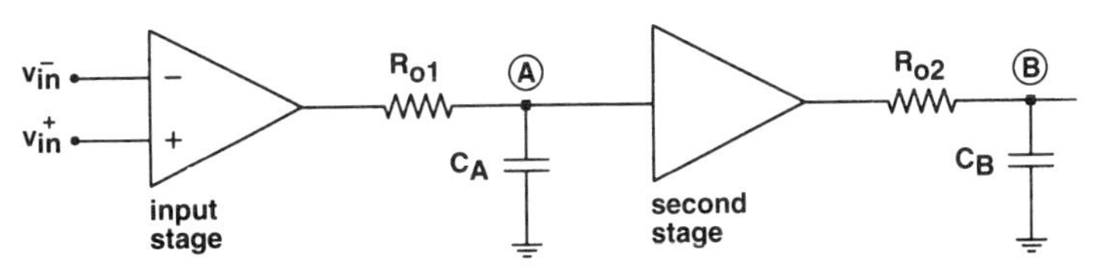

Figure 4.13. Block diagram showing the origin of the dominant poles.

evident from the figure that the transfer function of the amplifier $A_v(s) = V_{out}(s) / [V_{in}^+(s) - [V_{in}^-(s)]$ will contain the factors

$$\frac{1/sC_A}{R_{o1} + 1/sC_A} \frac{1/sC_B}{R_{o2} + 1/sC_B} = \frac{1}{(1 - s/s_A)(1 - s/s_B)}, \tag{4.23}$$

where the poles are $s_A = -1/R_{o1}C_A$ and $s_B = -1/R_{o2}C_B$. Since R_{o1} and R_{o2} are large, s_A and s_B will be close to the $j\omega$ axis in the s plane. Hence they will be the dominant poles of the amplifier. The effects of other poles will be noticeable only at very high frequencies.

If the op-amp is required to drive small internal capacitive leads only, the output source follower (Q_8, Q_9) may be eliminated and the output taken directly from node B. However, even for such capacitive loads, the maximum output current that can be sourced is limited by the current source Q_7.

For very high gain applications, the cascode differential amplifier of Fig. 4.8 can be used as the first stage of the op-amp. A two-stage op-amp with the cascode differential stage is shown in Fig. 4.14. Transistors Q_8 and Q_9 form a level shifter between the output of the first stage and the input of the second stage, to balance the dc level between the signal path. The gain of the first stage is given by Eq. (4.11), and the total gain is

$$A_v = g_{mi}R_{o1} \frac{g_{m6}}{g_{d6} + g_{d7}}, \tag{4.24}$$

where R_{o1} is given by Eq. (4.10). The frequency response given by Eq. (4.23) is still valid, with $s_A = -1/R_{o1}C_A$, where R_{o1} is replaced by its new value given by Eq. (4.10).

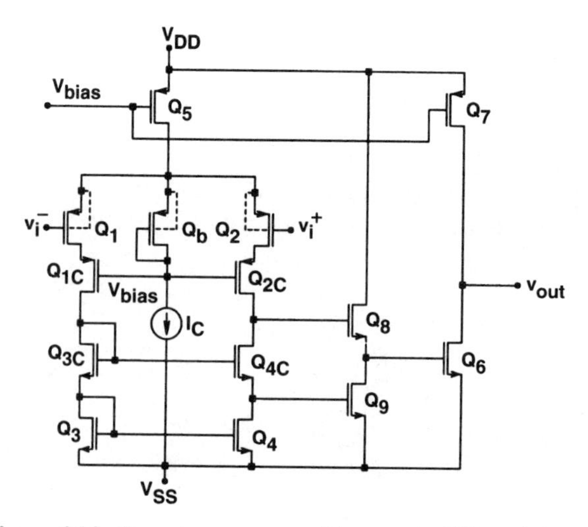

Figure 4.14. Two-stage op-amp with cascode differential stage.

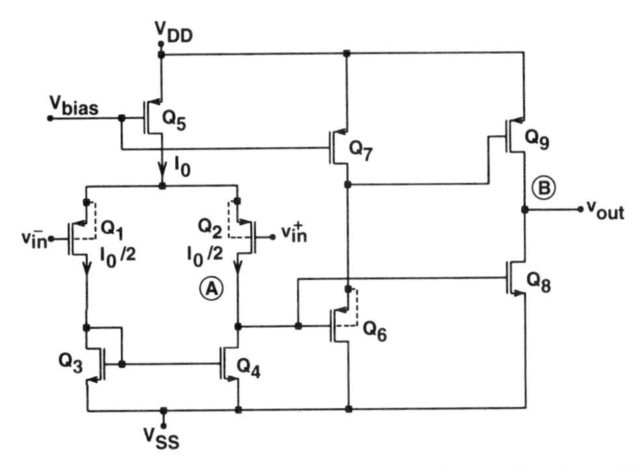

Figure 4.15. Improved uncompensated CMOS operational amplifier.

An improved CMOS op-amp [6] with increased output range and current drive capability is shown in Fig. 4.15. The output stage consists of devices Q_6 to Q_9, with Q_6 and Q_7 acting as a level shifter and Q_8 and Q_9 acting as a class B push-pull output stage. The dc biasing is designed so that Q_8 and Q_9 have equal-valued small gate-to-source dc biases. This maximizes the linear v_{out} range. The conceptual form of the CMOS gain stage with the level shifter is shown in Fig. 4.16.

The low-frequency small-signal gain of the second stage can be found from the equivalent circuit of Fig. 4.17. The node equation is

$$(g_{m8} + g_{m9})v_{\text{in}} + (g_{d8} + g_{d9})v_{\text{out}} = 0, \tag{4.25}$$

so that

$$A_{v2} = \frac{v_{\text{out}}}{v_{\text{in}}} = -\frac{g_{m8} + g_{m9}}{g_{d8} + g_{d9}}. \tag{4.26}$$

Since g_m can be 100 times larger than g_d, the gain is high. The low-frequency small-signal differential gain can easily be found from Eqs. (4.19) and (4.26):

$$A_v \simeq \frac{-g_{mi}}{g_{dl} + g_{di}} \times \frac{g_{m8} + g_{m9}}{g_{d8} + g_{d9}}. \tag{4.27}$$

Thus $|A_v|$ can be as high as 20,000. However, if the circuit has to drive a resistive load G_L, then $g_{d8} + g_{d9}$ is replaced by $g_{d8} + g_{d9} + G_L$, which normally reduces the gain significantly. Also, for a large load capacitance C_L, the pole of the compensated op-amp in a feedback arrangement resulting from the time constant $C_L/(g_{d8}$

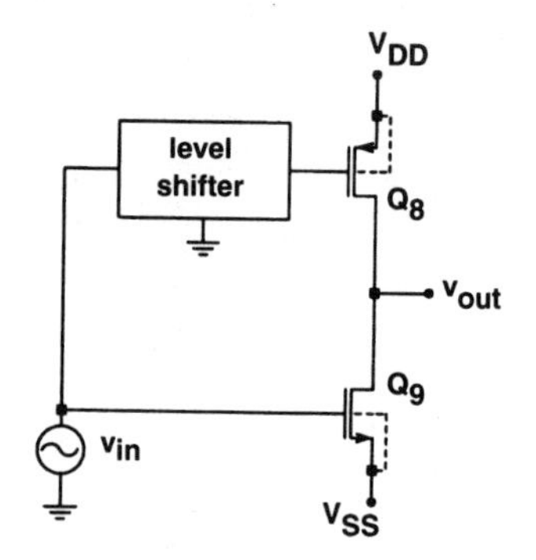

Figure 4.16. CMOS gain stage with level shifter.

$+ g_{d9})$ may move so close to the $j\omega$ axis that instability occurs. Hence this op-amp is again suited only for driving small-to-moderate-sized internal capacitive loads.

If the circuit of Fig. 4.15 is to be used to drive resistive loads, an output buffer stage must be added. This may be simply a source follower, similar to that in Fig. 3.33. However, better output current sourcing and sinking, and lower output impedance, can be obtained using more elaborate CMOS output buffers. These are discussed in Section 4.9.

Consider now the high-frequency behavior of the circuit of Fig. 4.15. As before, nodes A and B are at a high impedance level and are responsible for the dominant poles. The approximate equivalent circuit is shown in Fig. 4.18. Here, the input stage is represented by a simple Norton equivalent, obtained using Eqs. (4.19) and (4.21). Similarly, the Norton equivalent of the output stage can be found from Eqs. (4.26) and (4.22). As before, the transfer function contains a factor similar to that given in Eq. (4.23), where now $s_A = -(g_{d2} + g_{d4})/C_A$ and $s_B = -(g_{d8} + g_{d9})/C_L$. The overall transfer function is therefore

$$A_v(s) = \frac{V_{out}(s)}{V_{in}^+ - V_{in}^-} \approx \frac{g_{m1}}{g_{d2} + g_{d4}} \frac{g_{m8} + g_{m9}}{g_{d8} + g_{d9}} \frac{1}{(1 - s/s_A)(1 - s/s_B)} \cdot \quad (4.28)$$

$$= \frac{A_v(0)}{(1 - s/s_A)(1 - s/s_B)} \cdot$$

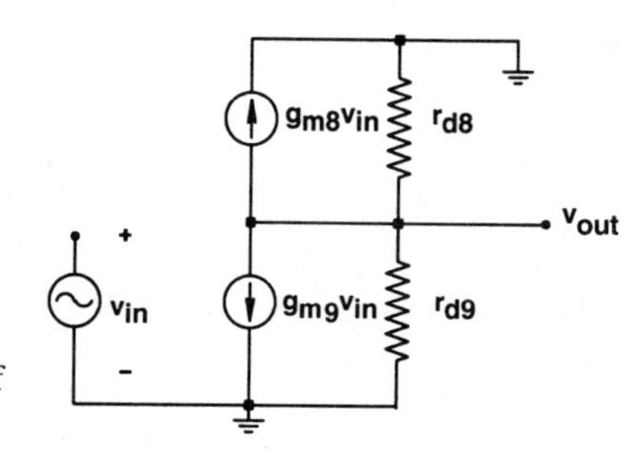

Figure 4.17. Small-signal equivalent circuit of the CMOS gain stage.

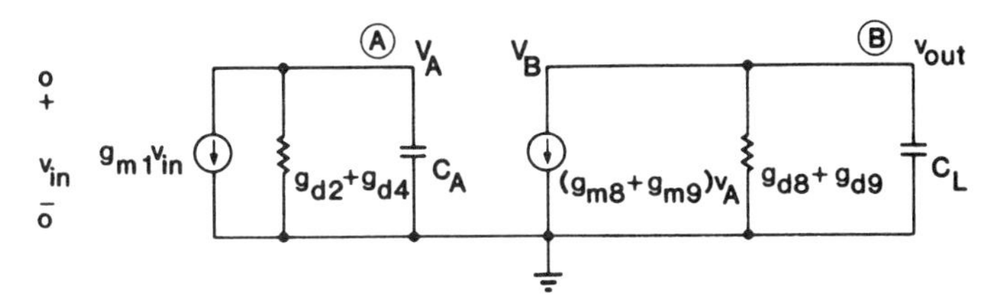

Figure 4.18. Two-stage representation of the CMOS operational amplifier of Fig. 4.15.

The frequency response is obtained by replacing s by $j\omega$. For low frequencies ($\omega \ll |s_A|, |s_B|$),

$$A_v(j\omega) \approx A_v(0) = \frac{g_{m1}}{g_{d2} + g_{d4}} \times \frac{g_{m8} + g_{m9}}{g_{d8} + g_{d9}}. \tag{4.29}$$

For high frequencies ($\omega \gg |s_A|, |s_B|$),

$$A_v(j\omega) \approx \frac{A_v(0)}{-\omega^2/s_A s_B} = -\frac{g_{m1}(g_{m8} + g_{m9})}{\omega^2 C_A C_L}. \tag{4.30}$$

Hence, for high frequencies, the amplifier inverts the input voltage. In switched-capacitor applications the op-amp always has a feedback capacitor C connected between its output and its inverting input terminals. A typical circuit is shown in Fig. 4.19. A sine-wave signal $V_{in}(j\omega)$ appearing at the inverting input terminal will thus be amplified by $-A_v(j\omega)$ and fed back to the input via capacitive divider C and C_1; here C_1 represents the overall capacitance of the input circuit driving the op-amp, including stray capacitances, and so on.

The op-amp and capacitor C and C_1 form a feedback loop, with a loop gain

$$A_L = -A_v(j\omega)\frac{C}{C + C_1}. \tag{4.31}$$

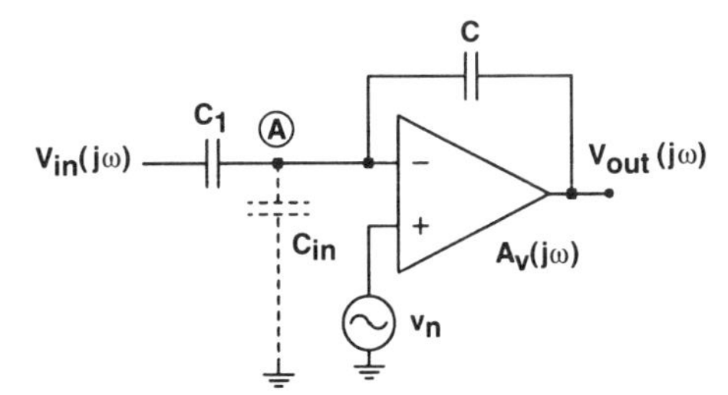

Figure 4.19. Operational amplifier with feedback capacitor C and input capacitor C_{in}.

In addition to the input voltage v_{in}, the circuit also contains the voltage v_n, representing the noise generated in (or coupled to) the op-amp. The circuit can be analyzed by using the node equation at node A:

$$C_1[V_{\text{in}}(j\omega - V_{\text{in}}^-(j\omega)] + C[V_{\text{out}}(j\omega) - V_{\text{in}}^-(j\omega)] = 0 \tag{4.32}$$

and the op-amp gain relation:

$$V_{\text{out}}(j\omega) = A_v(j\omega)[V_n(j\omega) - V_{\text{in}}^-(j\omega)]. \tag{4.33}$$

This gives

$$V_{\text{out}}(j\omega) = \frac{-C_1 V_{\text{in}}(j\omega) + (C + C_1)V_n(j\omega)}{C + (C + C_1)/A_v(j\omega)}. \tag{4.34}$$

The output voltage can become (in theory) infinite if the input signal or noise contains a sine-wave component with a frequency ω_1, such that

$$\begin{aligned} C + \frac{C + C_1}{A_v(j\omega_1)} &= 0, \\ A_v(j\omega_1) &= -1 - \frac{C_1}{C}. \end{aligned} \tag{4.35}$$

By Eq. (4.31) this corresponds to a loop gain of $A_L = 1$, a condition for oscillation.

At dc and very low frequencies Eq. (4.29) shows that $A_v(j\omega)$ is positive, and hence Eq. (4.35) cannot be valid. However, at high frequencies, by Eq. (4.30), $A_v(j\omega)$ becomes negative real, and at some ω_1 it may satisfy Eq. (4.35). When this occurs, the circuit will become unstable, and it will oscillate with a frequency ω_1. In theory, for our two-pole model, $A_v(j\omega)$ becomes negative real only for $\omega \to$ '; however, for large loop gain the circuit is only marginally stable for high frequencies, so that any additional small phase shift due to the high-frequency poles neglected in Fig. 4.18 may cause oscillation. Even if stability is retained, the transient response contains a lightly damped oscillation, which is unacceptable in most applications.

To prevent oscillation in feedback amplifiers, and to ensure a good transient response, an additional design step (called *frequency compensation*) is needed. It is based on the stability theory of feedback systems and is discussed briefly in the next section.

4.4. STABILITY AND COMPENSATION OF CMOS AMPLIFIERS

In Section 4.3 it was shown that the CMOS op-amp of Fig. 4.15 is only marginally stable when used in a feedback circuit. In this section the analysis of stability and the design steps required to ensure stable feedback op-amps are discussed.

A systematic investigation of stability can be based on the general block diagram of Fig. 4.20, which shows an op-amp in a negative feedback configuration. It is

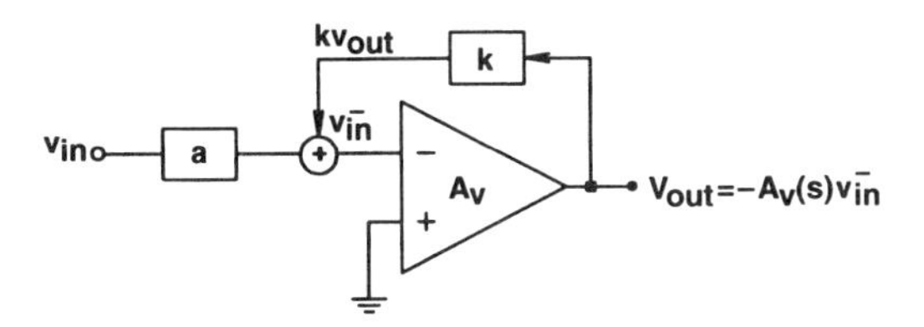

Figure 4.20. Operational amplifier with negative feedback.

assumed that k and a are positive constants, and $k \leq 1$. The voltage at the inverting input terminal is

$$V_{in}^{-} = aV_{in} + kV_{out} \tag{4.36}$$

and the output voltage is

$$V_{out} = A_v(s)V_{in}^{-}. \tag{4.37}$$

Hence the voltage gain is

$$A_{vf}(s) = \frac{V_{out}(s)}{V_{in}(s)} = \frac{-aA_v(s)}{kA_v(s) + 1}. \tag{4.38}$$

A_{vf} is often called the closed-loop gain, A_v is the open-loop gain of the system; and kA_v is the *loop gain.*

We assume next that all poles s_i of $A_v(s)$ are due to stray capacitances to ground in an otherwise resistive circuit. (This is an acceptable approximation if inductive effects are negligible, and all capacitances loading the high-impedance nodes are connected between voltages that are in phase or 180° out of phase of each other.) Then all s_i are negative real numbers, and $A_v(s)$ is in the form

$$A_{vf}(s) = \frac{K}{(s - s_1)(s - s_2)\cdots(s - s_n)}. \tag{4.39}$$

For $s = j\omega$, $A_v(j\omega)$ gives the frequency response of the op-amp. Its magnitude is

$$|A_v(j\omega)| = \frac{|K|}{\Pi_{i=1}^{n}(\omega^2 + |s_i|^2)^{1/2}} \tag{4.40}$$

and its phase is given by

$$\angle A_v(j\omega) = \angle K - \sum_{i=1}^{n} \tan^{-1}\frac{\omega}{|s_i|}. \tag{4.41}$$

Note that both $|A_v(j\omega)|$ and $\angle A_v(j\omega)$ are monotone-decreasing functions of ω.

The natural frequencies of the overall feedback system are the poles s_p of $A_{vf}(s)$, which by Eq. (4.38) satisfy the relation

$$kA_v(s_p) + 1 = 0. \qquad (4.42)$$

For stability, all s_p must be in the negative half of the s plane; that is, the real parts of all poles must be negative. Now assume that $\text{Re}[kA_v(j\omega)] > -1$ for all real values of ω. Then $kA_v(j\omega) \neq -1$, and hence no s_p can occur on the $j\omega$ axis; furthermore, it can easily be proven that if $A_v(s)$ has only poles with negative real parts, then under the stated assumption, so will $A_{vf}(s)$. The proof is implied in Problem 4.26. Thus the condition

$$\text{Re}[kA_v(j\omega)] > -1 \qquad \text{for all } \omega \qquad (4.43)$$

is *sufficient* to ensure stability. It is not, however, a *necessary* condition. Two other sufficient conditions for stability can also readily be stated. Let ω_{180} be the frequency at which the monotone decreasing phase of $kA_v(j\omega)$ reaches $-180°$; that is,

$$\angle kA_v(j\omega_{180}) = -180°. \qquad (4.44)$$

If now $|kA_v(j\omega_{180})| < 1$, Eq. (4.42) cannot hold on the $j\omega$ axis, and hence the circuit is stable. A measure of its stability is the *gain margin,* defined as

$$\text{GM} = \text{gain margin (in dB)} = 20 \log_{10}|kA_v(j\omega_{180})|. \qquad (4.45)$$

The gain margin must be negative for stability; the more negative it is, the larger the margin of stability of the circuit. Normally, a margin of at least 20 dB is desirable.*

Next, let $|kA_v(j\omega_{180})|$, which also decreases monotonically with ω, reach the value 1 (i.e., 0 dB) at the *unit-gain frequency* ω_0. Then, if the phase at ω_0 satisfies $\angle kA_v(j\omega_0) > -180°$, the system will be stable. The *phase margin* PM, defined as $\angle kA_v(j\omega_0) + 180°$, is a measure of its stability; the larger the phase margin, the more stable the circuit. Usually, at least a 60° (and preferably larger) margin is required. This will also give a desirable (i.e., nonringing) step response for the closed-loop amplifier. The overshoot, OS, of the step response of the feedback system decreases rapidly with increasing phase margin: for PM = 60°, OS = 8.7%; for PM = 70°, OS = 1.4%, and for PM = 75°, OS = 0.008%.

All the stability conditions above can readily be visualized and checked using *Bode plots.* These show $|kA_v(j\omega)|$ (in decibels) and $\angle kA_v(j\omega)$ (in degrees) as func-

* The gain margin is harder to control, and hence much less often used, than the phase margin described next.

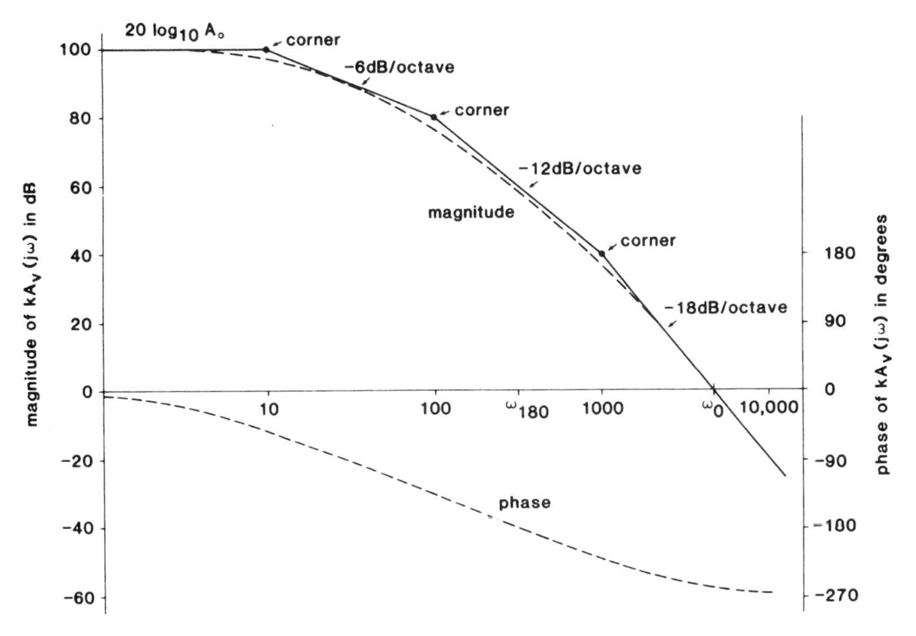

Figure 4.21. Bode plot for three real poles.

tions of ω on a logarithmic scale. Typical plots are shown in Fig. 4.21 (dashed curves) for the three-pole loop gain:

$$kA_v(j\omega) = \frac{A_0}{(1 - j\omega/s_1)(1 - j\omega/s_2)(1 - j\omega/s_3)}, \tag{4.46}$$

with $A_0 = 10^5$, $s_1 = -10$ rad/s, $s_2 = -10^2$ rad/s, and $s_3 = -10^3$ rad/s. Drawing the magnitude plot is simplified by using an asymptotic approximation to the logarithmic magnitude of the general term $a_i(j\omega) = 1/(1 - j\omega/s_1)$:

$$20 \log_{10}|a_i| = -20 \log_{10}\left|1 - \frac{j\omega}{s_i}\right| = -10 \log_{10}\left(1 + \frac{\omega^2}{|s_i|^2}\right). \tag{4.47}$$

Clearly, $20 \log_{10}|a_i| \approx 0$ for $|\omega| \ll |s_i|$, and $20 \log_{10}|a_i| \approx -20(\log_{10}\omega - \log_{10}|s_i|)$for $|\omega| \gg |s_i|$. Figure 4.22 illustrates the approximation of $|a_i|$ and also the phase $\angle a_i(j\omega)$ of $a_i(j\omega)$. An important conclusion which can be drawn from the figure is that for $|\omega| \gg |s_{i},|$, $20 \log_{10}|a_i|$ approaches a straight line with a slope of -6 dB/octave (i.e., decreases by 6 dB for each doubling of ω), while $\angle a_i$ approaches $-90°$ in this same region. In particular, $|\angle a_i| \approx 90°$ for $\omega > 5|s_i|$. Also, $|\angle a_i| \leq 30°$ for $\omega < 0.5|s_i|$; this fact will be used later.

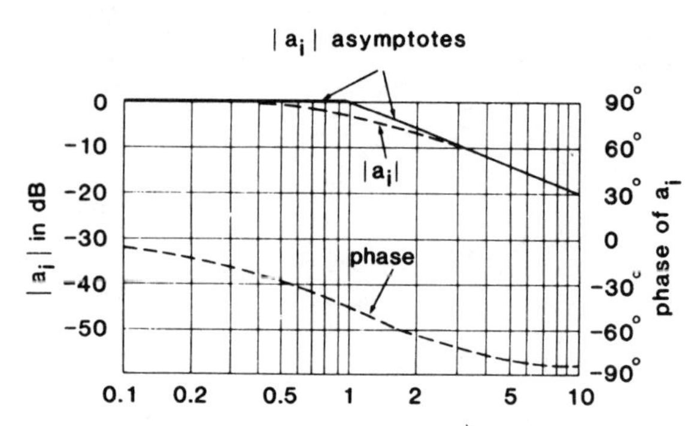

Figure 4.22. Gain and phase responses for a factor $a_i(j\omega) = (1 - j\omega/s_i)^{-1}$.

The logarithmic form of the loop gain satisfies

$$20 \log_{10} [kA_v (j\omega)] = 20 \log_{10} A_0 + \sum_{i=1}^{n} 20 \log_{10} [a_i(j\omega)]. \qquad (4.48)$$

Therefore, at the unity-gain frequency ω_0 the slope of the logarithmic loop gain versus logarithmic frequency is approximately $-6m$ dB/octave, while its angle is around $-90m$ degrees. Here, m is the number of those poles whose magnitude $|s_i|$ is less than ω_0. Clearly, for a substantial positive phase margin (say 60°, so that $\angle kA_v(j\omega_0) > 120°$), m should be less than 2. Ideally, m is 1 (i.e., there is only one pole satisfying $|s_i| < \omega_0$), and the other poles have much larger magnitudes than ω_0. Then the phase margin is close to 90°. (For $A_0 \gg 1$, $m = 0$ is impossible.)

Returning to the example of Fig. 4.21, the solid lines show the asymptotic approximation to the logarithmic magnitude of $kA_v(j\omega)$. The curves indicate that at the unity-gain frequency $\omega_0 \approx 4$ krad/s, the phase of $kA_v(j\omega)$ is about $-270°$. Hence the phase margin is negative, and the feedback system is potentially unstable.

The modification of $kA_v(s)$, which changes an unstable feedback system into a stable system, is called *frequency compensation.* Its purpose is usually to achieve the ideal situation described above; thus we aim to realize a loop gain that contains exactly one pole smaller in magnitude than ω_0, while all others are much larger. Since the feedback factor k can be anywhere in the $0 < k \leq 1$ range, and $k = 1$ represents the worst case (i.e., the largest ω_0 and hence the smallest phase margin), this will be assumed from here on. Note that $k \simeq 1$ corresponds to $C \gg C_{in}$ in Fig. 4.19 and $k = 1$ represents a short circuit between the output and the inverting input of the amplifier. Such a circuit is shown in Fig. 4.77 in connection with Problem 4.13.

It will next be shown how to carry out the compensation for the op-amp of Fig. 4.15. Referring to its equivalent small-signal representation (Fig. 4.18), we will first

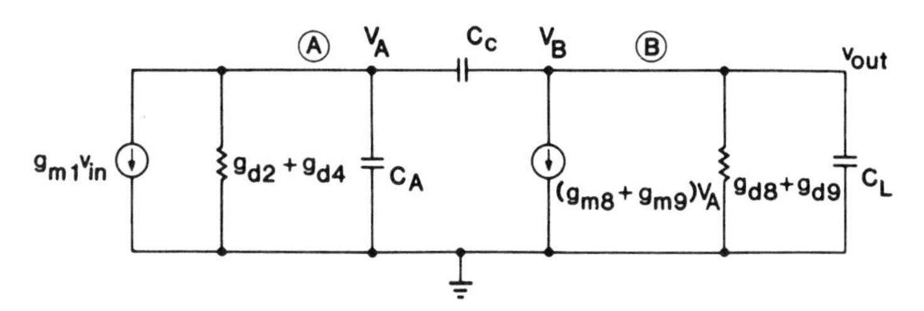

Figure 4.23. Two-stage representation of a CMOS operational amplifier with a pole-splitting capacitor C_c.

attempt to achieve compensation by connecting a compensating capacitor C_c between the high-impedance nodes A and B, as shown in Fig. 4.23. It is well known (see Ref. 2, Sec. 9.4) that for bipolar op-amps the addition of such capacitor moves the pole associated with node A to a much lower frequency, while that corresponding to node B becomes much larger. It is therefore often called a *pole-splitting capacitor* and (for bipolar op-amps) accomplishes the desired compensation. The situation is less favorable for MOS op-amps, as will be shown next. The node equations for nodes A and B in Fig. 4.23 are

$$g_{m1}V_{in} + (g_{d2} + g_{d4} + sC_A)V_A + sC_c(V_A - V_{out}) = 0 \quad (4.49)$$

and

$$sC_c(V_{out} - V_A) + (g_{m8} + g_{m9})V_A + (g_{d8} + g_{d9} + sC_L)V_{out} = 0. \quad (4.50)$$

Solving for V_{out}, the voltage gain

$$A_v(s) = \frac{A_0(1 - s/s_z)}{(1 - s/s_{p1})(1 - s/s_{p2})} \quad (4.51)$$

results. Hence the dc gain is

$$A_0 = \frac{g_{m1}(g_{m8} + g_{m9})}{(g_{d2} + g_{d4})(g_{d8} + g_{d9})} \quad (4.52)$$

and the zero is

$$s_z = \frac{g_{m8} + g_{m9}}{C_c}. \quad (4.53)$$

The calculation of the poles is simplified if it is a priori assumed that $|s_{p2}| \gg |s_{p2}|$ and that $g_{m8} + g_{m9} \gg g_{d2} + g_{d4}$ or $g_{d8} + g_{d9}$. Then, after some calculation [2, p. 519] (see Problem 4.6),

$$s_{p1} \approx \frac{(g_{d2} + g_{d4})(g_{d8} + g_{d9})}{(g_{m8} + g_{m9})C_c} = -\frac{g_{m1}}{A_0 C_c}, \tag{4.54}$$

$$s_{p2} \approx \frac{-(g_{m8} + g_{m9})}{C_A C_L + (C_A + C_L)C_c} = -\frac{-(g_{m8} + g_{m9})/C_A C_L}{1/C_c + 1/C_A + 1/C_L}, \tag{4.55}$$

where A_0 is the dc gain given in (4.52).

Physically, C_c (multiplied by the Miller effect) is added in parallel to C_A, thus reducing $|s_{p1}|$ by a very large (ca. 10^3) factor, while at the same time it increases the second pole frequency $|s_{p2}|$ via shunt feedback.

Clearly, $|s_{p1}|$ decreases, while $|s_{p2}|$ increases with increasing values of C_c. Thus C_c indeed splits the poles apart, as originally intended. Unfortunately, the desired compensation is nevertheless usually not achieved, due to the positive (right-half-plane) zero s_z. For the usual case of $1/C_c \ll 1/C_A + 1/C_L$, the inequalities

$$|s_{p2}| \approx \frac{g_{m8} + g_{m9}}{C_A + C_L} \gg s_z > |s_{p1}| \tag{4.56}$$

hold. The logarithmic magnitude of the factor $(1 - j\omega/s_z)$ is near zero for $|\omega| \ll s_z$, while it *increases* by about 6 dB/octave for $|\omega| \gg s_z$. The phase of the factor is $-\tan^{-1}(\omega/s_z)$; it *decreases* from 0 to $-90°$ as ω grows from zero to infinity. As a result, the plots shown in Fig. 4.24 are obtained. Clearly, at the unity-gain frequency ω_0 the phase is less than $-180°$. Hence in a feedback configuration the amplifier

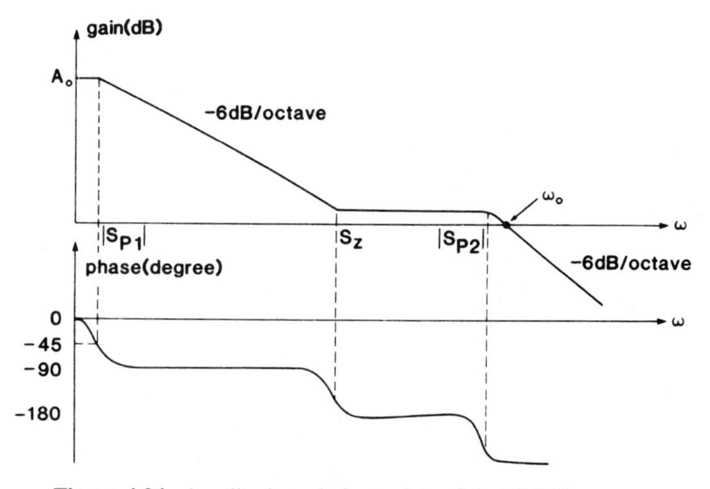

Figure 4.24. Amplitude and phase plots of the CMOS op-amp.

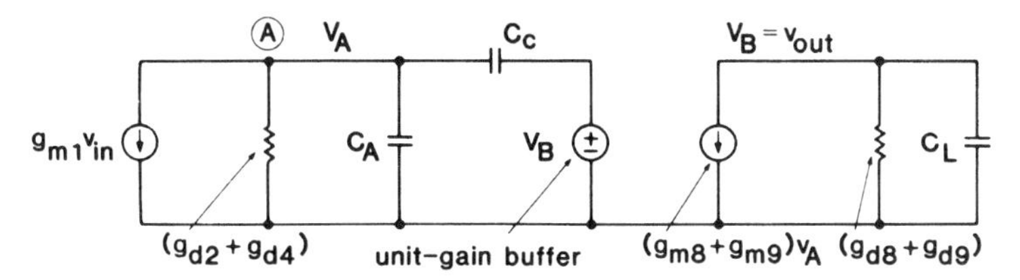

Figure 4.25. Unity-gain buffer arrangement used to eliminate the right-plane zero.

can become unstable. Note that if $g_{m8} + g_{m9}$ would be increased, $s_z/|s_{p1}|A_0$ would, by Eqs. (4.52) to (4.54), increase proportionally. It is clear from Fig. 4.24 that if $s_z/|s_{p1}|$ (in octaves) is greater than A_0 (in dB)/6, the unity-gain frequency ω_0 is less than s_z, and the phase margin is positive. Thus, for sufficiently high g_m values (such as are afforded by bipolar transistors), the inclusion of C_c accomplishes the desired stabilization. Unfortunately, the transconductance of MOSFETs is normally not high enough for the purpose, and other arrangements must be found to eliminate s_z.

One scheme for getting rid of s_z is to shift it to infinite frequency. Physically, the zero is due to the existence of two paths through which the signal can propagate from node A to node B. The first is through C_c, while the second is by way of the controlled source $(g_{m8} + g_{m9})v_A$. For $s = s_z$ the two signals from these paths cancel, and a transmission zero occurs. The zero can be shifted to infinite frequency by eliminating the feedforward path through C_c, at the cost of an extra unity-gain buffer (Fig. 4.25). A detailed analysis shows (Problem 4.7) that the numerator of $A_v(s)$ is now simply A_0, while the denominator remains nearly the same as in Eq. (4.51). A circuit implementing this scheme is shown in Fig. 4.26, where Q_{10}/Q_{11} form the buffer.

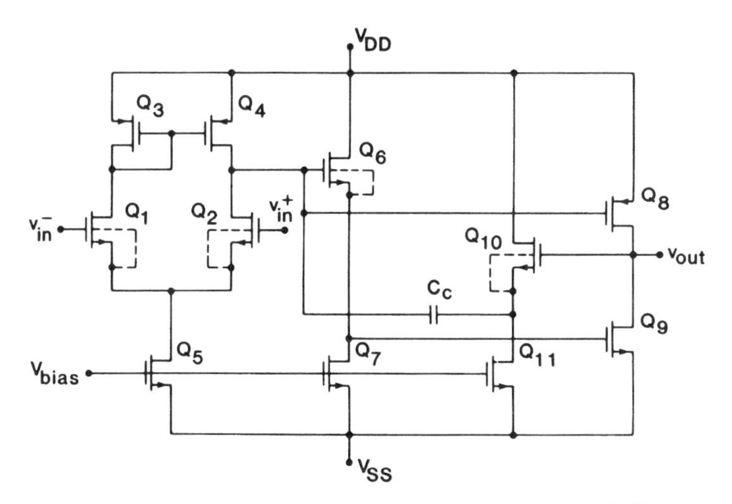

Figure 4.26. Internally compensated CMOS op-amp with unity-gain buffer used to avoid the right-half-plane zero.

An alternative (and simpler) scheme [7] can also be used. Consider the circuit shown in Fig. 2.27. Nodal analysis shows (Problem 4.8) that its transfer function is

$$A_v(s) = \frac{A_0(1 - s/s_z)}{(1 - s/s_{p1})(1 - s/s_{p2})(1 - s/s_{p3})}. \tag{4.57}$$

Here A_0, s_{p1}, and s_{p2} are (as before) given by Eqs. (4.52), (4.54), and (4.55), while now

$$s_z = -\frac{1}{[R_c - 1/(g_{m8} + g_{m9})]C_c}, \tag{4.58}$$

$$s_{p3} = -\frac{1}{R_c}\left(\frac{1}{C_c} + \frac{1}{C_A} + \frac{1}{C_L}\right). \tag{4.59}$$

As Eq. (4.58) shows, it is again possible for this circuit to shift s_z to infinity, if $R_c = 1/(g_{m8} + g_{m9})$ is chosen. Then, choosing a sufficiently large value for C_c can split the poles. To quantify this, it is reasonable to require that $|s_{p2}| > \omega_0$, the unity-gain frequency. For this choice, since in the frequency region between $|s_{p1}|$ and $|s_{p2}|$

$$A_v(j\omega) \simeq \frac{A_0}{j\omega / |s_{p1}|} \tag{4.60}$$

holds, the approximation $\omega_0 \approx A_0 |s_{p1}|$ can be used. Thus $|s_{p2}| > A_0|s_{p1}|$ may be specified. From Eq. (4.55), with $1/C_A \gg (1/C_c + 1/C_L)$, we require

$$\frac{g_{m8} + g_{m9}}{C_L} > \frac{g_{m1}}{C_c}, \tag{4.61}$$

so that a feedback capacitor satisfying

$$C_c > \frac{g_{m1}C_L}{g_{m8} + g_{m9}} \tag{4.62}$$

is needed. Since experience indicates that normally $C_c \sim C_L$ is a good choice, we require that $g_{m1} < g_{m8} + g_{m9}$. Another way of eliminating s_z for the circuit of Fig. 4.27 is by pole–zero cancellation. Choosing $s_z = s_{p2}$, from Eqs. (4.55) and (4.58),

$$R_c \simeq \frac{1 + (C_A + C_L)/C_c}{g_{m8} + g_{m9}} \tag{4.63}$$

is obtained. The resulting cancellation leaves the op-amp with a two-pole response.

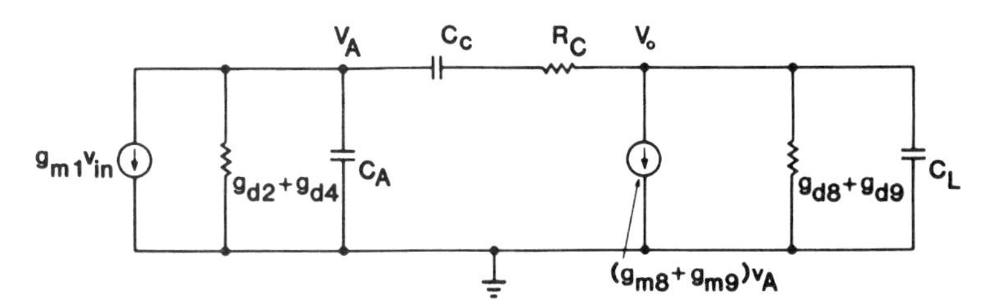

Figure 4.27. Small-signal equivalent circuit of CMOS op-amp with nulling resistor for compensation.

Compensation now requires that $|s_{p3}| > A_0|s_{p1}|$. Using Eqs. (4.55), (4.59), and (4.61), this condition can be rewritten in the form

$$C_c > C_c^0 \frac{1/C_c + 1/(C_A + C_L)}{1/C_c + 1/C_L + 1/C_A}. \tag{4.64}$$

Here C_c^0 is the bound given for C_c on the right-hand side of Eq. (4.62). The factor multiplying C_c^0 in Eq. (4.64) is usually much smaller than 1; hence, now a smaller C_c can be used. Its value can be obtained from the bound*

$$C_c > \frac{1}{2}\frac{g_{m1}/(g_{m8} + g_{m9}) - 1}{1/C_A + 1/C_L} + \left[\frac{1}{4}\left(\frac{g_{m1}/(g_{m8} + g_{m9}) - 1}{1/C_A + 1/C_L}\right)^2 + \frac{g_{m1}C_AC_L}{g_{m8} + g_{m9}}\right]^{1/2}. \tag{4.65}$$

The actual implementation of the scheme of Fig. 4.27 in the CMOS op-amp of Fig. 4.15 is shown in Fig. 4.28. The parallel-connected channels of the complementary transistors Q_{10} and Q_{11} form R_c. This push-pull arrangement helps to suppress even harmonics and thus improves the linearity of the resistor R_c.

Note that the condition $R_c = 1/(g_{m8} + g_{m9})$ is easily obtained by matching Q_{10} and Q_{11} to Q_9 and Q_{10}, respectively. Satisfying Eq. (4.63) is somewhat harder; however, the accuracy is not critical, and as explained above, this choice for R_c results in a smaller value for C_c.

The pole-splitting frequency compensation technique described so far is applicable to a two-stage topology where the output stage is a common-source gain stage preceded by a differential stage preamplifier. If the op-amp is loaded with a very small resistance, the gain of the output stage can become so small that the two-stage solution may not have enough dc gain. In this case, using the cascode load differential amplifier shown in Fig. 4.7 can increase the gain of the two-stage amplifier. Alternatively, the folded-cascode topology of Fig. 4.10 can also be used as the input stage

* In practice, it is usual to choose $C_c \approx C_L$. This choice satisfies the constraints of Eqs. (4.64) and (4.65) with a large margin.

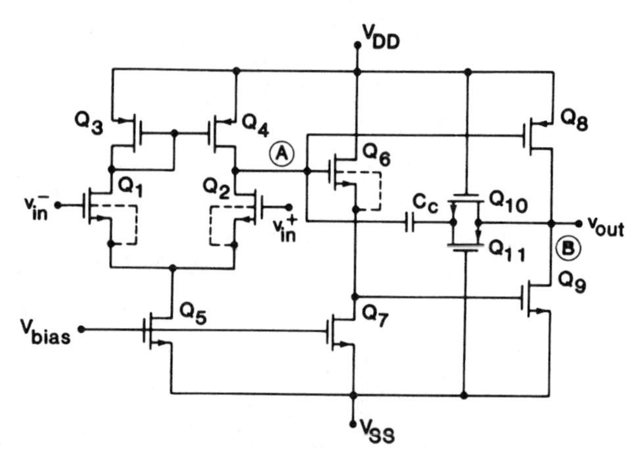

Figure 4.28. Improved internally compensated CMOS operational amplifier.

in which case the op-amp can be frequency stabilized with a pole-splitting capacitor that is connected between the two high-impedance nodes. Another method to enhance the op-amp gain is to use a multistage configuration. The simplest approach would be to insert a positive gain intermediate stage between the input and output stages. This is shown in simplified form in Fig. 4.29. The three-stage amplifier has three dominant poles, one at the output of each gain stage. The simple pole splitting does not remove the third pole, and to stabilize such a topology a nested Miller [8] compensation must be used. This compensation technique uses the two capacitors C_{c1} and C_{c2} shown in Fig. 4.29. C_{c1} is connected between the final output and the output of the intermediate stage. C_{c2} is connected between the final output and the output of the differential stage. Figure 4.30 shows the three-stage op-amp of Fig. 4.29 in more detail. It consists of a p-channel input differential pair Q_1–Q_4, followed

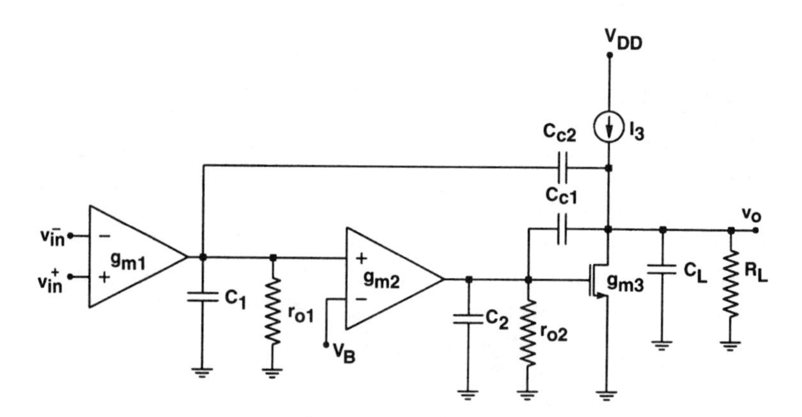

Figure 4.29. Nested Miller compensation scheme for a three-stage op-amp.

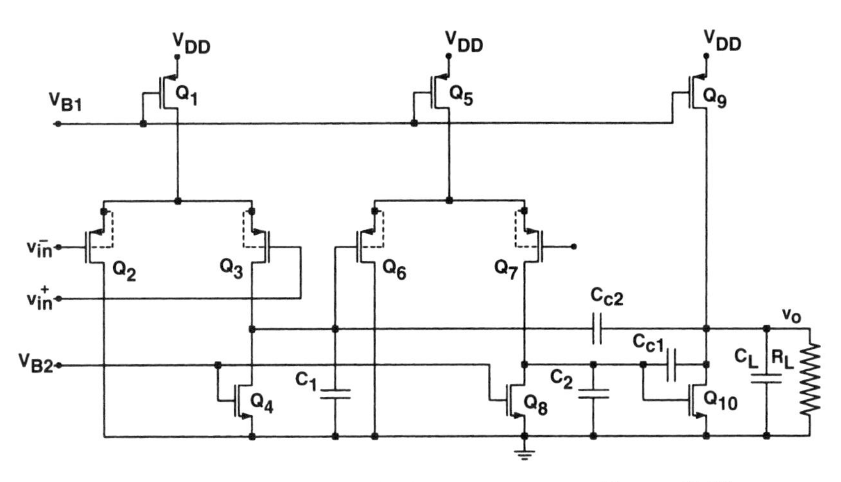

Figure 4.30. Simplified circuit diagram of a three-stage op-amp with nested Miller compensation.

by the differential pair Q_5–Q_8, that serves as the positive gain intermediate stage. The output stage is a common-source amplifier made of transistors Q_9 and Q_{10}. The op-amp is stabilized by capacitors C_{c1} and C_{c2}. The small-signal equivalent circuit of the three-stage amplifier using the nested Miller compensation scheme is shown in Fig. 4.31 [8].

The open-loop gain of the uncompensated op-amp has three dominant poles s_{p1}, s_{p2}, and s_{p3}. The location of the three poles are given by

$$s_{p1} = \frac{1}{C_1 r_{o1}}, \tag{4.66}$$

$$s_{p2} = \frac{1}{C_2 r_{o2}}, \tag{4.67}$$

$$s_{p3} = \frac{1}{C_L R_{\text{eq}}}, \tag{4.68}$$

where $R_{\text{eq}} = r_{o3} \| R_L$ is the equivalent output impedance of the third stage.

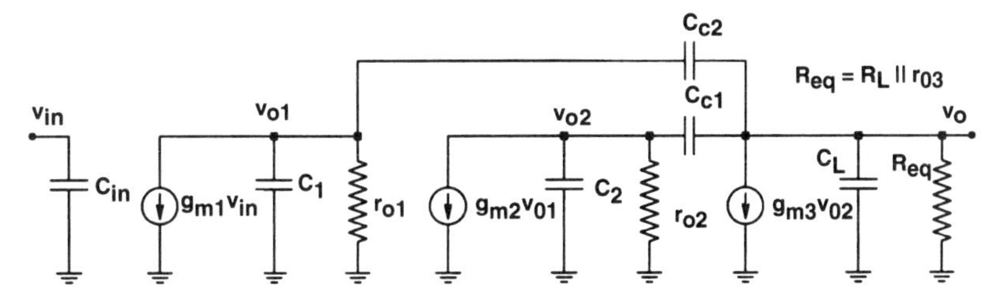

Figure 4.31. Small-signal equivalent circuit of the three-stage op-amp.

The transfer function of the uncompensated op-amp is

$$\frac{V_o(s)}{V_{in}(s)} = \frac{-A_0}{(s/s_{p1} + 1)(s/s_{p2} + 1)(s/s_{p3} + 1)}, \tag{4.69}$$

where A_0 is the dc open-loop gain given by

$$A_0 = g_{m1}r_{o1}\, g_{m2}r_{o2}g_{m3}R_{eq}. \tag{4.70}$$

The magnitude plot of the combined intermediate- and output-stage gain is shown in Fig. 4.32*a,* where $f_2 = 1/(2\pi\, r_{o2}C_2)$ and $f_3 = 1/(2\pi\, R_{eq}C_L)$ are the pole frequencies. The combination of the intermediate and output stages is compensated by the first Miller capacitor C_{c1}. The insertion of this capacitor splits the poles such that f_3 is shifted to a higher frequency, f_3', and f_2 to a lower frequency, f_2'. The new location of the poles is given by

$$s'_{p2} \simeq -\frac{1}{r_{o2}g_{m3}R_{eq}C_{c1}}, \tag{4.71}$$

$$s'_{p3} \simeq -\frac{g_{m3}}{C_L(1 + C_2/C_{c1}) + C_2}. \tag{4.72}$$

It is worth noting that the insertion of C_{c1} splits s_{p2} and s_{p3} to the same location, as was the case with the poles of the two-stage op-amp described earlier in this chapter. Also, the location of the pole s_{p1}, corresponding to the input stage, remains unaltered.

The frequency characteristic of the complete three-stage op-amp is shown in Fig. 4.32*b* where the third pole s_{p1} corresponding to the input stage has been added. The result of inserting the first compensation capacitor C_{c1} is a frequency response that contains the two dominant poles s_{p1} and s'_{p2}. These two poles can be split by inserting the second compensation capacitor C_{c2}, which shifts f_1 to a higher frequency, f_1', and f_2' to a lower frequency, f_2'' (dominant pole). The result is an op-amp with an open-loop response with one dominant pole at $s''_{p2}$and a magnitude response that has a straight 6-dB/octave roll-off from the dominant pole frequency f''_{p2} up to the unity-gain frequency.

Inserting the two nested Miller compensation capacitors C_{c1} and C_{c2}, as in the case of the two-stage op-amp, introduces right-half-plane zeros. Similar strategies described earlier in this section, such as the zero blocking technique or placing a resistor in series with the Miller capacitor to cancel the zero, can be used to eliminate the effect of the unwanted zeros.

Next, the stability conditions of the single-stage folded cascode op-amp shown in Fig. 4.10 will be considered. The approximate low-frequency equivalent circuit of this op-amp is shown in Fig. 4.11, and the overall frequency response is described by a first-order transfer function given by Eq. (4.18). Here $s_{p1} = -1/R_oC_L$ is the dominant pole, due to the load capacitance C_L in parallel to the output impedance R_o. Figure 4.33 illustrates the gain and phase response of the op-amp for two different

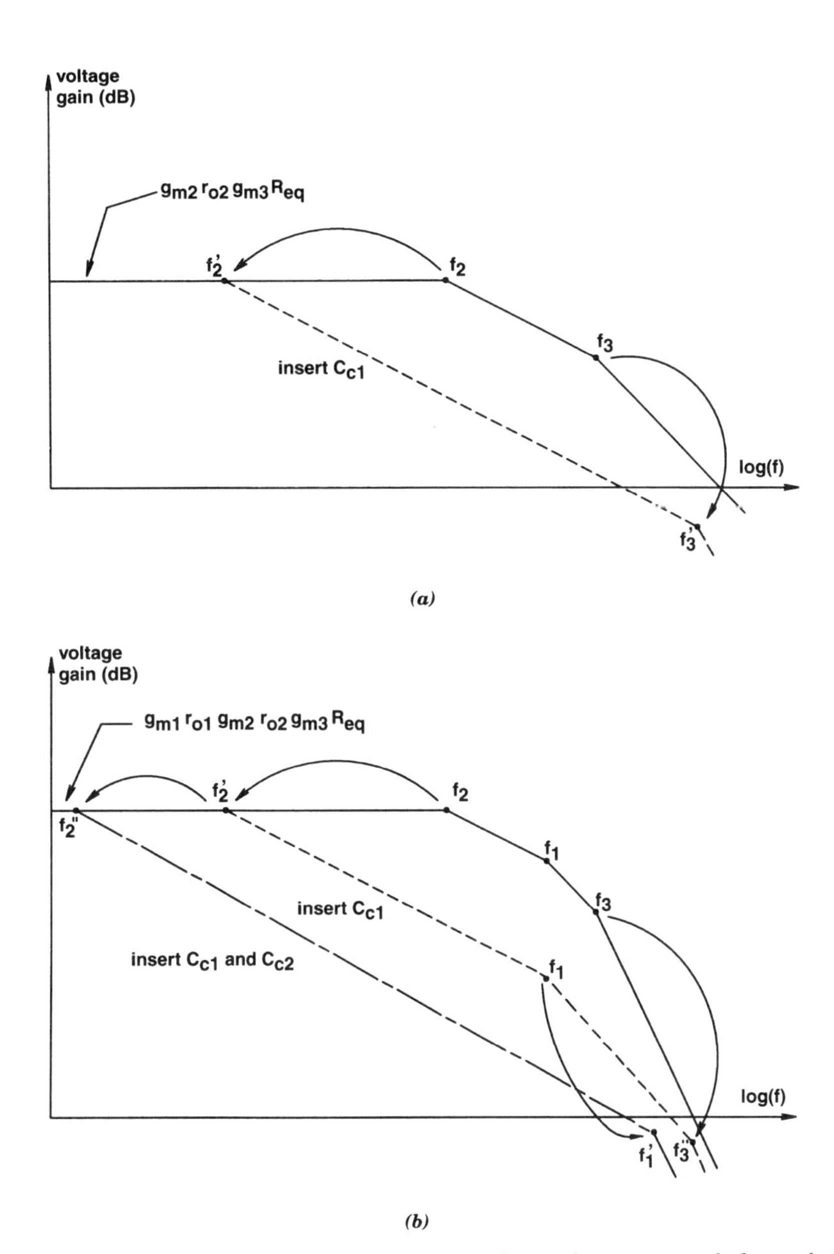

Figure 4.32. (*a*) Frequency response of the intermediate and output stages before and after inserting C_{c1}; (*b*) frequency response of the complete three-stage op-amp before and after inserting the nested Miller compensation capacitors C_{c1} and C_{c2}.

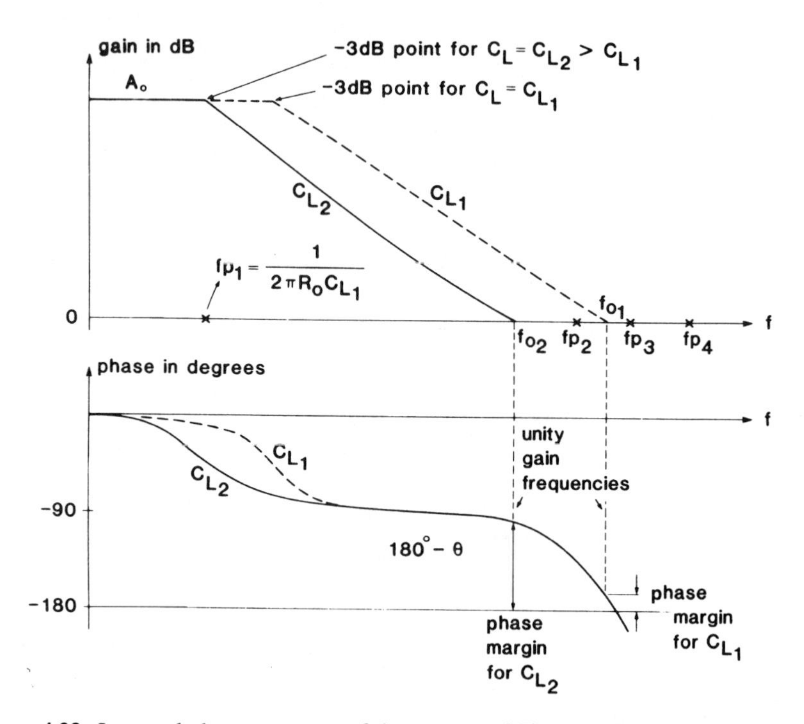

Figure 4.33. Loss and phase responses of the op-amp of Fig. 4.10 for two different values of the load capacitance C_L.

values of C_L. The contribution of the nondominant poles on the phase and amplitude responses is shown at higher frequencies. As the figure illustrates, the larger C_L, the greater the phase margin of the op-amp. This is the opposite of the conditions of the two- or three-stage op-amp, where C_L contributes to a nondominant high-frequency pole. There, increasing C_L reduces the distance between the dominant and nondominant poles, and thus decreases the phase margin. Thus the folded-cascode op-amp of Fig. 4.10 is particularly suitable for achieving wide and stable closed-loop bandwidths with large capacitive load, such as required in high-frequency switched-capacitor circuits.

In addition, the compensation in this circuit is achieved without coupling high-frequency noise from the power supplies to the output as in multistage op-amps. Hence the high-frequency PSRR can be high.

4.5. DYNAMIC RANGE OF CMOS OP-AMPS

Among the most important characteristics of an op-amp are the input-stage common-mode range (CMR) and the output-stage voltage swing. The input common-mode range specifies the range of the common-mode input voltage values such that the

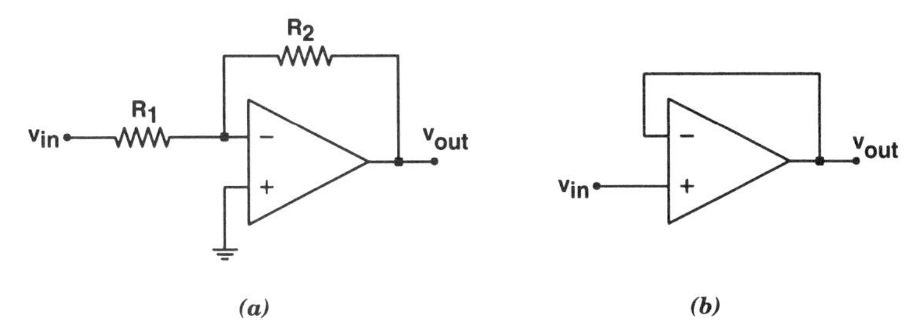

Figure 4.34. (*a*) Op-amp circuit without common-mode signal; (*b*) unity-gain op-amp configuration with common-mode signal.

differential stage continues to amplify the differential input voltage with approximately the same differential gain. The output voltage swing is the range over which the output voltage can vary without excessive distortion. Two possible configurations of an op-amp are shown in Fig. 4.34*a* and *b*. In Fig. 4.34*a* the op-amp is used with two external resistors as an inverting buffer. Since one input of the op-amp is connected to ground, the common-mode ac input is zero. In Fig. 4.34*b*, the op-amp is connected as a unity-gain buffer. All of the ac input signal is now applied as a common-mode input to the op-amp. While the output voltage swing is important for both cases, the input common-mode range is important only for the unity-gain buffer of Fig. 4.34*b* and is not important for the inverting buffer of Fig. 4.34*a*.

Figure 4.35 illustrates a *p*-channel-input CMOS differential stage. This stage will be used as an example to discuss the input common-mode range. The drain-to-source dc voltage of transistor Q_1 (and Q_2) is given by

$$V_{DS1} = V_{SS} + V_{GS3} - (V_{\text{in},\text{CM}} - V_{GS1}). \tag{4.73}$$

The minimum allowable common-mode input voltage occurs when Q_1 and Q_2 are

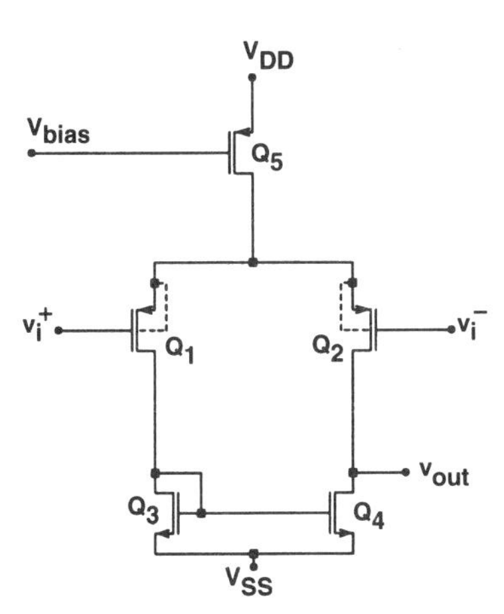

Figure 4.35. A *p*-channel input CMOS differential stage used to calculate common-mode range.

at the edge of their saturation regions. It can be obtained by setting $V_{DS1} = V_{Dsat1}$ in Eq. (4.73):

$$V_{\text{in,CM}} = V_{GS1} + V_{SS} + V_{GS3} - V_{Dsat1}. \tag{4.74}$$

Since $V_{GS1} = V_{Tp} + V_{Dsat1}$ and $V_{GS3} = V_{Dsat3} + V_{Tn}$,

$$V_{\text{in,CM}}^{\text{min}} = V_{SS} + V_{Dsat3} + V_{Tn} - |V_{Tp}|. \tag{4.75}$$

Since $V_{Tp} \leq 0$ and $V_{Tn} - |V_{Tp}| \approx 0$, the minimum common-mode input voltage is approximately equal to V_{SS} plus the drain-to-source saturation voltage of transistor Q_3.

A similar analysis can be performed to determine the highest common-mode input voltage. As the input voltage is increased, the drain-to-source voltage of Q_5 is reduced. The maximum common-mode voltage is achieved when Q_5 is about to leave the saturation region, or $V_{DS5} = V_{Dsat5}$. The drain-to-source of Q_5 is given by

$$V_{DS5} = V_{\text{in,CM}} - V_{GS1} - V_{DD}. \tag{4.76}$$

$V_{\text{in,CM}}^{\text{max}}$ is obtained by setting $V_{DS5} = V_{Dsat5}$:

$$\begin{aligned} V_{Dsat5} &= (V_{\text{in}})_{\text{max}} - V_{Dsat1} - V_{Tp} - V_{DD}, \\ V_{\text{in,CM}}^{\text{max}} &= V_{DD} + V_{Dsat5} + V_{Dsat1} + V_{Tp}. \end{aligned} \tag{4.77}$$

Since V_{Dsat5}, V_{Dsat1}, and V_{Tp} are all negative, we have

$$V_{\text{in,CM}}^{\text{max}} = V_{DD} - |V_{Dsat5}| - |V_{Dsat1}| - |V_{Tp}|. \tag{4.78}$$

Combining Eqs. (4.75) and (4.78), the input common-mode range is found:

$$V_{SS} + V_{Dsat3} + V_{Tn} - |V_{Tp}| \leq V_{\text{in,CM}} \leq V_{DD} - |V_{Dsat5}| - |V_{Dsat1}| - |V_{Tp}|. \tag{4.79}$$

Note that typical values are $V_{Tp} = -0.8$ V and $|V_{Dsat1}| = |V_{Dsat5}| = 0.3$ V. So from Eq. (4.79) it is clear that while the p-channel input differential stage of Fig. 4.35 has a reasonably good negative common-mode swing, the positive common-mode swing is poor and is limited to at least 1.4 to 1.6 V below the positive power

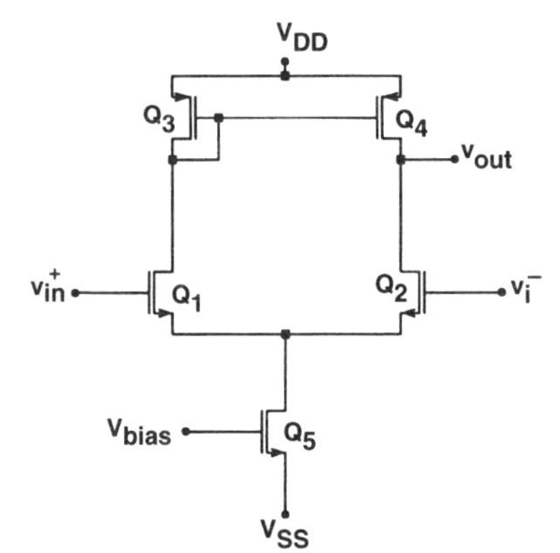

Figure 4.36. An *n*-channel input CMOS differential stage.

supply voltage. A similar analysis can be carried out for the differential stage with *n*-channel inputs, shown in Fig. 4.36. The common-mode range can be derived as

$$V_{SS} + V_{Dsat1} + V_{Dsat2} + V_{Tn} + \leq V_{\text{in,CM}} \leq V_{DD} - |V_{Dsat3}| + V_{Tn} - |V_{Tp}|. \tag{4.80}$$

For this case the positive input common-mode limit is approximately one $|V_{Dsat3}|$ below the positive supply voltage for $V_{Tn} - |V_{Tp}| \approx 0$. The negative limit of the input common-mode voltage is 1.4 to 1.6 V above the negative supply voltage. The input common-mode ranges of the two differential stages of Figs. 4.35 and 4.36 are complementary. While the *p*-channel input differential stage has good negative and poor positive input common-mode swing, the *n*-channel input has the complementary range. Op-amps with wide positive and negative input common-mode ranges can therefore be obtained using a combination of *p*- and *n*-channel differential stages. They are discussed in Section 4.10.

The output voltage swing of the two-stage op-amp is discussed next. Such an op-amp with a *p*-channel differential input is shown in Fig. 4.37. The output voltage swing is limited by the requirement that transistors Q_6 and Q_7 must remain in the saturation region. It can be easily shown that this results in the condition

$$V_{SS} + V_{Dsat6} \leq V_{\text{out}} \leq V_{DD} - |V_{Dsat7}|. \tag{4.81}$$

If the output swings beyond the range specified by Eq. (4.81), transistors Q_6 and Q_7 will leave the saturation region, reducing the gain of the output stage. Further increase of the output voltage will be limited by the power supply voltages.

A single-stage folded-cascode op-amp with *p*-channel input devices and an improved biasing scheme was discussed earlier and was shown in Fig. 4.10. First,

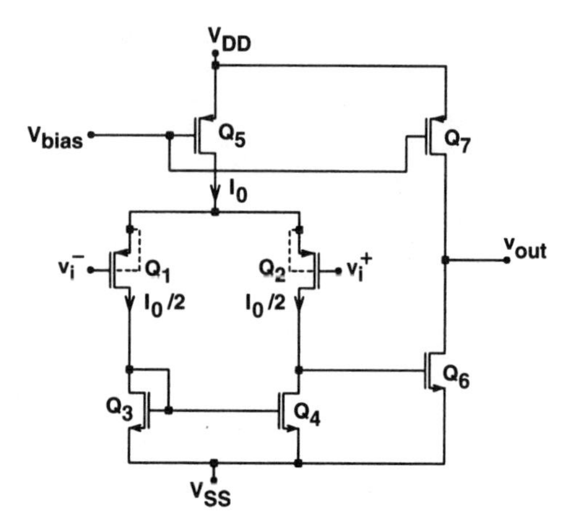

Figure 4.37. Two-stage CMOS op-amp.

consider the lower limit of the input common-mode range. With the improved biasing scheme, transistors Q_5 and Q_6 are biased slightly above the saturation region, so

$$V_{DS5} = V_{DS6} \simeq V_{D\text{sat}}. \tag{4.82}$$

As before, the drain-to-source voltage of Q_1 is given by

$$V_{DS1} = V_{SS} + V_{DS5} - (V_{\text{in,CM}} - V_{GS1}). \tag{4.83}$$

Setting $V_{DS5} = V_{D\text{sat5}}$ and $V_{GS1} = V_{D\text{sat5}} + V_{Tp}$ in Eq. (4.83), we have

$$V_{DS1} = V_{SS} + V_{D\text{sat5}} - (V_{\text{in,CM}} - V_{D\text{sat1}} - V_{Tp}). \tag{4.84}$$

For Q_1 at the edge of saturation, the minimum value of V_{DS1} is $V_{D\text{sat1}}$:

$$V_{D\text{sat1}} = V_{SS} + V_{D\text{sat5}} - V_{\text{in,CM}}^{\text{min}} + V_{D\text{sat1}} + V_{Tp}. \tag{4.85}$$

Rearranging yields

$$V_{\text{in,CM}}^{\text{min}} = V_{SS} + V_{D\text{sat5}} + V_{Tp}. \tag{4.86}$$

Since $V_{TP} < 0$ and $V_{D\text{sat5}} > 0$, we can rewrite Eq. (4.86) as

$$V_{\text{in,CM}}^{\text{min}} = V_{SS} + V_{D\text{sat5}} - |V_{Tp}|. \tag{4.87}$$

Normally, $V_{D\text{sat5}} < |V_{Tp}|$. Hence $V_{\text{in}}^{\text{min}} < V_{SS}$, and the input common-mode signal can go below V_{SS}.

To find the maximum input common-mode voltage, we note that the performance of the circuit is similar to that of the differential stage shown in Fig. 4.35. This leads to

$$V_{\text{in,CM}}^{\max} = V_{DD} + V_{D\text{sat}7} + V_{D\text{sat}1} + V_{Tp}. \tag{4.88}$$

Since $V_{D\text{sat}1}$, $V_{D\text{sat}7}$ and V_{Tp} are all negative numbers, Eq. (4.88) gives

$$V_{\text{in,CM}}^{\max} = V_{DD} - |V_{D\text{sat}7}| - |V_{D\text{sat}1}| - |V_{Tp}|. \tag{4.89}$$

Combining Eqs. (4.87) and (4.89) the input common-mode range can be written as

$$V_{SS} + V_{D\text{sat}5} - |V_{Tp}| \leq V_{\text{in,CM}} \leq V_{DD} - |V_{D\text{sat}7}| - |V_{D\text{sat}1}| - |V_{Tp}|. \tag{4.90}$$

As Eq. (4.90) shows, the input stage of the single-stage folded-cascode op-amp of Fig. 4.10 has an excellent lower limit for its common-mode range, lower than the negative supply voltage. The upper limit of the common-mode range is, however, by as much as 1.4 to 1.6 V below the positive supply voltage. A complementary cascode op-amp with n-channel input devices will be characterized by excellent positive input common-mode range (which includes the positive supply voltage) and a minimum input common-mode limit that is 1.4 to 1.6 V above the most negative supply voltage.

The op-amp of Fig. 4.10 uses an improved biasing scheme such that both Q_6 and Q_4 are biased at the edge of saturation:

$$\begin{aligned} V_{DS6} &= V_{D\text{sat}6}, \\ V_{DS4} &= V_{D\text{sat}4}. \end{aligned} \tag{4.91}$$

The maximum output voltage swing was derived earlier and is given by Eq. (4.16). From that equation, the output voltage swing is limited to a range that is at least $2V_{D\text{sat}}$ above V_{SS} and $2V_{D\text{sat}}$ below V_{DD}. Of course, v_{out} can swing beyond the range described in (4.16); however, as the output crosses the specified upper (lower) limit, first transistors Q_{2c} (Q_{4c}) leave the saturation region, and (as illustrated in Fig. 3.14) the output impedance drops, resulting in a reduction in the overall gain. Further, increase (decrease) of v_{out} causes Q_4 (Q_6) to leave the saturation region and results in drastic reduction in the output impedance and hence in the gain. In this region the op-amp has very little differential gain and the output signal will be severely distorted.

In summary, the single-stage folded-cascode op-amp of Fig. 4.10 has an excellent negative-input common-mode range but a poor positive common-mode range. It has

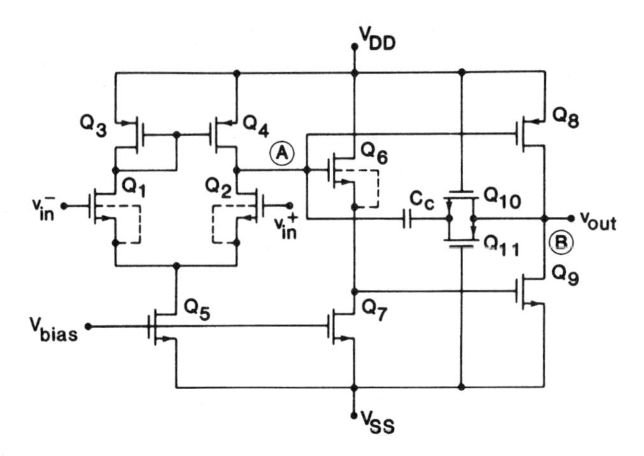

Figure 4.38. Internally compensated CMOS operational amplifier.

a reasonably wide output voltage swing; the output voltage can reach to within 0.5 to 1 V of the supply voltages without serious distortion or drop in gain.

4.6. FREQUENCY RESPONSE, TRANSIENT RESPONSE, AND SLEW RATE OF COMPENSATED CMOS OP-AMPS [9,10]

Next, an approximating frequency- and time-domain analysis of the compensated CMOS op-amp of Fig. 4.38 will be given. For small input signals v_{in} the transistors will operate in their saturation regions, and their small-signal models can be used. Then, for moderate frequencies (i.e., for $|s_{p1}| \ll \omega \ll |s_{p2}|$) the input stage Q_1 to Q_5 can be replaced by a frequency-independent voltage-controlled current source, while subsequent stages, Q_6 to Q_{11}, can be replaced by a frequency-independent amplifier with the feedback capacitor C_c connected between its input and output terminals (Fig. 4.39). The model is valid as long as the signal frequencies are much larger than $|s_{p1}|$ but are negligibly small compared to the magnitude of the high-frequency pole s_{p2}. From Fig. 4.39, $V_{out}(s) = g_{mi}V_{in}(s)/sC_c$, so that the high-frequency gain is given by $A_v(j\omega) = V_{out}(j\omega)/V_{in}(j\omega) = g_{mi}/j\omega C_c$. *The unity-gain frequency is thus* $\omega_0 = g_{mi}/C_c$. For $|s_{p2}| \gg \omega_0$, the phase of A_v at ω_0 will thus be close to 90°. This can be obtained by choosing C_c sufficiently large.

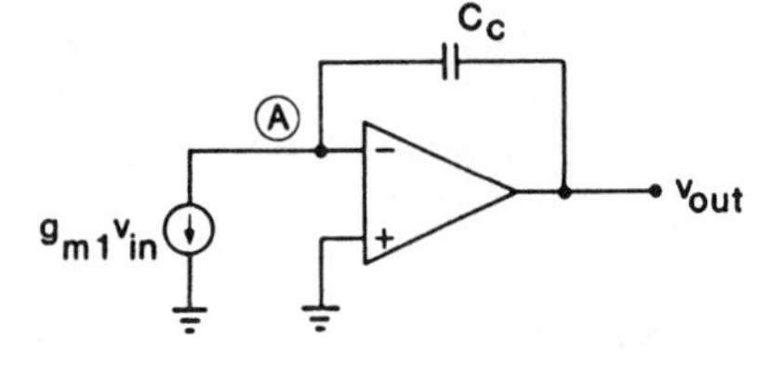

Figure 4.39. Small-signal model of the CMOS op-amp used to calculate its frequency response.

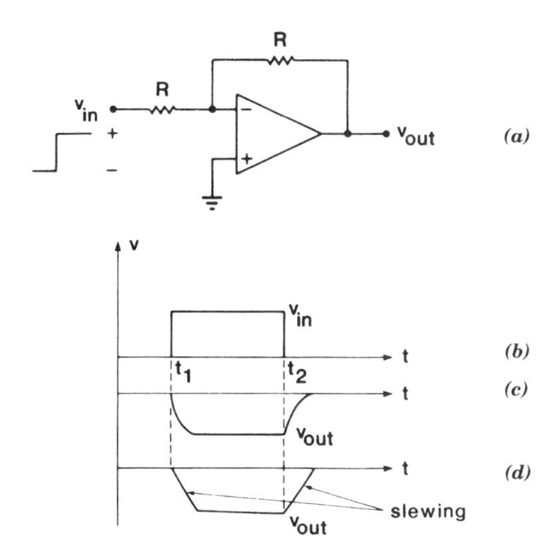

Figure 4.40. Slewing response of the CMOS op-amp connected in the inverting mode: (*a*) circuit; (*b*) input signal; (*c*) small-signal output waveform; (*d*) large-signal output waveform.

Consider next the voltage inverter shown in Fig. 4.40*a*. Assume again that the amplifier is compensated, so that its voltage gain can be approximated by

$$A_v(s) = \frac{A_0}{1 - s/s_{p1}}. \tag{4.92}$$

Hence for an input step $v_{in}(t) = V_1 u(t)$, the output voltage is in the form

$$v_{out}(t) = -V_1 u(t) \frac{A_0}{A_0 + 2} [1 - e^{-(A_0/2+1)[s_{p1}]t}] \tag{4.93}$$

(Problem 4.11). Thus, for a square input voltage (Fig. 4.40*b*) the exponentially varying waveform of Fig. 4.40*c* should occur at the output.* If the amplitude V_1 is small (say, much less than 1 V), this is in fact what happens. If, however, the input voltage is large (e.g., $V_1 = 5$ V), the experimentally observable output voltage is of the form shown in Fig. 4.40*d*. The nearly linear (rather than exponential) rise and fall of $v_{out}(t)$ is called *slewing,* and the nearly constant slope dv_{out}/dt of the curve is called the *slew rate.* Slewing is a nonlinear (large-signal) phenomenon, and hence it must be analyzed in terms of the large-signal model of the op-amp shown in Fig.

* The time constant is $t_0 = 2/(A_0 |s_{p1}|) = 2/\omega_0$.

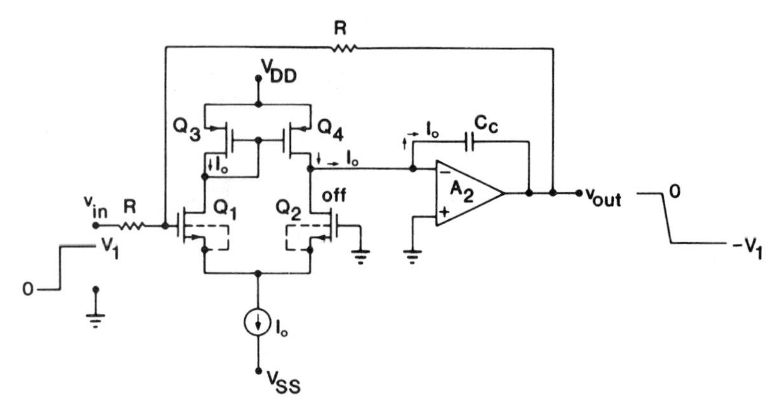

Figure 4.41. Large-signal model for calculating the slew rate of a CMOS op-amp in the inverting mode.

4.41. Prior to the arrival of the input step, $v_{in} = 0$, and the currents in Q_1 and Q_2 are both equal to $I_o/2$. After the large step occurs at the input, Q_1 conducts more current and cuts off Q_2. Hence the current conducted by Q_1 and Q_3 is now I_o (Fig. 4.41). Since Q_3 and Q_4 form a current mirror, the current in Q_4 (which charges C_c) is also I_o. Assuming that the output stage A_2 can sink the current I_o, the slew rate is

$$S_r = \left|\frac{dV_{out}}{dt}\right| = \left|-\frac{1}{C_c}\frac{dQ_c}{dt}\right| = \frac{I_o}{C_c}, \tag{4.94}$$

where Q_c is the charge in C_c. Here $C_c = g_{mi}/\omega_0$, where [from Eq. (2.18)] the transconductance of the input stage is

$$g_{mi} = 2\sqrt{\frac{I_o}{2}k'\frac{W}{L}} \tag{4.95}$$

and ω_0 is the unity-gain frequency of the op-amp. Combining these relations, we obtain

$$S_r = \frac{I_o\omega_0}{g_{mi}} = \omega_0\sqrt{\frac{I_o}{2k'(W/L)}}. \tag{4.96}$$

Thus the slew rate can be increased by *increasing* the unity-gain bandwidth and the bias current of the input stage, and by *decreasing* the W/L ratio of the input transistors.

It should be noted that the transconductance of MOSFETs is much lower than that of bipolar devices. This is ordinarily a major disadvantage; however, it results in significantly higher slew rates for MOS op-amps than for bipolar ones for a given unity-gain bandwidth since C_c can be smaller.

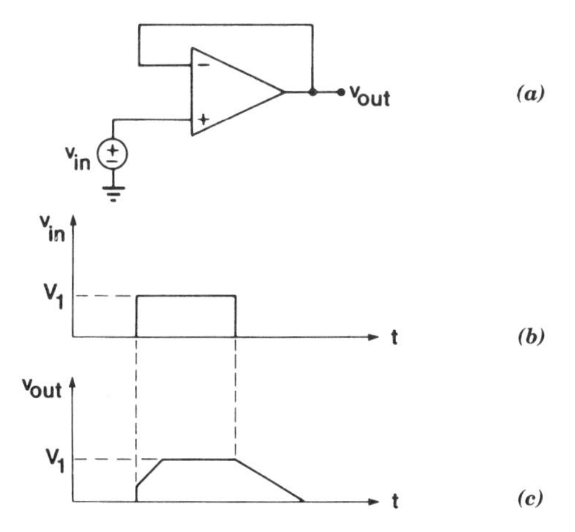

Figure 4.42. Slewing in a voltage follower: (*a*) op-amp used as a voltage follower; (*b*) large input signal; (*c*) output response.

The negative slewing of v_{out} continues until it reaches $-V_1$. At that time the gate voltage of Q_1 [which, due to the two resistors R, equals $(v_{in} + v_{out})/2$] reaches zero voltage. Hence at that time the quiescent bias conditions are restored, and Q_1 to Q_4 all carry a current $I_o/2$. Therefore, the charging of C_c and the decrease of the output voltage cease.

The complementary process takes place when v_{in} drops back to zero at $t = t_2$. Now Q_1 cuts off, since $v_{out} = -V_1$ still holds and hence its gate voltage drops to $-V_1/2$. Thus Q_3 and Q_4 cut off, and C_c is discharged through Q_2 with a current I_o, provided that A_2 can source at least the same current. The slew rate of v_{out} is hence again I_o/C_c. The process stops when v_{out} (and hence the gate voltage of Q_1) reaches zero voltage.

In Fig. 4.41, the op-amp operates in the inverting mode. Figure 4.42*a* illustrates the use of the op-amp as a unity-gain voltage follower. Figure 4.42*b* shows an input pulse waveform; Fig. 4.42*c* shows the corresponding output response under large-signal conditions. As the diagram shows, the rising edge contains a positive step followed by a fast slewing rise, while the falling edge is a relatively slow linear slope.

The behavior of the rising edge can be understood by considering the equivalent circuit shown in Fig. 4.43. In the circuit, the stray capacitance C_w across the input-stage current source I_o is included. Note that C_w is quite large in CMOS op-amps where the common sources of the input devices Q_1 and Q_2 are connected to the *p*-well, since this creates a large capacitance between the source and the substrate.

A large input signal $v_{in}(t) = V_1 u(t)$ turns Q_2 fully on. Therefore its source voltage v_w rises and hence Q_1 and Q_3 are turned off. Thus Q_2 carries the full current $I_o + i_w$, where $i_w(t)$ is the current through C_w. Since normally the combined impedance

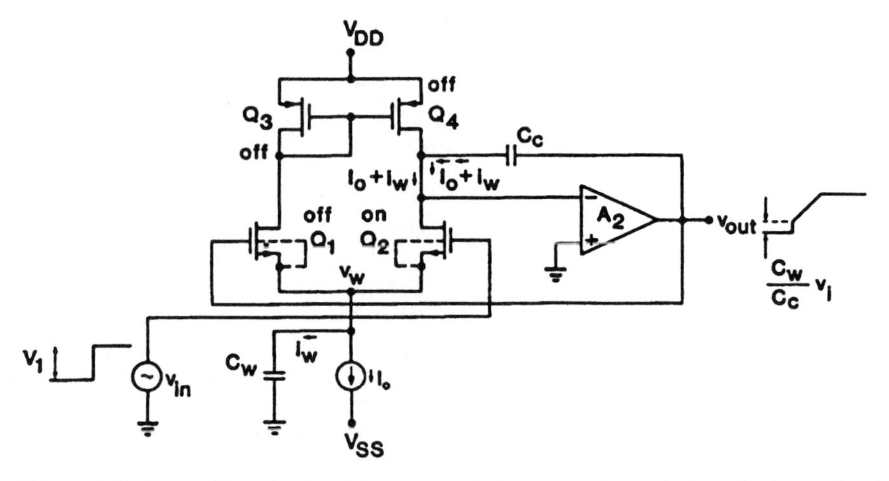

Figure 4.43. Equivalent circuit of the voltage follower used to calculate the large-signal behavior for positive inputs.

of C_w and the current source I_o is much larger than the driving impedance $(1/g_{m2})$ of Q_2, the incremental source voltage is $v_w(t) \approx v_{\text{in}}(t)$. Hence

$$i_w(t) = C_w \frac{dv_w(t)}{dt} \approx C_w \frac{dv_{\text{in}}(t)}{dt}, \tag{4.97}$$

which is the impulse function $V_1 C_w \delta(t)$. The output voltage satisfies

$$\begin{aligned} v_{\text{out}}(t) &= \frac{1}{C_c} \int_0^t (I_o + i_w)\, dt \\ &= \frac{I_o}{C_c} + \frac{C_w}{C_c} \int_0^t \frac{dv_{\text{in}}}{dt}\, dt = \frac{I_o}{C_c} t + \frac{C_w}{C_c} V_l u(t). \end{aligned} \tag{4.98}$$

The first term represents the linear rise, with a slew rate I_o/C_c, while the second represents the small pedestal seen at the beginning of the rising edge.

For a negative step, the equivalent circuit of Fig. 4.44 applies. Now the input signal turns Q_2 off, and Q_1, Q_3, and Q_4 all carry the current $I_o - i_w$. Considering next the two capacitors C_c and C_w, we note that C_c is connected between $v_{\text{out}}(t)$ and (virtual) ground, while C_w is connected between v_w and (true) ground. Now $v_w(t)$ follows the gate voltage $v_{\text{out}}(t)$ of Q_1, and hence $v_w \approx v_{\text{out}}$, so that

$$\frac{dv_{\text{out}}}{dt} = -\frac{I_o - i_w}{C_c} = -\frac{i_w}{C_w}. \tag{4.99}$$

Therefore, $i_w = I_o C_w/(C_c + C_w)$ and

$$\frac{dv_{\text{out}}}{dt} = -\frac{I_o}{C_c + C_w}. \tag{4.100}$$

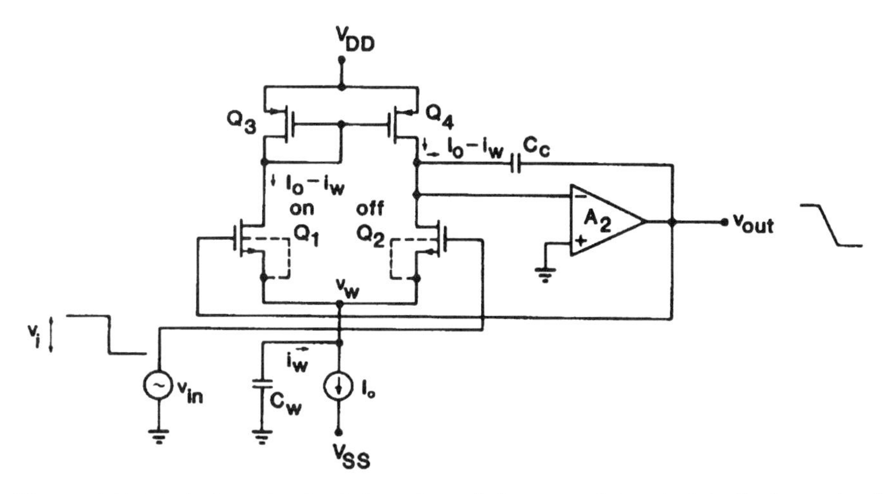

Figure 4.44. Equivalent circuit of the voltage follower used to calculate the large-signal behavior for negative inputs

Thus the negative slew rate is reduced by the presence of C_w, from I_o/C_c to $I_o/(C_c + C_w)$, that is, by a factor $1 + C_w/C_c$.

4.7. NOISE PERFORMANCE OF CMOS OP-AMPS

Noise represents a fundamental limitation of the performance of MOS op-amps: the equivalent noise voltage may be several times greater than a comparable bipolar amplifier. The noise performance of an MOS op-amp is due to both thermal and $1/f$ noise sources. The dominating noise source depends on the frequency range of interest. At low frequencies the $1/f$ noise dominates, whereas at high frequencies the thermal noise is more important and the $1/f$ noise can be ignored. Hence it is important to analyze the causes of noise and the possible measures that can reduce it. As an example, the noise of a two-stage CMOS op-amp will be analyzed and the noise contribution of each transistor to the total input referred noise will be presented. A similar analysis can be carried out for folded cascode or other types of op-amps.

Figure 4.45 shows an uncompensated CMOS op-amp, with the noise generated by each device Q_i represented symbolically by an equivalent voltage source v_{ni} connected to its gate.* (The calculation of the gate-referred noise voltages v_{ni} was described briefly in Section 2.7.) We can next combine the noise sources v_{n1} to v_{n4} in the differential input stage into a single equivalent source v_{nd} connected to the input of an otherwise noiseless input stage, as shown in Fig. 4.46. (Note that the noise of Q_5 is a common-mode signal and is hence suppressed by the CMRR of the

* Such a source indicates that a noise current $g_{mi}v_{ni}$ flows in Q_i.

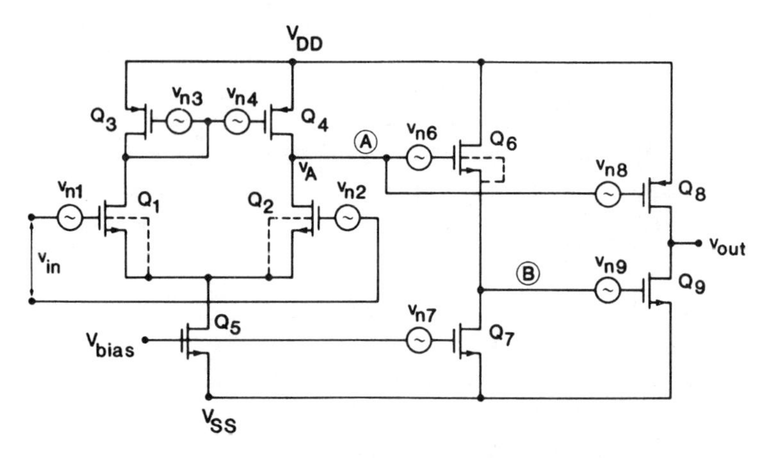

Figure 4.45. Noise sources in a CMOS operational amplifier.

op-amp; it is therefore omitted in Fig. 4.45.) The voltage gain from the noise sources v_{n1} and v_{n2} to the output node A of the input stage can be calculated using its low-frequency equivalent circuit. This gives

$$A_d = \frac{v_A}{v_{n1}} = \frac{v_A}{v_{n2}} = \frac{g_{m1}}{g_{d2} + g_{d4}}. \tag{4.101}$$

This is the same as the differential signal gain of the stage. Similarly, the gain between sources v_{n3} and v_{n4} and node A can be calculated. Physically, the noise source v_{n3} introduces a noise current $g_{m3}v_{n3}$ into Q_3, which is mirrored in Q_4. Hence v_{n3} causes currents of Q_3 and Q_4 to change by $g_{m3}v_{n3}$, and thus v_A by $g_{m3}v_{n3}/(g_{d2} + g_{d4})$. The effect of v_{n4} is similar. The gain is therefore

$$A_v = \frac{v_A}{v_{n3}} = \frac{v_A}{v_{n4}} = \frac{g_{m3}}{g_{d2} + g_{d4}}. \tag{4.102}$$

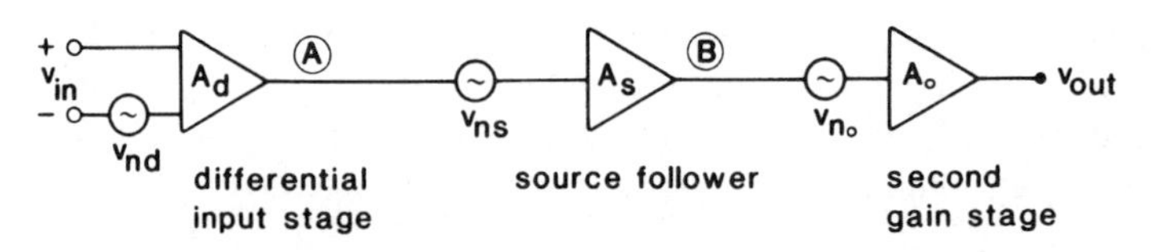

Figure 4.46. Block diagram of a three-stage CMOS operational amplifier with noise sources.

Since these sources are all uncorrelated, they result in a mean-square voltage

$$\overline{v_A^2} = A_d^2(\overline{v_{n1}^2} + \overline{v_{n2}^2}) + A_v^2\left(\overline{v_{n3}^2} + \overline{v_{n4}^2}\right) \tag{4.103}$$

at node A. Hence the equivalent input noise voltage $v_{nd} = v_A/A_d$ has the mean-square value

$$\overline{v_{nd}^2} = \overline{v_{n1}^2} + \overline{v_{n2}^2} + \left(\frac{g_{m4}}{g_{m1}}\right)^2 \left(\overline{v_{n3}^2} + \overline{v_{n4}^2}\right). \tag{4.104}$$

Hence, to minimize $\overline{v_{nd}^2}$, clearly v_{n1} and v_{n2} should be small and $g_{m4} \ll g_{m1}$. The former, by the discussions of Section 2.7, requires that the area (W/L) and transconductance (g_m) of Q_1 and Q_2 be large. To obtain large g_m, the bias current and W/L ratio should be large—this, however, requires large devices and high power dissipation.

The noise contribution of the load devices can be reduced, as (4.104) shows, by making their transconductances as small as their biasing conditions permit. This can be achieved by increasing their lengths L. Thus, assuming that the areas of the input and load devices are given, the W/L ratios of the input devices Q_1 and Q_2 should be as *large*, and those of the load devices Q_3 and Q_4 as *small* as other considerations permit. Also, it has been found experimentally [11] that the rms equivalent $1/f$ noise voltage v_n is about three times larger for an n-channel device than for a p-channel device. Since in Eq. (4.104) $(g_{m4}/g_{m1})^2 \ll 1$, it is hence advantageous to use p-channel input devices with n-channel loads, rather than the other way around, as shown in Fig. 4.45. Applying all these principles, the equivalent input noise voltage v_{nd} can be reduced appreciably [11].

Similarly, the noise sources of the source follower (Q_6, Q_7) can be replaced by an equivalent source v_{ns} (Fig. 4.46). From the low-frequency small-signal equivalent circuit,

$$\overline{v_{ns}^2} = \overline{v_{n6}^2} + \left(\frac{g_{m7}}{g_{m6}}\right)^2 \overline{v_{n7}^2}. \tag{4.105}$$

Referring $\overline{v^2{}_{ns}}$ back to the input of the op-amp, the total equivalent input noise voltage becomes

$$\overline{v_n^2} \simeq \overline{v_{nd}^2} + \frac{\overline{v_{ns}^2}}{A_d^2} = \overline{v_{n1}^2} + \overline{v_{n2}^2} + \left(\frac{g_{m4}}{g_{m1}}\right)^2 \left(\overline{v_{n3}^2} + \overline{v_{n4}^2}\right) + \frac{\overline{v_{n6}^2} + (g_{m7}/g_{m6})^2\overline{v_{n7}^2}}{A_d^2}. \tag{4.106}$$

For low frequencies where $A_d^2 \gg 1$, the effect of v_{ns} is negligible; however, at higher frequencies this will no longer be true. Since Q_6 and Q_7 are used as a level shifter, the gate–source voltage drop of Q_6 must be large. By Eq. (2.9) this will be achieved for a given i_{D6} if $k_6 = k'(W/L)_6$ is small. Hence Q_6 is a long thin device, and $(g_{m7}/g_{m6})^2 \gg 1$. At frequencies where $|A_d(\omega)| \approx g_{m7}/g_{m6}$, the effect of v_{n7} is comparable to that of v_{n1} and v_{n2}. Hence care must be taken in the design of Q_7 to make it a low-noise device.

The effect of the noise sources v_{n8} and v_{n9} can be analyzed similarly and can be represented by an equivalent source v_{n0}. However, they usually do not affect the total equivalent input noise voltage significantly.

Normally, all v_{ni} contain a $1/f$ noise component that dominates it at low frequencies. Hence the equivalent input noise voltage is greatest at low frequencies (below 1 kHz), where $|A_d(\omega)| \gg 1$. Thus the input devices Q_1 and Q_2 tend to be the dominant noise sources, and their optimization is the key to low-noise design.

Using chopper-stabilized differential configuration, the low-frequency $1/f$ noise of the op-amp can be canceled, and a large (over 100 dB) dynamic range obtained for an integrated MOS low-pass filter. For wide-band operational amplifiers and a low clock frequency, aliasing can increase the effect of the high-frequency noise to the point where it overwhelms the $1/f$ noise. Hence the unity-gain frequency ω_0 should be kept as low as is permitted by the application at hand.

4.8. FULLY DIFFERENTIAL OP-AMPS

In cases when power supply and substrate noise rejection is an important consideration, the use of fully differential (balanced) signal paths may be advantageous. In such circuits the input voltages are symmetrical with respect to the common-mode input voltage V_{cmi}, and the output voltages are symmetrical with respect to the common-mode output voltage V_{cmo}. This allows the designer to choose the values of input and output common-mode voltages (V_{cmi} and V_{cmo}) independently, for optimum performance. Although for maximum swing, V_{cmo} should be equal to half the total supply voltage, the same may not be the case for V_{cmi}. This makes the design of fully differential circuits more complicated and the required chip area 50 to 100% larger than the single-ended realization of the same network. However, there are many compensating advantages in terms of noise immunity. In fully differential op-amps, power supply and substrate noise appear as common-mode signals and are hence rejected by the circuit. In addition, as will be shown, the effective output voltage swing is doubled by the balanced op-amp configuration, while the input circuit (and hence most of the noise) remains the same as for the single-ended op-amps.

Additional advantages also exist. Figure 4.47 shows the circuit of a fully differential switched-capacitor (SC) integrator. In this circuit the switches illustrated schematically in the figure introduce a clock-feedthrough noise into the circuit. This can be minimized by the differential configuration, since (just as the power supply noise) it will appear as a common-mode signal. The symmetry of the circuit should be fully preserved in the physical layout to obtain good rejection of common-mode signals even in the presence of stray elements and nonidealities. The differential configuration also eliminates systematic offset voltages.

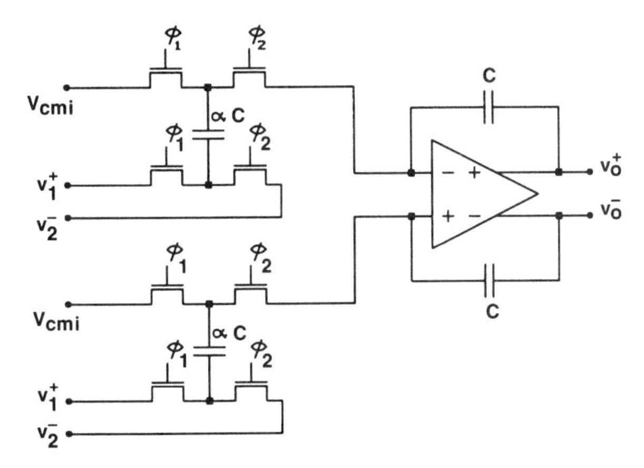

Figure 4.47. Fully differential switched-capacitor integrator.

The noise rejection properties of the fully differential circuits in actual implementation are not as effective as the theory predicts. This is partly because the noise coupled from the power supplies or substrate is not fully symmetrical. Also the clock-feedthrough noise from switches has a voltage-dependent component that couples to one signal path more than the other. However, by using careful and symmetrical layout methodologies it is almost guaranteed that the noise rejection properties of fully differential circuits are far superior than those of single-ended designs.

The circuit diagram of a fully differential single-stage folded-cascode op-amp is shown in Fig. 4.48. This circuit is obtained by modifying the op-amp of Fig. 4.10 and replacing the p-channel current mirrors with two cascode current sources Q_3,

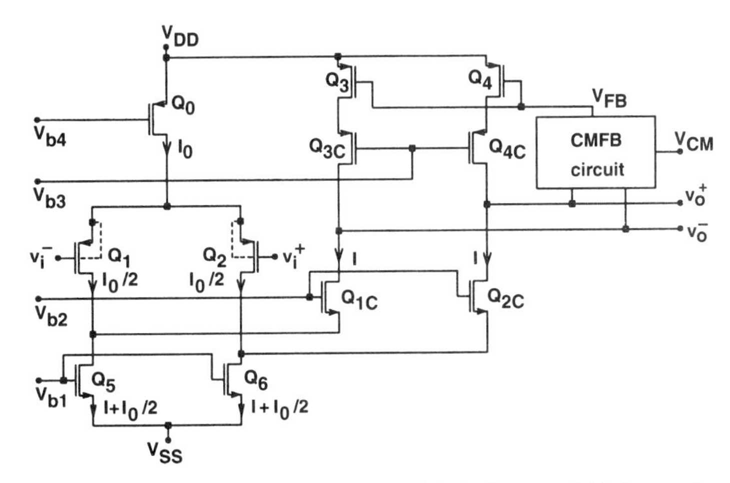

Figure 4.48. Circuit diagram of a fully differential single-stage folded-cascode op-amp.

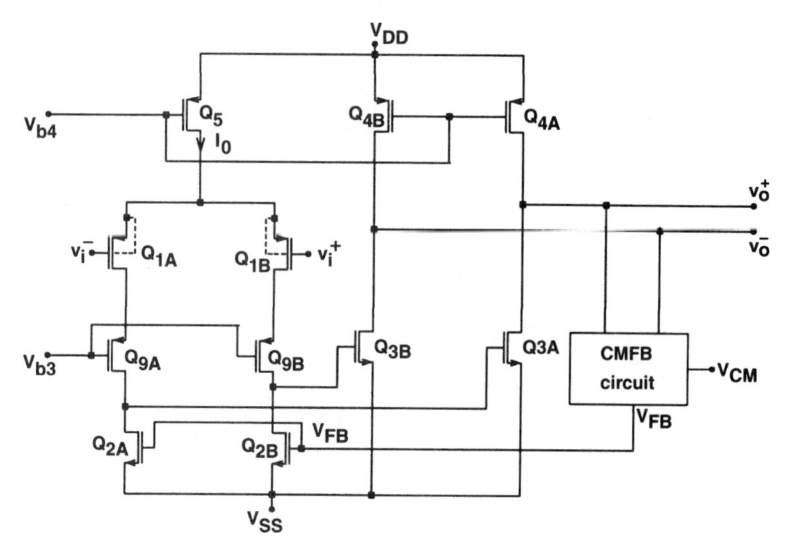

Figure 4.49. Circuit diagram of a two-stage fully differential op-amp.

Q_{3c} and Q_4, Q_{4c}. Figure 4.49 shows an alternative two-stage fully differential op-amp [12]. The differential input stage consists of transistors Q_{1A}, Q_{1B}, Q_{2A}, Q_{2B}, Q_{9A}, Q_{9B}, and Q_5. The common-gate devices Q_{9A} and Q_{9B} have been added to increase the gain of the operational amplifier and to reduce the differential input capacitance. The two differential output stages are formed with the two common-source amplifiers, consisting of transistors Q_{3A}, Q_{4A}, Q_{3B}, and Q_{4B}. A common-mode feedback (CMFB) circuit has been added to both op-amps. The CMFB circuit takes its inputs from the differential output of the op-amp and provides a common-mode feedback signal. This is necessary, since in a fully differential op-amp the common-mode output voltage must be internally forced to ground or to some other reference potential. By contrast, in a single-ended op-amp one of the input terminals is usually grounded and the other becomes virtual ground due to an externally applied negative feedback. This stabilizes the common-mode voltages at both input and output terminals.

One of the main drawbacks associated with the fully differential op-amp is the need for the CMFB circuit. Besides requiring extra area and power, the CMFB circuit limits the output swing, increases noise, and slows down the op-amp. The design of a good CMFB circuit is one of the most complicated parts of the fully differential op-amp design. There are two major design approaches for the CMFB circuits, the switched-capacitor approach [13] and the continuous-time approach [12,14]. The switched-capacitor approach is normally used in switched-capacitor circuits, while the continuous-time approach is used in non-sampled-data applications.

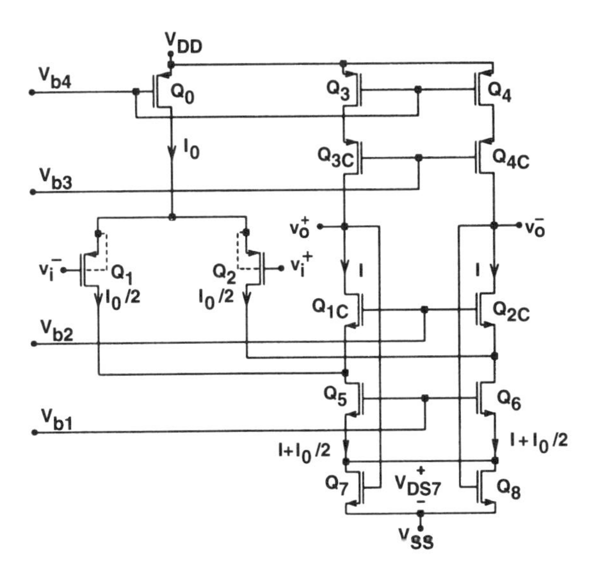

Figure 4.50. Fully differential folded-cascode op-amp with continuous-time CMFB circuit.

Figure 4.50 shows the single-stage fully differential folded-cascode op-amp of Fig. 4.48 with a continuous-time CMFB circuit added [15,16]. The common-mode feedback operates the following way. Since the gate voltages of Q_5 and Q_6 are fixed at V_{b1} and their currents are $I + I_o/2$, their source voltages are also stabilized. This fixes the drain-to-source voltages v_{DS7} and v_{DS8} of Q_7 and Q_8. The value of V_{b1} is chosen such that $|v_{DS7}| \ll V_{D\text{sat}}$, so that both Q_7 and Q_8 operate in their linear (ohmic) regions. Their aspect ratios $(W/L)_7 = (W/L)_8$ are chosen such that in equilibrium the common-mode output voltage $v_{\text{out},c} = (v_o^+ + v_o^-)/2$ has some desired value (usually, ground potential). If the common-mode voltage $v_{\text{out},c}$ would now drop for any reason, the resistance of Q_7 and Q_8 increases. This reduces $|v_{GS5}|$ and $|v_{GS6}|$, and since the current in Q_5 and Q_6 remains unchanged, it forces $|v_{DS5}|$ and $|v_{DS6}|$ to increase. Thus the drain voltages of Q_5 and Q_6 rise. This, by the argument just presented, reduces $|v_{GS1c}|$ and $|v_{GS2c}|$ and thus increases their drain voltages, which are v_o^+ and v_o^-. The common-mode voltage $v_{\text{out},c}$ is thus increased. The gain of the negative feedback loop is readily seen to be $g_{m7}r_{d7}g_{m5}r_{d5}g_{m1c}r_{d1c}$, which can be very high. This feedback loop also stabilizes $v_{\text{out},c}$ against transistor parameter variations arising from fabrication imperfections.

Since Q_7 and Q_8 operate in their linear (ohmic) regions, their drain currents are linear functions of their gate voltages. Thus it can readily be shown (Problem 4.16) that a *differential* voltage $\pm v$ at the output terminals does not affect the overall drain–source resistance of the parallel combination of Q_7 and Q_8. Thus the common-mode output voltage does not change if a differential input signal is applied; this is, of course, a desirable feature.

Using small-signal analysis, it can be shown (Problem 4.17) that the differential gain of the stage is

$$A_D = 2g_{m1}R_o, \tag{4.107}$$

where R_o is the output impedance at either output node:

$$\begin{aligned} R_o = R_o^+ &= \frac{1}{(g_{d1} + g_{d5})/g_{m1c}r_{d1c} + g_{d3}/g_{m3c}r_{d3c}} \\ = R_o^- &= \frac{1}{(g_{d2} + g_{d6})/g_{m2c}r_{d2c} + g_{d4}/g_{m4c}r_{d4c}} . \end{aligned} \tag{4.108}$$

Since the circuit is a folded cascode, it does not require a level shifter. Also, since the desired output is a differential signal, no differential-to-single-ended conversion is required. Thus the nondominant poles introduced by these stages do not appear. The only high-impedance nodes are the output terminals, and the corresponding dominant poles are those due to the time constants $R_o^+ C_L^+$ and $R_o^- C_L^-$, where C_L^+ and C_L^- are the load capacitances at the output terminals. To achieve compensation, the dominant poles can hence be shifted to lower frequencies by increasing C_L^+ and C_L^-. Since no internal compensation is required, the op-amp can have a fast settling time and is hence well suited for the implementation of high-frequency switched-capacitor filters.

Since the output impedances of the circuit can be made very high, the dc differential gain A_D can be large, comparable to that of a basic two-stage op-amp. A possible bias chain circuit for the op-amp of Fig. 4.50 is shown in Fig. 4.51. Choosing $I = I_o/2$, the aspect ratios can be found as

$$\begin{aligned} (W/L)_3 &= (W/L)_4 = \frac{(W/L)_0}{2} = \frac{(W/L)_{18}}{2}, \\ (W/L)_7 &= (W/L)_8 = (W/L)_{14}, \\ (W/L)_5 &= (W/L)_6 = (W/L)_{15}, \\ (W/L)_{1c} &= (W/L)_{2c} = \frac{(W/L)_{16}}{2}, \\ (W/L)_{3c} &= (W/L)_{4c} = \frac{(W/L)_{17}}{2}. \end{aligned} \tag{4.109}$$

$(W/L)_{19}$ should be chosen (in conjunction with the other aspect ratios in the bias chain) to set I_o to its desired value. For this bias circuit with the aspect ratios given above, the dc currents of Q_7, Q_8, and Q_{14} are equal. Also, their dc drain voltages are approximately the same. Hence their gate-to-source dc bias voltages satisfy

$$V_{GS7} = V_{GS8} \simeq V_{GS14} = V_{cm} - V_{SS}. \tag{4.110}$$

This shows that $V_{G7} = V_{G8} \simeq V_{G14} = V_{cm}$. Thus the output common-mode voltage is equal to V_{cm} when this bias circuit is designed using Eq. (4.109).

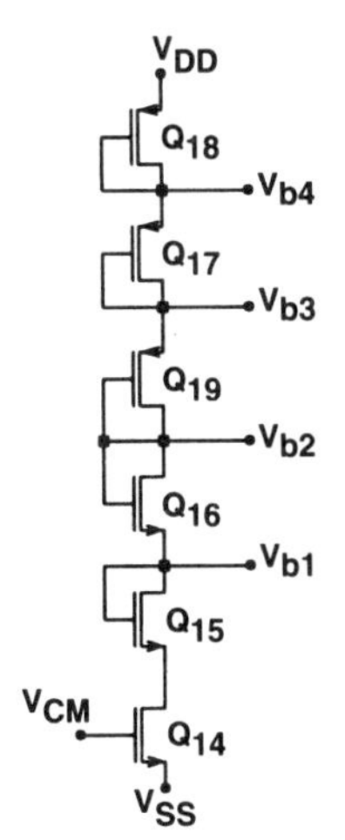

Figure 4.51. Bias chain for the fully differential op-amp of Fig. 4.50.

An alternative form of continuous-time common-mode feedback circuit is shown in Fig. 4.52 [17, pp. 287–291]. This circuit can be used to provide common-mode feedback for the folded-cascode differential op-amp of Fig. 4.48. The feedback voltage V_{FB} will bias transistors Q_3 and Q_4. In the absence of a differential voltage ($v_o^+ = v_o^-$) transistors Q_8, Q_9, Q_{12}, and Q_{13} will carry currents equal to $I/2$ and Q_{15} will have a current equal to I. The current of Q_{15} will be mirrored into Q_3 and Q_4, which will set the output current of the differential op-amp. The common-mode feedback therefore forces Q_8 and Q_{13} to have the same gate–source voltages as Q_9 and Q_{12}, forcing the common-mode output of the op-amp to V_{cm}. In the presence of a differential voltage, as long as (v_o^- has a value that is exactly the negative of

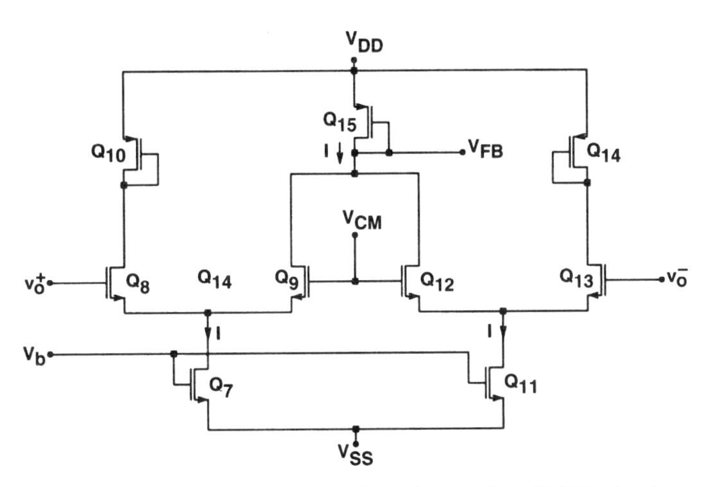

Figure 4.52. Alternative form of continuous-time CMFB circuit.

v_o^+ (with respect to V_{cm}), the current in Q_8 will increase (decrease) while the current in Q_{13} will decrease (increase), keeping the current in Q_{15} unchanged. The common-mode voltage will be kept close to V_{cm} as long as the differential voltage is not so large that the transistors in the common-mode feedback circuit (Q_8, Q_{13}) turn off. The sizes of the transistors in the differential pair can be selected in such a way that their gate–source voltages are maximized, hence extending the operating range. The input range, however, is a major limitation of this circuit.

Other than the limited input differential range, the common-mode feedback circuits described so far have two additional drawbacks. First the circuit that detects the output common-mode signal has a nonlinear characteristic. Second, the open-loop gain of the common-mode feedback may not be sufficiently large, due to its inherent limitations. The first problem can be avoided by using linear common-mode detectors such as a pair of identical resistors or the corresponding switched-capacitor equivalent for sampled data circuits. To reduce the effect of the second problem, the output common-mode feedback circuit should have a dc gain and bandwidth as large as the respective differential-mode circuitry. This can be accomplished by having the differential and output common-mode paths share as much circuitry as possible, thus treating both signals as equally as possible [12]. These concepts can be applied to the common-mode feedback circuit of the two-stage fully differential op-amp of Fig. 4.49. The complete circuit diagram of the op-amp is shown in Fig. 4.53. The common-mode feedback signal V_c is formed with two equal-valued resistors that are connected between the two differential outputs and is given by $V_c = (v_o^+ + v_o^-)/2$. The common-mode feedback circuit consisting of transistors Q_{6A}, Q_{6B}, Q_{6C}, Q_8, and Q_7 is merged with the differential-mode circuitry at the

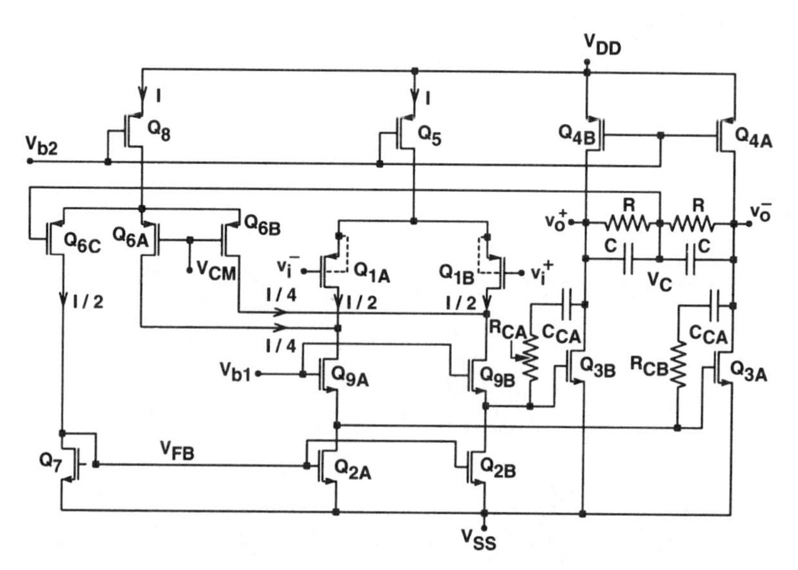

Figure 4.53. Schematic diagram of a two-stage fully differential op-amp with a common-mode feedback circuit.

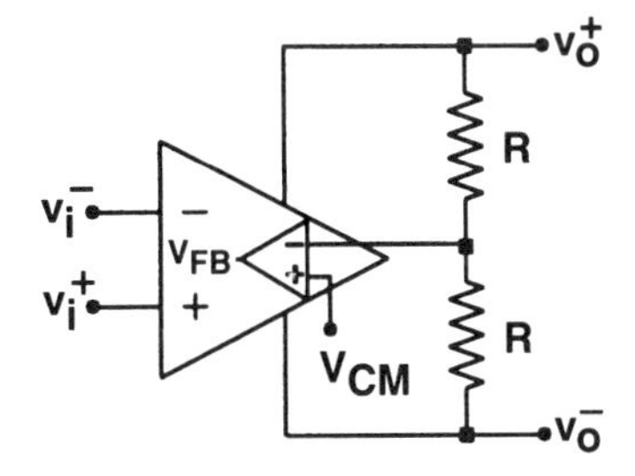

Figure 4.54. Conceptual representation of the resistive divider common-mode signal detector of Fig. 4.53.

very front end of the operational amplifier. Therefore, an equally amplified common-mode feedback signal and differential-mode input signal are combined as currents into the loads Q_{2A} and Q_{2B}. From there on to the outputs, the signals share the same circuitry, including the compensation networks consisting of R_{CA}, C_{CA}, R_{CB}, and C_{CB}. One potential drawback of this circuit is the resistors in the common-mode signal detector, which loads the op-amp differential output stages and hence degrades the amplifier dc gain. This effect can be reduced by designing the output stage transistors Q_{3A}, Q_{3B}, Q_{4A}, and Q_{4B} to have large W/L ratios and large currents, hence increasing their transconductances.

The use of resistive divider common-mode signal detector is shown conceptually in Fig. 4.54. In sampled analog systems such as switched-capacitor circuits, the same method used for processing the differential signals can be used for the common-mode detector circuit. Figure 4.55 shows the symbolic representation of the differential switched-capacitor integrator of Fig. 4.47 with a switched-capacitor common-mode detector. In this circuit a pair of integrating capacitors C_{c1} and C_{c2} and a pair of switched-capacitor resistors $\alpha_1 C_{c1}$ and $\alpha_1 C_{c2}$ implements the common-mode signal detector. This circuit is equivalent to a parallel RC circuit which takes the average

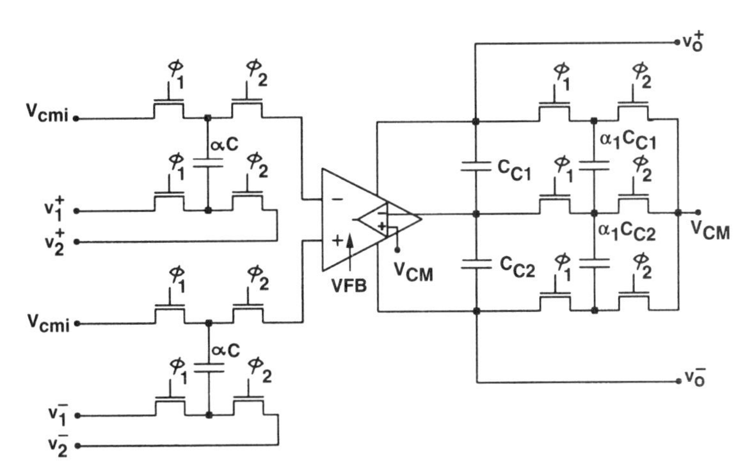

Figure 4.55. Fully differential integrator with switched-capacitor common-mode feedback circuit.

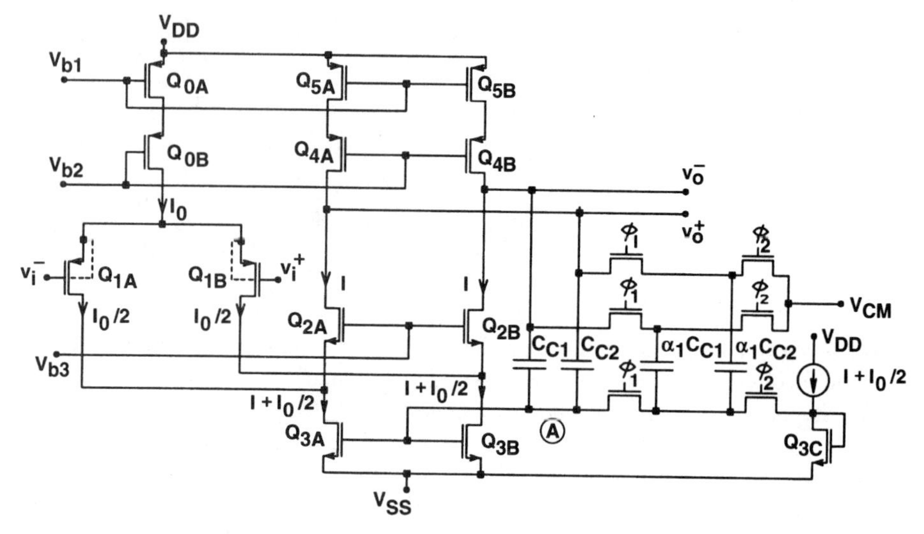

Figure 4.56. Fully differential folded-cascode operational amplifier with switched-capacitor common-mode feedback.

of the two voltages v_o^+ and v_o^-. For proper operation, the time constant of the equivalent RC should be much faster than the one in the differential signal path [13,18]. Normally, the sizes of $\alpha_1 C_{c1}$ and $\alpha_1 C_{c2}$ are between one-fourth and one-tenth of that of the nonswitched capacitors.

The circuit diagram of a fully differential folded-cascode operational amplifier with switched-capacitor common-mode feedback is shown in Fig. 4.56. The common-mode output level of the amplifier is maintained by the switched-capacitor feedback circuitry. Capacitors C_1 and C_2 in Fig. 4.56 have equal values and form a voltage divider. A bias voltage is generated across these capacitors based on the average voltage of the differential outputs and is used to control the gates of the NMOS transistors in the output stage (node A). The dc voltages across C_{c1} and C_{c2} are established by the switched-capacitors $\alpha_1 C_{c1}$ and $\alpha_1 C_{c2}$. These capacitors are first charged between the desired output common-mode voltage and a fixed bias voltage and are subsequently thrown in parallel to C_{c1} and C_{c2}. The fixed bias voltage is equal to the desired voltage used to bias the output-stage current sources. Only changes in the common-mode output are coupled to node A, which returns the common-mode output voltage to the desired level through negative feedback. During the period that the switches controlled by ϕ_1 are on, corrective charges are transferred to C_{c1} and C_{c2} through the switched capacitors to prevent any drift in the common-mode output voltage. The switched-capacitor common-mode feedback circuit has a wide output voltage swing and is the preferred choice in applications where the op-amp is used in a fully differential circuit involving switched capacitors.

4.9. CMOS OUTPUT STAGES

The main objective of the output stage of an operational amplifier is to be able to drive a load consisting of a large capacitance (up to several nanofarads) and/or a small resistance (down to 50 Ω or less) with an acceptably low level of signal distortion. It is also desirable to have a large output voltage range, preferably from rail to rail. To achieve the extended voltage swing, the output transistors should be connected in a common-source configuration. In fact, in CMOS operational amplifier design practice a push-pull stage is often used as an output stage, as shown in Fig. 4.57. The push-pull stage consists of two complementary common-source transistors Q_1 and Q_2, allowing rail-to-rail output voltage swing. The gates of the two output transistors are normally driven by two in-phase ac signals separated by a dc voltage [19,20]. When the input signals are above their corresponding dc values, the drain current of the NMOS device will be larger than the drain current of the PMOS transistor, and hence the output stage pulls a current from the load. If, on the other hand, the input signals are below their dc values, the output stage sources more current than it sinks and thus it pushes a current into the load.

Another important feature of the output stage is the efficiency, which requires a high ratio between the maximum signal current that can be delivered to the load and the quiescent current of the output stage. To achieve this requirement, a class B biasing scheme can be used. Because an output stage using this type of biasing will provide a large output current with a quiescent current that is approximately zero. The drawback, however, is that output stages with class B biasing introduce a large crossover distortion. The distortion can be reduced by using a class A biasing scheme. However, the maximum output current of a class A biased output stage is equal to its quiescent current, which results in poor power efficiency for a rail-to-rail output signal.

A compromise can be achieved between crossover distortion and quiescent power dissipation by using an output stage that is biased between class A and class B. This is called the class AB biasing scheme. In the push-pull output stage of Fig. 4.57, the class AB biasing scheme can be accomplished by keeping the voltage between

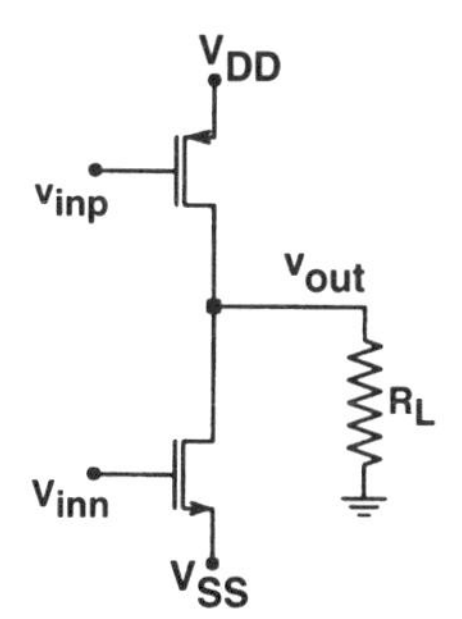

Figure 4.57. Push-pull CMOS output stage with rail-to-rail output swing.

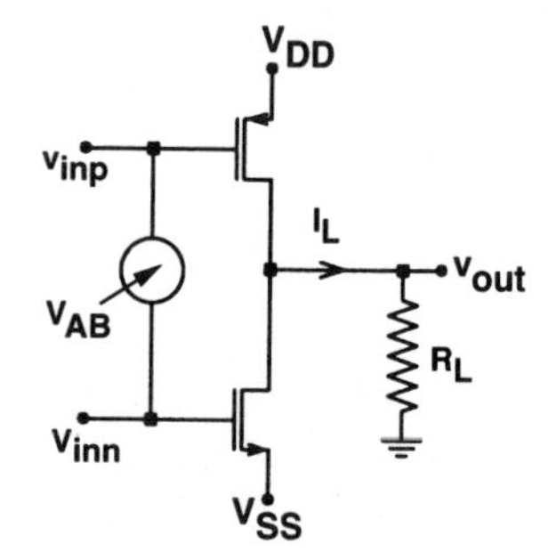

Figure 4.58. Push-pull CMOS output stage with class AB biasing.

the gates of the output transistors constant. This principle is shown in Fig. 4.58. To make the quiescent current and the relation between the push and pull currents independent of the supply voltage and process variations, the voltage source V_{AB} in Fig. 4.58 has to track these parameters. Figure 4.59 shows the desired class AB transfer function where the output transistors are biased with a small quiescent current, which improves the crossover distortion compared to a class B biased output stage. Also shown is the maximum output current, which is much larger than the quiescent current and increases the power efficiency compared to a class A biased output stage. To further reduce the crossover distortion, the transistor that is not delivering the output current should be biased with a small amount of residual current. This current will eliminate the turn-on delay of the nonactive output device, hence reducing the crossover distortion [21]. This minimum current is represented by I_{min} in Fig. 4.59.

Two other important parameters of the push-pull rail-to-rail output stage of Fig. 4.58 are the output voltage range and the maximum output current that is supplied to the load. To determine the output voltage range, first assume that the input signal voltage in Fig. 4.58 is increasing. This will cause the NMOS transistor to pull more current from the load, and thus the output voltage decreases. This process continues until the NMOS device ends up in the triode region and the output voltage becomes

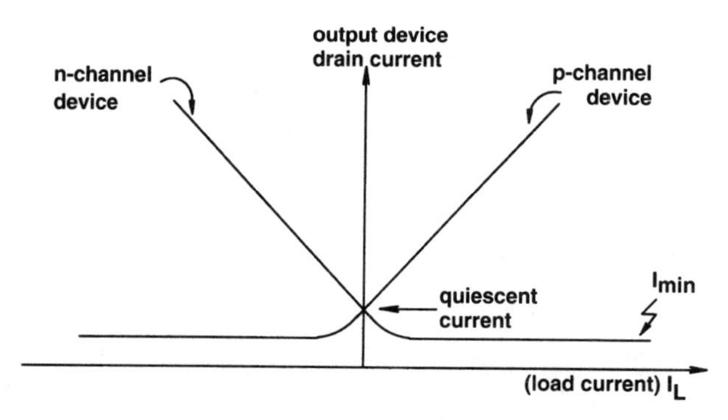

Figure 4.59. Output-stage current for class AB biasing.

Figure 4.60. Multistage low-output-impedance op-amp.

limited. The same happens for the PMOS device when the input signal decreases. The output voltage swing can be extended by maximizing the gate-to-source voltage swing and by choosing the largest possible W/L ratio for the output devices. The allowable gate-to-source voltage drive and the dimensions of the output transistors also determine the maximum output current of the output stage. In conclusion, an adequately designed class AB output stage should allow the gate-to-source voltage of the output transistors to get as close to the supply rails as possible.

Operational amplifiers that are required to drive a heavy load at the output (especially a small resistance) use a multistage structure, as shown in Fig. 4.60 [22]. The first stage is typically a transconductance preamplifier, which provides differential input and a large gain. The output stage is a class AB biased push pull circuit that provides low output impedance and a large current-driving capability. In the following a variety of output stages are presented and discussed.

The first circuit considered is shown in Fig. 4.61 [22]. For a zero applied differential input signal ($v_{in}^{+} - v_{in}^{-} = 0$), the two matched current sources I, made of devices Q_{10} and Q_{11}, uniquely define the circuit quiescent current level. For simplicity let us assume that the four NMOS input transistors Q_1, Q_2, Q_5, and Q_6 and the four PMOS input transistors Q_3, Q_4, Q_7, and Q_8 are identical devices. Then $V_{GS1} = V_{GS2} = V_{GS5} = V_{GS6}$ and $V_{GS3} = V_{GS4} = V_{GS7} = V_{GS8}$, which results in $I_1 = I_2 = I$. Under this condition the current in output devices Q_{14} and Q_{15} is also determined by the current source devices Q_{10} and Q_{11} and the W/L ratio of the two current mirrors Q_{12}, Q_{14} and Q_{13}, Q_{15}. It follows, therefore, that the class AB biasing scheme

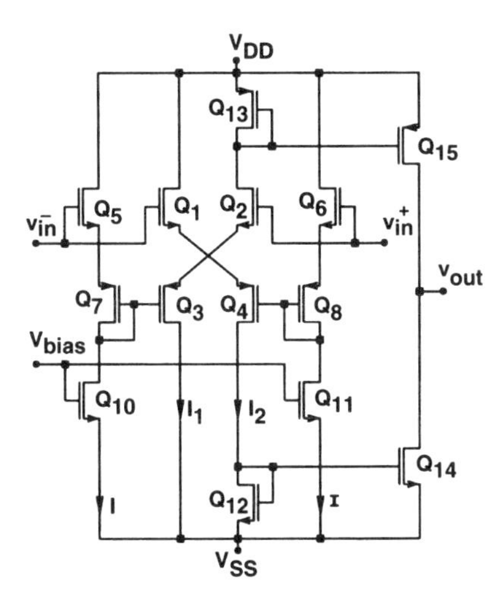

Figure 4.61. Schematic of class AB amplifier.

makes the quiescent power consumption of the circuit be controlled precisely by the two matched current sources in the input stage.

When the differential input signal is positive ($v_{in}^{+} - v_{in}^{-} > 0$), the voltage drop across Q_2 and Q_3 is increased by $\Delta v_{in} = v_{in}^{+} - v_{in}^{-}$, while the drop across Q_1 and Q_4 is decreased by the same amount. Consequently, the current through Q_{13} is increased while the current in Q_{12} is reduced close to zero, and a current much larger than the quiescent level, determined by the W/L ratio of Q_{13} and Q_{15}, is available through Q_{15} to be delivered to the load.

The peak value for currents I_1 and I_2 and hence the output current is a function of the applied input voltage. In practice, however, as the current level increases, the sum of the voltage drops across Q_1, Q_4, and Q_{12} (Q_3, Q_2, and Q_{13}) also increases, until it is equal to the total supply voltage. At this point devices Q_1 and Q_4 (Q_2 and Q_3) or both enter the linear region of operation, and the current level becomes practically constant independent of the applied input voltage.

The circuit of Fig. 4.61 can be used as a stand-alone single-stage op-amp. It is also suitable for a differential output op-amp since it can produce a complementary output simply by adding two additional current mirrors symmetrical with respect to Q_{12},Q_{14} and Q_{13},Q_{15}. Figure 4.62 shows the entire fully differential amplifier schematic where cascode transistors can be added to the output stage to increase the output impedance and hence the gain [14]. A common-mode feedback should be added to V_{cm} in Fig. 4.62 for proper operation. The amplifier of Fig. 4.61 can also be used as the output section for a two-stage amplifier. Figure 4.63 shows the schematic of a two-stage low-output-impedance op-amp, where the first-stage transconductance amplifier is made of a simple folded-cascode differential amplifier. Note that in this configuration one input of the differential output stage is connected to an appropriate dc potential (V_{bias4}), such as halfway between the two supplies. It is

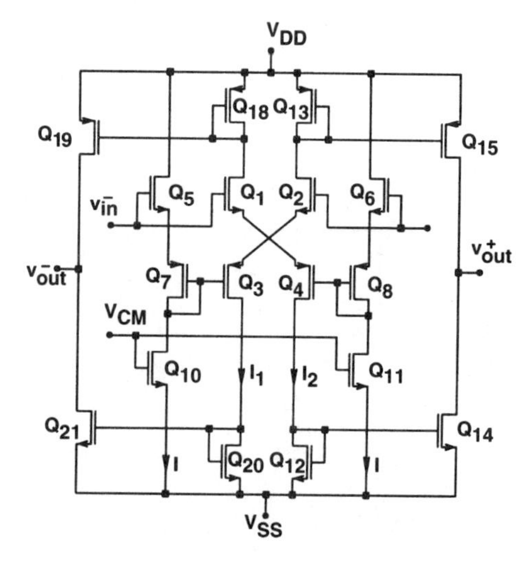

Figure 4.62. Simplified schematic of fully differential op-amp.

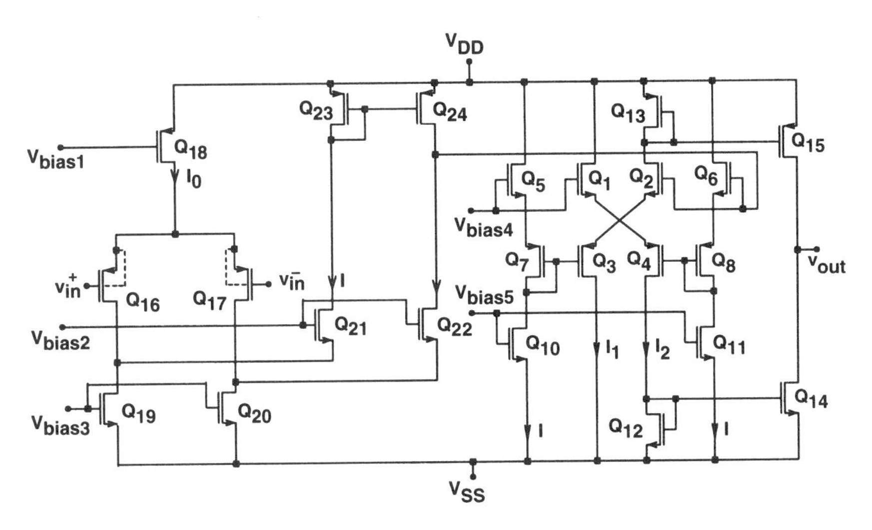

Figure 4.63. Two-stage low-output-impedance op-amp.

also possible to connect the differential input of the output stage to the differential output of a fully differential folded-cascode transconductance amplifier. This is shown in Fig. 4.64, where a common-mode feedback is used to stabilize the differential output of the first stage [14]. This technique is discussed in more detail later in the section.

A simplified version of the class AB output stage of Fig. 4.61 is shown in Fig. 4.65*a* [23]. Compared to the amplifier of Fig. 4.61, the positive input and all the

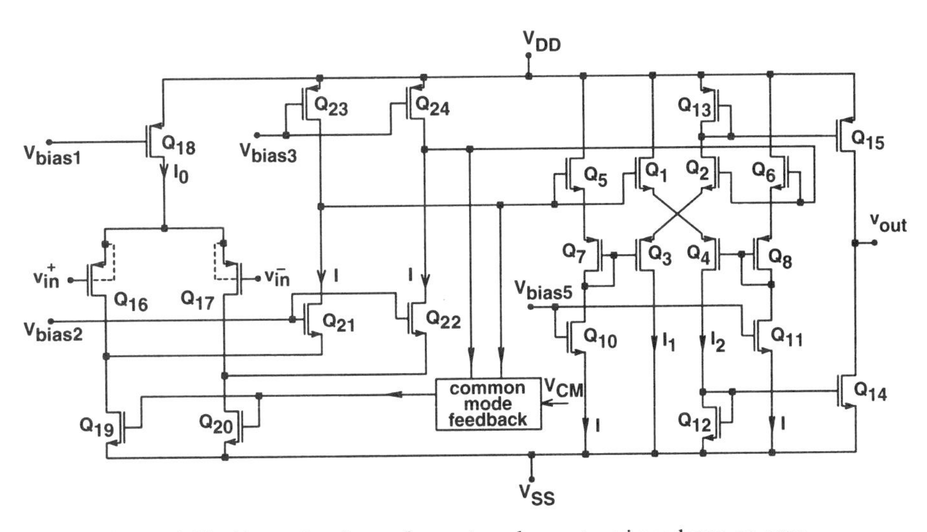

Figure 4.64. Alternative form of two-stage low-output-impedance op-amp.

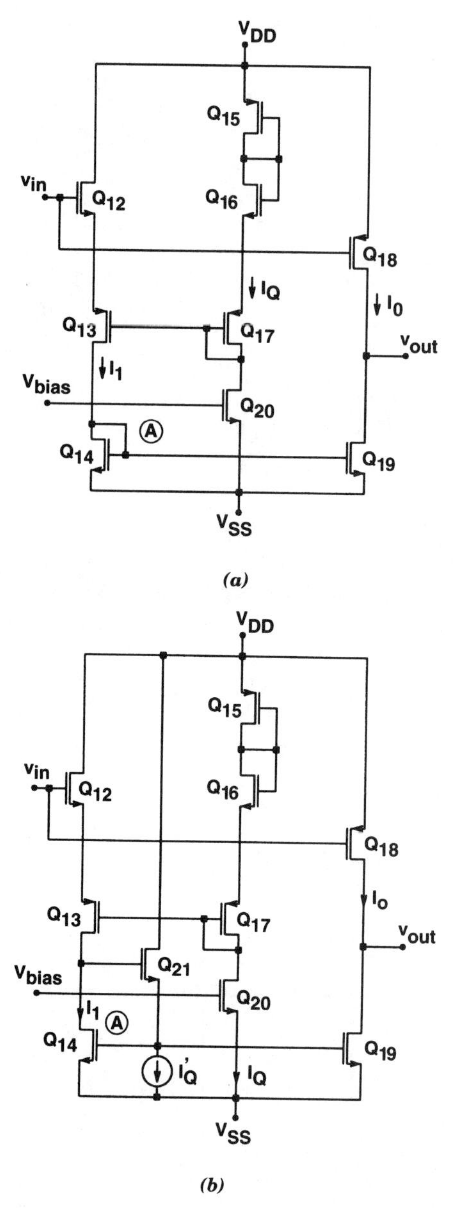

Figure 4.65. (*a*) Class AB output stage; (*b*) wideband class AB output stage.

corresponding devices are eliminated. The input v_{in} is intentionally biased one V_{GS} below the positive supply and it directly drives the gate of the p-channel output device. The n-channel output transistor is driven by the output of the common-gate amplifier, consisting of devices Q_{12}, Q_{13}, and Q_{14}. Assuming that

$$(W/L)_{12} = p(W/L)_{16}, \tag{4.111}$$

$$(W/L)_{13} = p(W/L)_{17}, \tag{4.112}$$

$$(W/L)_{18} = mp(W/L)_{15}, \tag{4.113}$$

$$(W/L)_{19} = m(W/L)_{14}, \tag{4.114}$$

then in quiescent condition,

$$V_{GS18} + V_{GS12} + V_{GS13} = V_{GS15} + V_{GS16} + V_{GS17} \tag{4.115}$$

and

$$I_1 = pI_Q,$$

$$I_o = mpI_Q.$$

Also, from Eqs. (4.111)–(4.114) and Eq. (4.115) it is clear that in quiescent conditions

$$V_{GS13} = V_{GS17}, \tag{4.116}$$

$$V_{GS12} = V_{GS16}, \tag{4.117}$$

$$V_{GS18} = V_{GS15}. \tag{4.118}$$

Under these conditions, the bias voltage of v_{in} is set to $V_{DD} - V_{GS15}$. The input voltage v_{in} in the negative direction can swing from $V_{DD} - V_{GS18}$ all the way down to V_{SS}, causing Q_{18} to provide a large sourcing current. The maximum sourcing current is given by

$$(I_o^+)_{\max} = k_n'\left(\frac{W}{L}\right)_{18}(V_{DD} - V_{SS} - V_T)^2. \tag{4.119}$$

In the positive direction, v_{in} can swing from $V_{DD} - V_{GS18}$ to V_{DD}. The maximum sinking current is given by

$$(I_o^+)_{\max} = \frac{mk_{12}k_{13}}{(\sqrt{k_{12}} + \sqrt{k_{13}})^2}\left(\sqrt{\frac{I_Q}{k_{15}}} + \sqrt{\frac{I_Q}{k_{16}}} + \sqrt{\frac{I_Q}{k_{17}}} + V_{Tn}\right)^2, \tag{4.120}$$

where $k_i = k_i'\,(W/L)$.

Referring to Fig. 4.65*a*, the operation of the circuit is as follows. To source an output current, v_{in} goes low and drives the gate of Q_{18} more negative, which increases its drain current. At the same time, the output of the positive-gain common-gate amplifier consisting of devices Q_{12}, Q_{13}, and Q_{14} also goes low, reducing the current of Q_{19}. To sink current, v_{in} goes high; it shuts off Q_{18} by reducing its gate drive and also pulls the gate of Q_{19} high, which in turn increases its drain current. The maximum source and sink currents are given by Eqs. (4.119) and (4.120).

To drive small resistive loads, the devices used in the output stage are very large. Since the transistor parasitic capacitances increase much faster than its transconductance, the frequency response of the output stage and therefore the entire op-amp deteriorates. More specifically, as a result of a dramatic increase in C_{GS19}, the nondominant pole at node A moves closer to the dominant pole and hence makes the frequency compensation very difficult. An alternative wideband circuit is shown in Fig. 4.65*b*, where a source follower Q_{21} has been added to bias node A, and the impedance has been reduced from $1/g_{m14}$ in Fig. 4.65*a* to $1/g_{m21}$ in Fig. 4.65*b*.

The complete circuit schematic of the two-stage low-output-impedance op-amp that uses the output stage of Fig. 4.65*b* is shown in Fig. 4.66. The input transconductance stage is made of a folded-cascode differential amplifier. The bias current is set by resistor R_B and the three diode-connected devices Q_{31} to Q_{33}, which are connected in series between V_{DD} and the negative supply. The frequency response of the circuit exhibits several real poles. It has two dominant poles, which are due to the output impedance and the corresponding load capacitances of the folded-cascode differential amplifier and the output stage. The remaining nondominant poles are at nodes A to H, which are located at much higher frequencies than are the dominant poles. This is because the impedance levels at all these nodes are determined by the $1/g_m$ value of a large MOS device, which is much smaller than the output impedance at the two nodes that determine the two dominant poles. The op-amp therefore can be compensated using the Miller capacitance C_c in series with a zero-nulling NMOS device operating in triode region, which is connected between the two high-impedance nodes.

Another approach for a class AB push-pull output stage is illustrated in Fig. 4.67 [24]. Here the two large common-source output transistors Q_1 and Q_2 are driven by two error amplifiers, A_1 and A_2. The feedback loop around A_1 (A_2) and Q_1 (Q_2) ensures low output impedance. In this configuration the error amplifiers must satisfy a number of requirements for proper operation of the output stage. First, since the error amplifiers have an input-referred dc offset voltage on the order of several millivolts, they must have a reduced value of open-loop gain on the order of 10. Otherwise, in the case of a large gain, the input referred offset voltage, when referred to the output, will cause unacceptable variations of the quiescent currents in Q_1 and Q_2. A second requirement is that the error amplifiers must have a rail-to-rail output and input swing so that they can provide good drive capability for the output transistors. Finally, they must be broadband, to prevent crossover distortion. This may become difficult to achieve, however, due to stability constraints.

An alternative form of the output stage of Fig. 4.67 can be achieved by considering the CMOS class AB complementary source-follower stage of Fig. 4.68, which is a

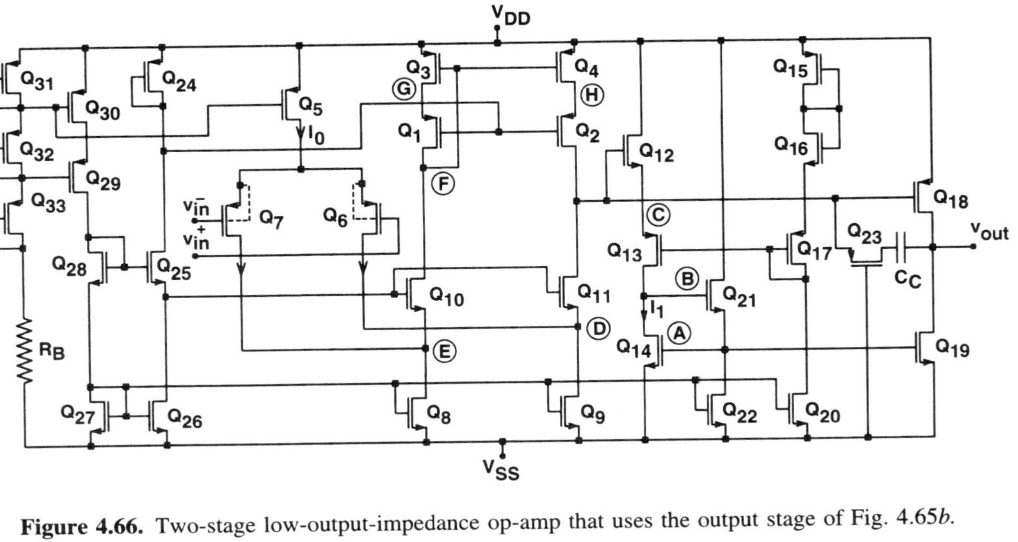

Figure 4.66. Two-stage low-output-impedance op-amp that uses the output stage of Fig. 4.65*b*.

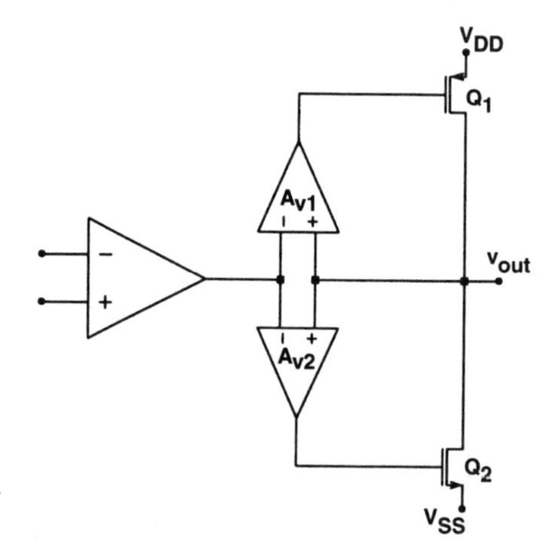

Figure 4.67. Block diagram of CMOS op-amp with class AB push-pull output stage.

direct analog of its bipolar counterpart. In the circuit, Q_1 and Q_4 form a gain stage, while Q_5 and Q_6 drive the load R_L and Q_2 and Q_3 provide a voltage drop between the gates of Q_3 and Q_6 to reduce crossover distortion. The sizes of Q_2 and Q_3 are chosen such that the gate-to-source voltages of Q_5 and Q_6 are slightly larger than their threshold voltages. The primary drawback of this circuit is that the gate-to-source voltage of the output transistors limits the output voltage swing. The maximum output voltage for $R_L \rightarrow \infty$ is $V_{DD} - V_{T5}$ and the minimum is $V_{SS} - |V_{T6}|$, where V_{T5} (V_{T6}) is the threshold voltage of Q_5 (Q_6). If R_L draws current from, say, Q_5, the device must provide a drain current v_{out}/R_L and hence needs a gate-to-source voltage $V_{GS} = V_{T5} + \sqrt{v_{\text{out}}/(k_5 R_L)}$. This increases rapidly with decreasing R_L and hence represents an important limitation on the achievable positive output voltage swing. Similar considerations hold, of course, for negative swings, due to the necessary V_{GS6}.

Many of the problems of the output stages of Fig. 4.67 and 4.68 are easily solved

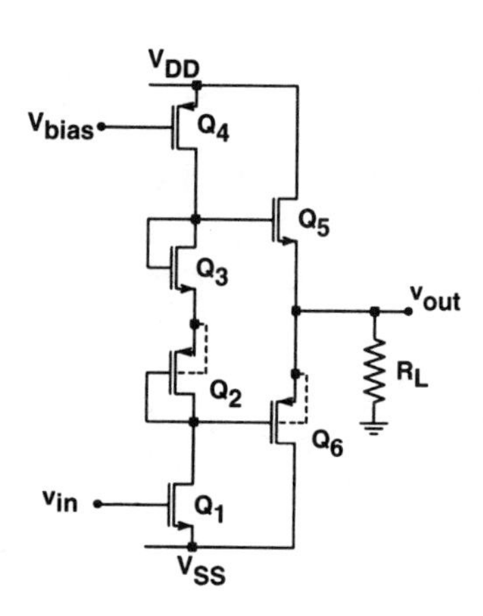

Figure 4.68. CMOS class AB push-pull output stage based on the traditional bipolar implementation.

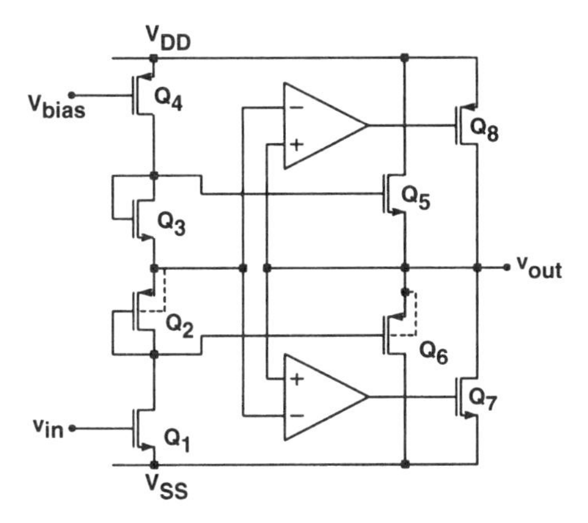

Figure 4.69. Combined class AB and class B output stage.

by merging the two output stages as shown in Fig. 4.69. The problem of quiescent current control is solved by deliberately introducing an offset voltage into A_1 and A_2 in such a way that transistors Q_7 and Q_8 are not carrying any current in the quiescent state. Transistors Q_5 and Q_6 therefore control the quiescent output current. The quiescent output current will be proportional to the current through Q_2 and Q_3 and is a function of the W/L ratios of Q_5 to Q_3 and Q_6 to Q_2. In steady state the class AB circuit consisting of transistors Q_5 and Q_6 carries the entire output current. The class B output stage, consisting of op-amps A_1 and A_2, has a very large current-driving capability but is kept off in quiescent conditions. When driving small resistive loads, the class B output stage takes over operation of the stage.

Besides their usefulness in quiescent current control, transistors Q_5 and Q_6 provide a high-frequency feedforward path from the input to the output of the push-pull stage and reduce the excess phase shift introduced by op-amps A_1 and A_2. The op-amps still require some nominal phase compensation to make them stable in the closed-loop unity-gain mode. The frequency characteristic of the overall amplifier is determined largely by Q_5 and Q_6 rather than that of the composite output stages.

Another class AB biased push-pull output stage is shown in Fig. 4.70 in block diagram form [14]. In this circuit the output stage, consisting of devices Q_1 and Q_2 is preceded by a fully differential preamplified stage A_1. The output stage has several important properties: It has low standby power dissipation, which can be controlled by a supply-independent current source, it has a good current-driving capability, and it has a simple configuration, so it avoids additional parasitic poles. The output stage consists of devices Q_1 to Q_4, and the quiescent current level at the output stage is controlled by the common-mode voltage of the preamplifier stage. Here since the output stage is driven by a fully differential-output preamplifier stage, an additional common-mode feedback (CMF) is necessary to set the dc voltage values of the

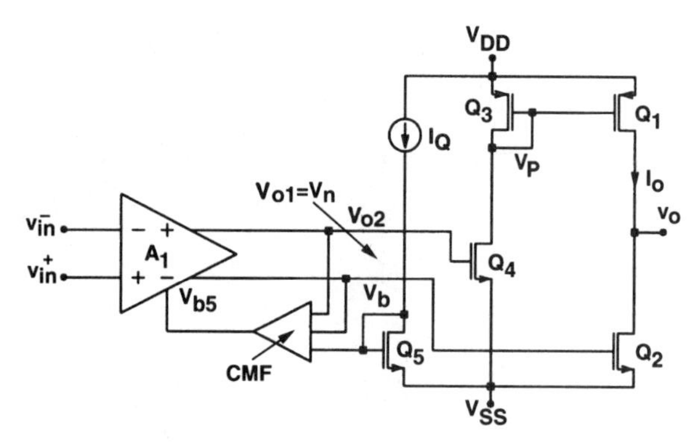

Figure 4.70. Class AB push-pull output stage with common-mode feedback.

differential output (i.e., $v_{o1} = v_{o2} = V_n = V_b$). Therefore, assuming that $(W/L)_1/(W/L)_3 = (W/L)_2/(W/L)_4$, the current I_o in the output devices is given by

$$I_o = \frac{(W/L)_2}{(W/L)_5} I_Q \tag{4.121}$$

which can be made supply independent.

For this circuit, assuming that the input stage does not impose a limit on the output stage, when it needs to sink the maximum available current, $v_{o1} = V_n$ can swing all the way up to V_{DD}, resulting in a rail-to-rail gate-to-source voltage drive for Q_2. To source current, v_{o2} swings to V_{DD}, forcing Q_4 into the linear region. This in turn causes V_p to move close to V_{SS}, resulting in a large V_{GS} for Q_1. Because of the complementary nature of the differential outputs of the input stages when one of the outputs is close to V_{DD}, the other one is close to V_{SS}. As a result, the output stage delivers a push-pull drive to the output transistors Q_1 and Q_2 in such a way that when one of the devices is heavily conducting, the other one is turned off. Also, since the V_{GS} drives of the output transistors can be as high as the full supply voltage, they can supply a large amount of output current with relatively small device sizes.

The circuit diagram of the complete op-amp is shown in Fig. 4.71. Transistors Q_{40} to Q_{47}, along with resistor R_B, constitute the biasing section. Transistors Q_{14} to Q_{25} realize the differential output folded-cascode preamplifier stage. Transistors Q_{43} and Q_{46} establish bias voltages for the high-swing cascode current sources. Transistors Q_5 to Q_{13}, which establish a bias voltage equal to V_b at the output of the differential stage, achieve the continuous-time common-mode feedback. In the conceptual block diagram representation of the preamplifier stage shown in Fig. 4.70, the common-mode feedback sets the voltage values of the two differential outputs as

$$v_{o1} = v_{o2} = V_n = V_b. \tag{4.122}$$

In Fig. 4.71 the common-mode feedback is realized by devices Q_8 to Q_{13}, while Q_5 and the current source Q_6 and Q_7 generate the bias voltage $V_b = V_{GS5}$. Devices Q_{12}

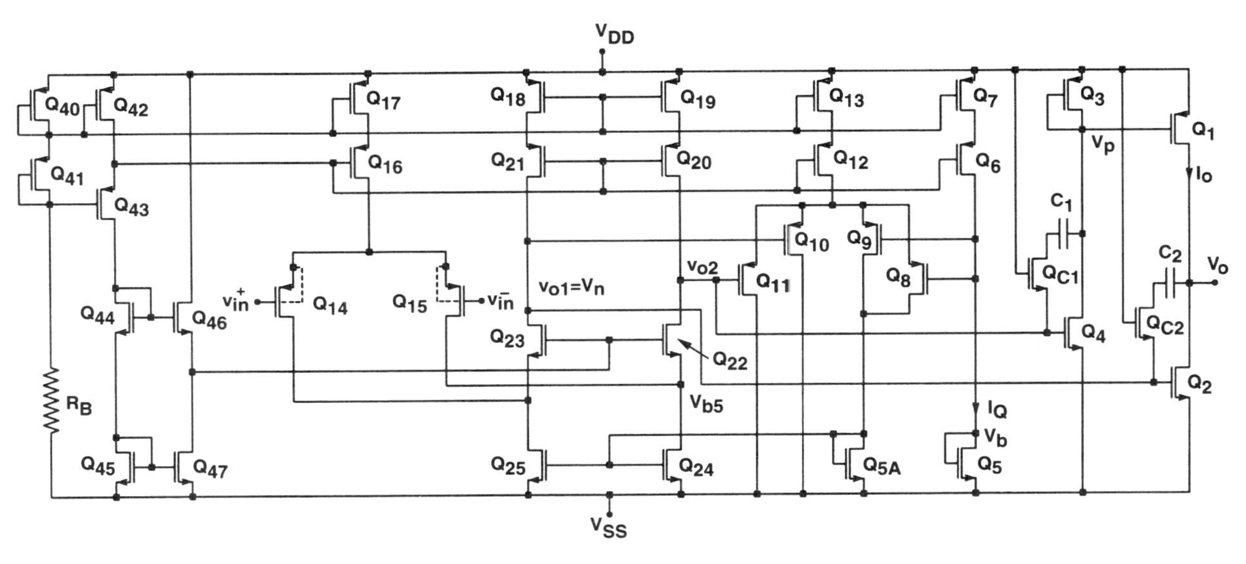

Figure 4.71. Circuit schematic of op-amp with push-pull output stage.

and Q_{13} form a current source that supplies a fixed current to the four source-coupled devices Q_8 to Q_{11}. The W/L ratio of Q_8 to Q_{11} are equal; therefore, the tail current is divided equally among the four devices, resulting in equal gate-to-source voltages, so

$$V_{GS8} = V_{GS9} = V_{GS10} = V_{GS11}. \tag{4.123}$$

Since the gate potentials of all four devices are equal, the differential output of the preamplifier stage will be biased to V_b. The output stage, consisting of devices Q_1 to Q_4, is similar to the circuit shown in Fig. 4.70, where the quiescent output current is given by Eq. (4.121).

The open-loop gain of the op-amp is calculated as $A = g_{m14}r_o g_{m2}R_L$, where r_o is the output impedance of the preamplifier stage and R_L is the load of the output stage. The op-amp can be compensated using two Miller capacitors C_1 and C_2 along with their zero-nulling MOS resistors, Q_{C1} and Q_{C2}. Assuming widely spaced poles, the dominant pole is calculated as

$$s_{pD} = \frac{1}{r_o(C_{M1} + C_{M2})}, \tag{4.124}$$

where $C_{M1} = (1 + g_{m4}/g_{m3})(C_1/2)$ and $C_{M2} = g_{m2}R_LC_2$ [14]. The load capacitance C_L, along with the other parasitic capacitances, generate poles and zeros that are above the unity-gain frequency. The common-mode feedback circuit creates a pole–zero doublet which is an order of magnitude below the unity-gain frequency. These are given by

$$s_{p1} = -\frac{g_{m10}}{C_{M1}C_{M2}/(C_{M1} + C_{M2})}, \tag{4.125}$$

$$s_{z1} = -\frac{2g_{m10}}{C_{M1} + C_{M3}}, \tag{4.126}$$

where $C_{M3} = (1 + 2g_{m10}R_2)C_2$ and R_2 is the resistance of the zero-nulling MOS device Q_{C2}. Since s_{p1} and s_{z1} are close to each other, the pole–zero doublet does not cause any instability in the frequency response.

The class AB push-pull output stages presented thus far use elaborate techniques to solve the level-shifting problem between the two signals that drive the output devices and set their quiescent bias currents. A simple, yet very powerful technique that is based on the push-pull stage of Fig. 4.58 is shown in Fig. 4.72 [25]. The circuit consists of the push-pull output stage made of devices Q_1 and Q_4. In this circuit, transistors Q_5 and Q_6 form a typical gain stage where the input signal alters the relative conduction levels of the common-gate devices. The class AB bias circuit sets up the two loops, Q_1,Q_3 and Q_2,Q_4, that fix the voltage drop between the gates of the output devices.

Referring to Fig. 4.72, the quiescent conditions of the output stage are established

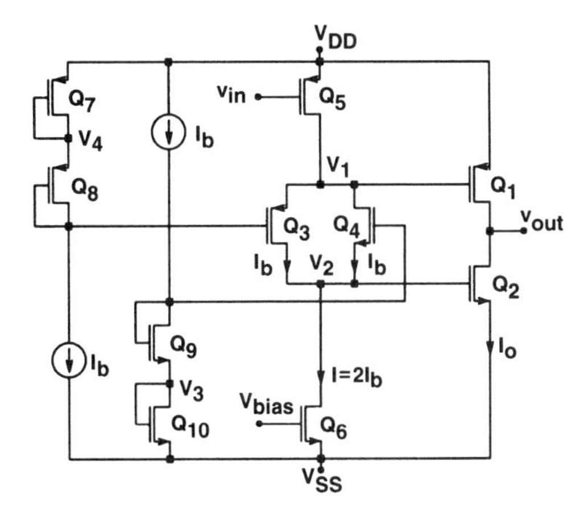

Figure 4.72. Rail-to-rail output stage with common-gate level shifters.

as follows. The complementary currents I_{b1} and I_{b2} ($I_{b1} = I_{b2} = I_b$) from the bias generator flow into complementary stacks of diode-connected transistors Q_7,Q_8 and Q_9,Q_{10} whose drain potentials are used to bias the gates of the common-gate transistors Q_3 and Q_4. In steady state the current $I = 2I_b$ through transistor Q_6 is equally divided between devices Q_3, and Q_4 so that each one carries a current equal to I_b. Assuming that

$$(W/L)_4 = (W/L)_9, \tag{4.127}$$

$$(W/L)_3 = (W/L)_8, \tag{4.128}$$

and since Q_3,Q_8 and Q_4,Q_9 carry the same drain currents I_b, they will have equal gate-to-source voltages and we have $V_2 = V_3$ and $V_1 = V_4$. As a result Q_2,Q_{10} and Q_1,Q_7 will also have equal gate-to-source voltages and the steady-state output current is given by

$$I_o = I_b \frac{(W/L)_1}{(W/L)_7} = I_b \frac{(W/L)_2}{(W/L)_{10}}. \tag{4.129}$$

When the output stage is driven to sink a large load current, v_{in} goes low and pulls v_1 and v_2 up to a high level close to V_{DD}. Under this condition Q_4 is completely shut off and Q_3 carries the full current of $I = 2I_b$ from Q_5. The source of Q_3 rises to its maximum point, thus cutting back on the conduction of Q_1. The drain of the common-gate device Q_3 also goes high, hence pulling the gate of Q_2 high, making it heavily conductive. Under the conditions of strong sourcing, v_{in} goes high, causing v_1 and v_2 to be pulled low. In this case Q_3 is completely shut off and Q_4 carries the full current of $I = 2I_b$. The source of Q_4 pulls the gate of Q_1 low, hence making it highly conductive.

A complete differential op-amp employing the output stage of Fig. 4.72 is shown in Fig. 4.73. A folded-cascode input stage is used, providing high open-loop gain. The biasing circuit consisting of resistor R_B and transistors Q_{13} to Q_{21} is designed such that all devices are biased at the onset of their saturation regions, hence maximizing the output voltage swing while providing high gain. The common-gate level-shifter circuit is inserted in the output stage of the folded-cascode differential stage. The op-amp of Fig. 4.73, with an output stage that uses common-gate level shifter to bias the push-pull output devices, is very compact and power efficient and is very suitable for driving small resistive loads.

4.10. OP-AMPS WITH RAIL-TO-RAIL INPUT COMMON-MODE RANGE

In this chapter two generations of CMOS op-amps were presented. The first generation of op-amps was limited to transconductance amplifiers. They had modest performance and were able to drive only capacitive loads. In addition to high-performance transconductance amplifiers, the second generation of these op-amps were able to drive resistive as well as capacitive loads with a level of performance which is compatible to that of their bipolar counterparts. In addition to having a high-performance output stage, third-generation CMOS op-amps have a wide-input common-mode range, which includes the positive and negative power supply rails. The wide-input common-mode range is important for low-voltage applications where the op-amp is connected in a unity-gain buffer configuration.

In an earlier section it was shown that op-amps with p-channel differential stage input devices such as the one in Fig. 4.35 have an input common-mode range that includes the negative power supply. The positive common-mode range is, however, limited to 1.4 to 1.6 V below the positive supply. Op-amps with n-channel input devices have the opposite properties. They have a positive-input common-mode range that includes the positive power supply, whereas the negative range is 1.4 to 1.6 V above the negative supply. For applications where the common-mode input range goes beyond or at least includes both power supply rails, such as the unity-gain buffer shown in Fig. 4.36*b*, the op-amp requires an input stage with an NMOS and PMOS differential pair connected in parallel. Several op-amp structures with rail-to-rail common-mode input stage are available [14,26]. One example of such an input stage is shown in Fig. 4.74. It has a folded-cascode differential-output structure with common-mode feedback.

The input stage of the op-amp in Fig. 4.74 is constructed from the parallel connection of p- and n-channel differential stages made of transistors Q_6 to Q_9 and Q_{26} to Q_{29}, respectively. The differential-stage tail current sources are formed from devices Q_6,Q_7 and Q_{28},Q_{29}. They use a cascode structure to increase the output impedance and hence the CMRR, as suggested by Eq. (3.80). High common-mode rejection is important when the op-amp is connected as a unity-gain buffer. Transistors Q_{40} to Q_{47} form a simple bias circuit. The two p-channel transistors, Q_{40} and Q_{41}, combined with resistor R_B, generate the reference current. This current is mirrored into transistors Q_{42} to Q_{46}, which generate the bias voltages for the high-swing cascode current

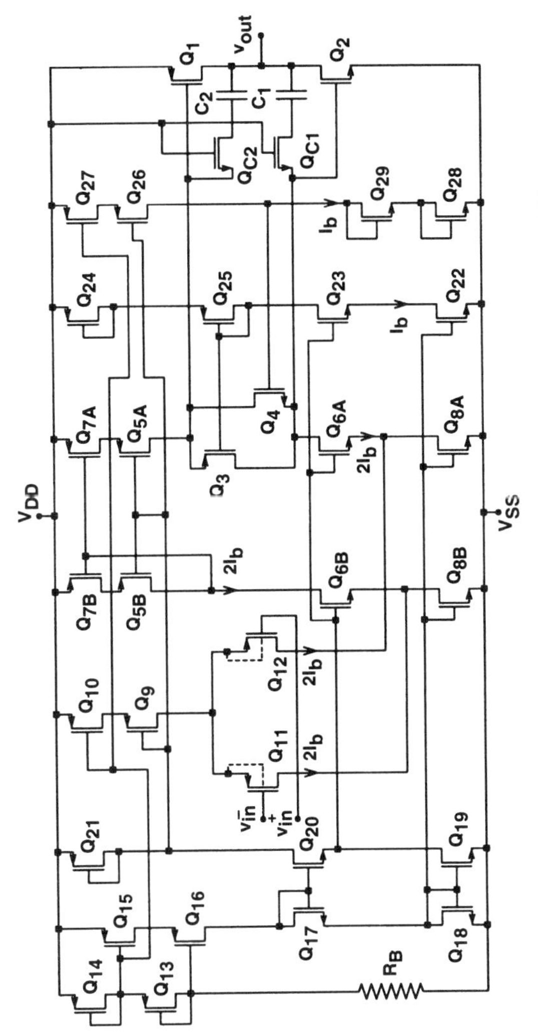

Figure 4.73. Rail-to-rail op-amp using the output stage of Fig. 4.72.

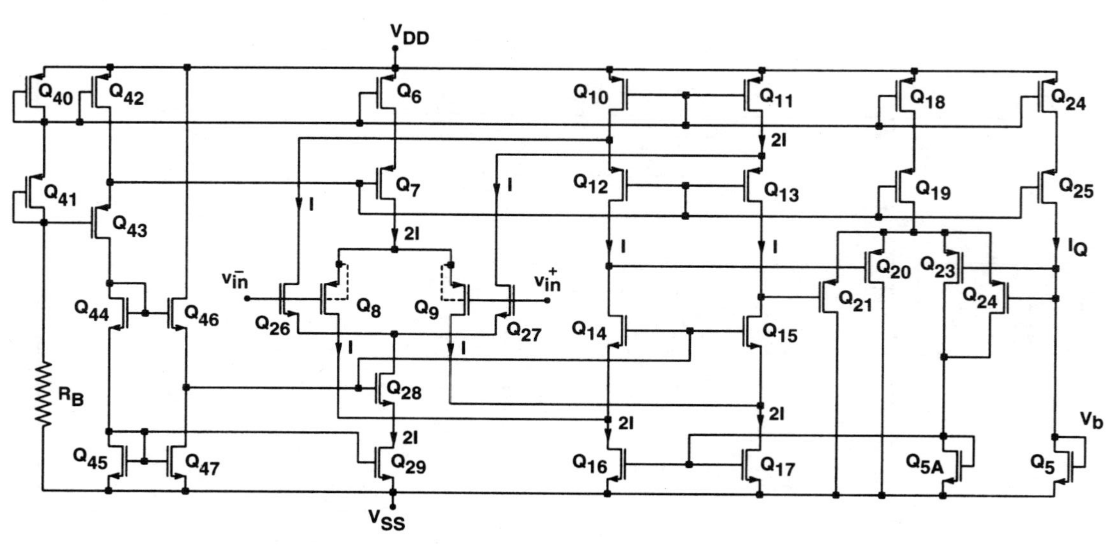

Figure 4.74. Rail-to-rail differential-input/differential-output stage with common-mode feedback.

sources. The input stage uses a folded-cascode structure made of transistors Q_{10} to Q_{17}. An additional common-mode feedback (CMF) circuit sets the dc voltage values of the two differential outputs. The configuration of the common-mode feedback circuit is based on the conceptual block diagram shown in Fig. 4.70. The operation of the common-mode feedback circuit was described in Section 4.9.

One major shortcoming of the differential stage of Fig. 4.74 is the fact that a current imbalance occurs in the load device when the common-mode input voltage approaches V_{DD} or V_{SS}. Consider the case when the common-mode voltage approaches V_{DD}. Then as the input devices Q_8 and Q_9 turn off the current components I that enter the drain of the devices, Q_{16} and Q_{17} become zero. This current imbalance causes the drains of Q_{16} and Q_{17} to snap to V_{SS} and the op-amp ceases operating in the linear region. To remedy this problem, the circuit is modified by adding transistors Q_{30} to Q_{37} (Fig. 4.75) to the input stage and dividing the currents in the devices Q_{10} and Q_{11} as well as Q_{16} and Q_{17} into two components: one that is constant and the other that is input dependent. The input-dependent current component becomes zero when their corresponding input pair devices turn off [14,27]. The complete op-amp is shown in Fig. 4.75. Now the sum of currents through the pairs Q_{10A}–Q_{10B}, Q_{11A}–Q_{11B}, Q_{16A}–Q_{16B}, and Q_{17A}–Q_{17B} are normally equal to $2I$, while the currents in Q_{12} to Q_{15} are equal to I. When the input common-mode voltage approaches V_{DD} (or V_{SS}), devices Q_8, Q_9, and Q_{32} (or Q_{26}, Q_{27}, and Q_{36}) cut off and cause the currents through devices Q_{16A} and Q_{17A} (or Q_{10A} and Q_{11A}) to drop to zero. As a result, output voltages v_{o1} and v_{o2} remain stable over the complete common-mode input voltage range.

The output stage for the op-amp of Fig. 4.75 is formed from transistors Q_1 to Q_4. The common-mode feedback circuit sets the quiescent current I_o in the output devices to

$$I_o = \frac{(W/L)_2}{(W/L)_5} I_Q. \tag{4.130}$$

The class AB push-pull output stage is based on the configuration of Fig. 4.70; its operation was described in Section 4.9.

Frequency compensation for the op-amp is achieved using two Miller capacitors along with their corresponding zero-nulling MOS resistors. The open-loop gain is calculated to be $A = (g_{m8} + g_{m26}) r_o g_{m2} R_L$, where r_o is the output impedance of the input stage and R_L is the load resistor.

One major drawback of the rail-to-rail op-amp shown in Fig. 4.75 is that the transconductance (g_m) of the input devices varies by a factor of 2 over the input common-mode range. This large variation in g_m prevents the op-amp from having an optimum frequency compensation over the entire operating range. To keep the transconductance constant, the g_m value of the lower and upper parts of the input range should be increased by a factor of 2. This is because in the middle of the common-mode range the g_m value of the input stage is the sum of the g_m values of the p- and n-channel devices. In the lower common-mode range the n-channel devices turn off and the g_m value of the input stage is reduced to the g_m value of the

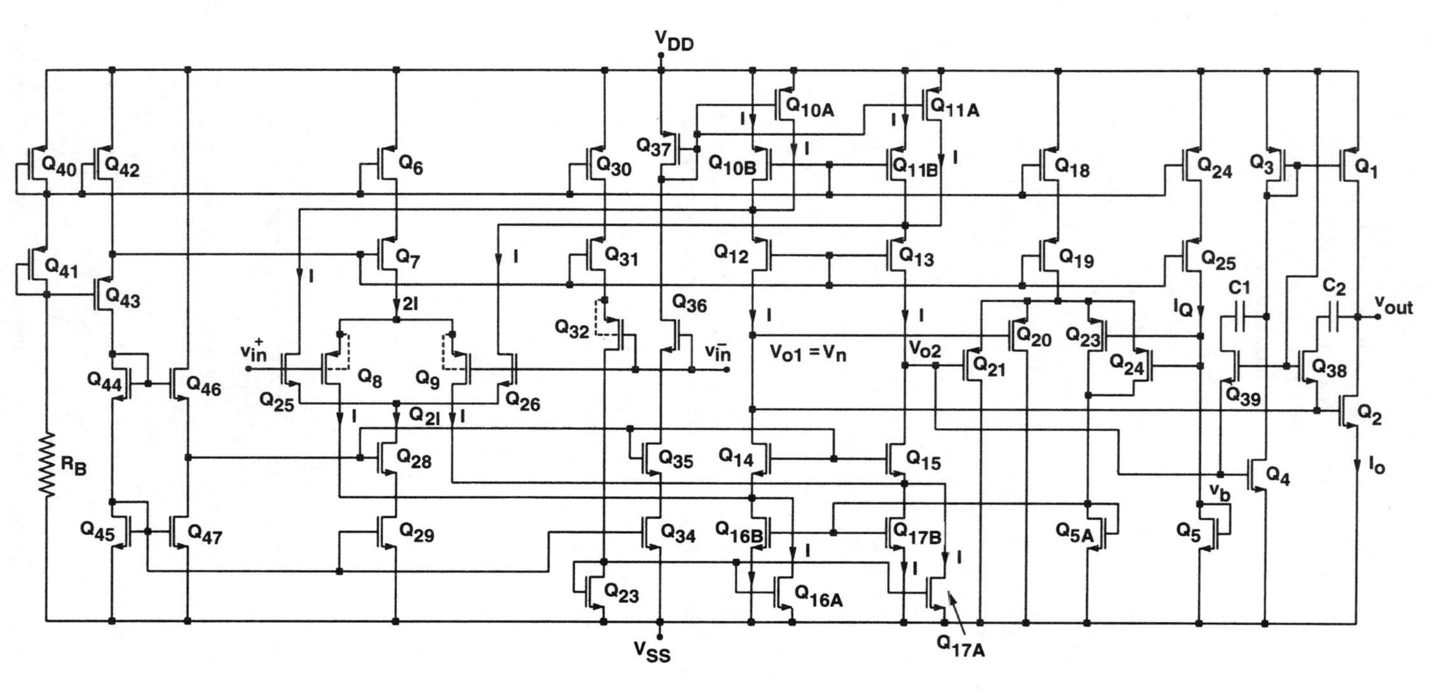

Figure 4.75. Complete rail-to-rail input-stage op-amp.

p-channel devices. Clearly, the opposite takes place in the upper common-mode range. The g_m values of the p- and n-channel input devices are given by

$$(g_m)_p = \sqrt{\mu_p C_{ox}\left(\frac{W}{L}\right)_p I_o}, \qquad (g_m)_n = \sqrt{\mu_n C_{ox}\left(\frac{W}{L}\right)_n I_o}. \tag{4.131}$$

From the equations above, it can be observed that to maintain a constant g_m value, the following condition should be satisfied between the W/L ratios of the p- and n-channel input devices:

$$\frac{(W/L)_p}{(W/L)_n} = \frac{\mu_n}{\mu_p} \tag{4.132}$$

As suggested by Eq. (4.131), the g_m value of the MOS device is proportional to the square root of its drain current. Therefore, while the p- and n-channel input devices should satisfy the condition of Eq. (4.132), to maintain a constant g_m value, the tail currents of the input stages should also vary by a factor of 4 over the input common-mode range. In other words, in the middle of the common-mode range, the p- and n-channel input stages would have tail currents that are equal to I_o. In the upper and lower parts of the input range, the tail currents of the p- and n-channel differential stages should be increased to $4I_o$, respectively. This will maintain an approximately constant g_m for the input stage over the entire input common-mode range.

This principle is applied to the circuit shown in Fig. 4.76 [20]. A rail-to-rail input stage is shown, where p- and n-channel differential pairs are placed in parallel.

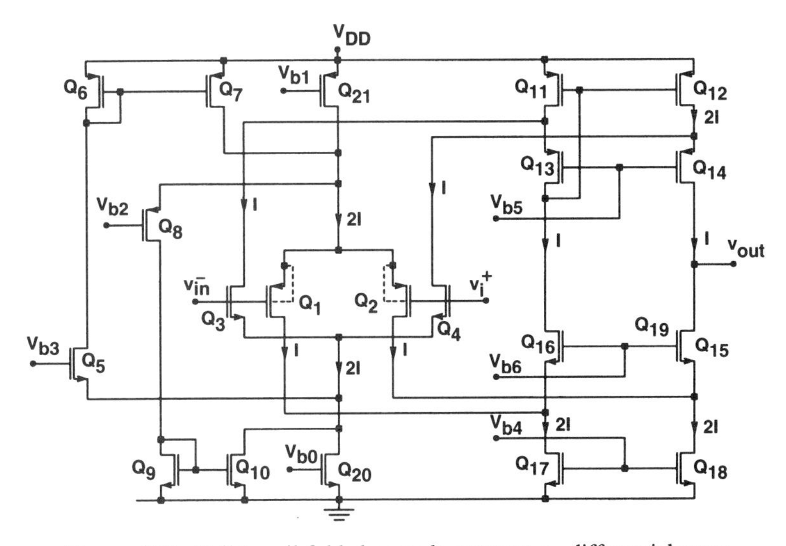

Figure 4.76. Rail-to-rail folded-cascode constant-g_m differential stage.

Similar to the op-amp of Fig. 4.75, the stage is able to reach the positive and negative supply rails through the n-channel (Q_3–Q_4) and p-channel (Q_1–Q_2) input pairs. The constant-g_m property is achieved by the addition of the current switches Q_5 and Q_8 and the two current mirrors Q_6–Q_7 and Q_9–Q_{10}, each with a gain of 3. To better understand the constant-g_m control circuit, the input common-mode range will be divided into three regions.

When the input common-mode goes below $V_{b3} = (V_{GS})_{Q3} + (V_{D\text{sat}})_{Q20}$, the n-channel input devices will start turning off and the p-channel input pair will be operational. In this case the n-channel current switch Q_5 will turn on while the p-channel current switch Q_8 is off. The current of transistor Q_5 is multiplied by a factor of 3 and is added to the tail current of the p-channel differential stage. The tail current is therefore increased by a factor of 4, which results in doubling the input g_m.

In the middle of the range, the input voltage is greater than $V_{b3} = (V_{GS})_{Q3} + (V_{D\text{sat}})_{Q20}$ but less than $V_{b2} = V_{DD} + |(V_{GS})_{Q1}| + |(V_{D\text{sat}})_{Q21}|$. Now the p- and n-channel current switches, Q_5 and Q_8 are both off, with both input pairs operational. The result is that the g_m of the input stage is equal to the sum of the g_m of the p- and n-channel differential pairs.

Finally, when the input common-mode range exceeds $V_{b2} = V_{DD} + |(V_{GS})_{Q1}| + |(V_{D\text{sat}})_{Q21}|$, the p-channel input stage will start turning off, while the n-channel pair will still be operational. The p-channel current switch will turn on, and the current of transistor Q_8, after being multiplied by a factor of 3, will be added to the tail current of the n-channel input pair. The result is that the tail current of the n-channel pair is multiplied by a factor of 4, which increases the effective input g_m by a factor of 2.

The cascode current mirrors Q_{11} to Q_{14} and the folded-cascode devices Q_{15} and Q_{16} form the single-ended output stage. The biasing schemes shown in Fig. 4.10 can be used to maximize the output voltage range. The op-amp of Fig. 4.76 can be used as a stand-alone single-stage transconductance amplifier with a rail-to-rail input range that can be used to drive capacitive loads; or with the addition of one of the high-performance output stages described in Section 4.9, it can be used as a general-purpose op-amp that is able to drive resistive as well as capacitive loads.

PROBLEMS

4.1. Prove Eqs. (4.3) to (4.7) for the circuit of Fig. 4.3. How much is the dynamic range for **(a)** the op-amp alone and **(b)** the feedback amplifier if $V_{CC} = 10$ V, $A = 10^3$, $\sqrt{\overline{v_n^2}} = 50\ \mu$V, and $R_2 = 10R_1$?

4.2. Show that the load conductance represented by Q_3 in Fig. 4.6 is $g_l = g_{m3} + g_{d3}$.

4.3. Show that the small-signal output impedance of the differential stage of Fig. 4.7 is given by Eq. (4.10).

4.4. Show that the small-signal output impedance of the single-stage folded-cascode op-amp of Fig. 4.9 is given by Eq. (4.12).

4.5. Prove Eq. (4.14) for the high-swing folded-cascode op-amp of Fig. 4.10.

4.6. Prove Eq. (4.55) for the poles of the circuit of Fig. 4.23. [*Hints:* Calculate $A_v(S)$ from Eqs. (4.49) and (4.50). Write its denominator as

$$\left(1 - \frac{s}{s_{p1}}\right)\left(1 - \frac{s}{s_{p2}}\right) \simeq 1 - \frac{s}{s_{p1}} + \frac{s^2}{s_{p1}s_{p2}}.$$

Find s_{p1} and s_{p2}; use $(g_{m8} + g_{m9})/(g_{d2} + g_{d4}) \gg 1$ and $(g_{m8} + g_{m9})/(g_{d8} + g_{d9}) \gg 1$.]

4.7. Prove that the zeros of $A_v(s)$ for the circuit of Fig. 4.25 are at $s \rightarrow \infty$, while its poles are the same as for the circuit of Fig. 4.23.

4.8. Derive Eqs. (4.57) to (4.59) for the circuit of Fig. 4.27. Why is $A_v(s)$ now a third-order function, whereas for the circuit of Fig. 4.23 it was only second order?

4.9. Prove Eq. (4.70) for the three-stage op-amp of Fig. 4.30.

4.10. Prove Eqs. (4.71) and (4.72) for the small-signal equivalent circuit of Fig. 4.31.

4.11. Show that Eq. (4.93) gives the small-signal output voltage of the circuit of Fig. 4.40*a* if $v_{in} = V_1 u(t)$ and the op-amp transfer function is given by Eq. (4.92).

4.12. Using the low-frequency small-signal models for devices Q_1 to Q_4, show that the voltage-gain relations of Eqs. (4.101) and (4.102) hold for the noisy input stage shown in Fig. 4.45.

4.13. The circuit of Fig. 4.77 can be used to measure the unity-gain bandwidth ω_0 of an op-amp. Show that ω_0 is the frequency at which $V_{out}(\omega) = V_{in}(\omega)/\sqrt{2}$; that is, the voltage gain is 3 dB below its dc value.

4.14. For the circuit of Problem 4.13, let the open-loop gain of the op-amp have a phase margin of 60° at the unity-gain bandwidth ω_0. How much is the phase shift between V_{out} and V_{in} at ω_0?

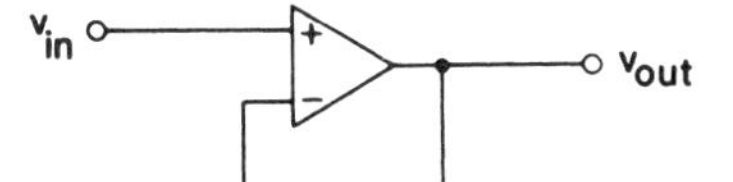

Figure 4.77. Op-amp in unity-gain configuration (Problem 4.13).

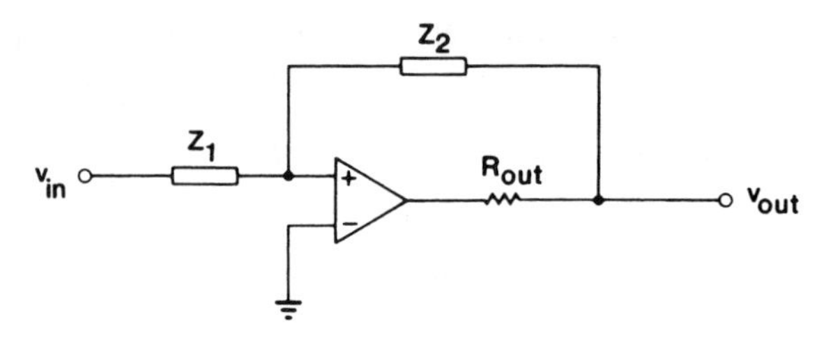

Figure 4.78. Op-amp with nonzero output impedance (Problem 4.15).

4.15. Show that in the circuit of Fig. 4.78, the effective output impedance is $R_{out}(-A_c + 1)/A$, where $A_c \simeq -Z_2/Z_1$ is the closed-loop gain of the stage. Assume that $A \gg 1$ and $R_{out}/A \ll |Z_1|$ *and* $|Z_2|$.

4.16. Let the output voltages of the differential op-amp of Fig. 4.50 change by a differential amount so that $v_{out}^+ \rightarrow v_{out}^+ + v$ and $v_{out}^- \rightarrow v_{out}^- - v$. Show that the sum of the drain currents of Q_3 and Q_4 remains unchanged. [*Hint:* Assume that Q_3 and Q_4 are in their linear regions, and hence you can use Eq. (2.7).]

4.17. Prove relations (4.107) and (4.108), giving the gain and output impedances, respectively, of the differential op-amp of Fig. 4.50.

4.18. Show that the aspect ratios of the bias circuit of Fig. 4.51 are given by Eq. (4.109) if we choose $I = I_o/2$ in Fig. 4.50, $I_{bias} = I_o$ as the bias chain current, and zero dc common-mode output voltage.

4.19. Show that (if necessary) a differential output op-amp can be constructed from two single-ended-output op-amps using the circuit [15] of Fig. 4.79.

4.20. Show that for the output stage of Fig. 4.65 the maximum sourcing and sinking currents are given by Eqs. (4.119) and (4.120), respectively.

4.21. Find the output impedance of the circuit of Fig. 4.68 (*Hint:* Set $v_{in} = 0$ and connect the output terminal to a test source. Calculate the current through the source.)

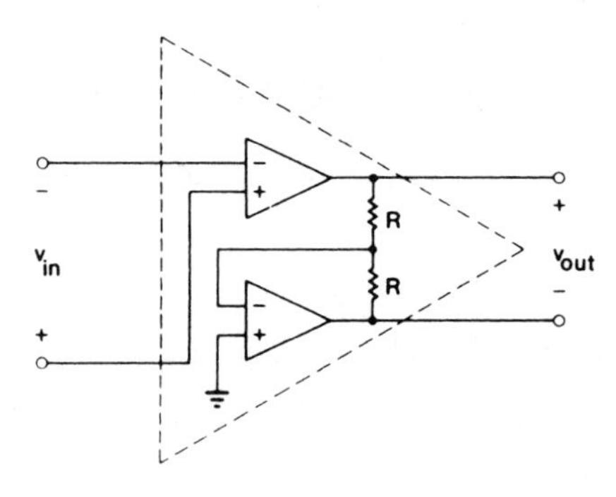

Figure 4.79. Simplified equivalent circuit of a differential output op-amp (Problem 4.19).

4.22. Prove Eq. (4.121) for the output stage of Fig. 4.70.

4.23. Show that the dominant pole of the compensated op-amp of Fig. 4.71 is given by Eq. (4.124).

4.24. Prove that the common-mode feedback circuit of Fig. 4.71 creates a pole–zero doublet given by Eqs. (4.125) and (4.126).

4.25. Prove Eq. (4.130) for the op-amp of Fig. 4.75.

4.26. Prove that if (1) $Av(s)$ is a rational function with its poles having negative real parts and (2) $\mathrm{Re}[kA_v(j\omega)] > -1$ for all ω, the s values satisfying $kA_v(s) = -1$ all have negative real parts. [*Hint:* By the maximum-modulus theorem of complex functions, if $kAv(s)$ has no poles in the right half of the s plane, its real part in the same region has its minimum value on the $j\omega$ axis.]

4.27. Show that the input common-mode limit for the differential stage of Fig. 4.36 is given by Eq. (4.80).

4.28. Show that the maximum sourcing and sinking currents for the Fig. 4.65*a* is given by Eqs. (4.119) and (4.120).

4.29. Show that the dominant pole of the op-amp of Fig. 4.71 is given by Eq. (4.124).

4.30. Prove that the common-mode feedback circuit in the op-amp of Fig. 4.71 creates a pole–zero doublet given by Eqs. (4.125) and (4.126).

REFERENCES

1. D. J. Hamilton and W. G. Howard, *Basic Integrated Circuit Engineering,* McGraw-Hill, New York, 1975.
2. P. R. Gray and R. G. Meyer, *Analysis and Design of Analog Integrated Circuits,* Wiley, New York, 1977.
3. K. C. Hsieh, P. R. Gray, D. Senderowicz, and D.G. Messerschmitt, *IEEE J. Solid-State Circuits, SC-16* (6), 708–715 (1981).
4. P. R. Gray and R. G. Meyer, MOS operational amplifier design—a tutorial overview, *IEEE J. Solid-State Circuits, SC-13* (3), 285–294 (1978).
5. R. J. Widler, *IEEE J. Solid-State Circuits, SC-13* (4), 184–191 (1969).
6. R. Gregorian and W. E. Nicholson, Jr., *IEEE J. Solid-State Circuits, SC-14* (6), 970–980 (1979).
7. P. R. Gray, Basic MOS operational amplifier design—an overview, in *Analog MOS Integrated Circuits:* Part II, IEEE Press, New York, 1980.
8. J. H. Huijsing and D. Lineborger, *IEEE J. Solid-State Circuits, SC-20* (6), 1144–1150 (1985).
9. P. R. Gray and R. G. Meyer, *Analysis and Design of Analog Integrated Circuits,* Wiley, New York, 1993.

10. J. E. Solomon, *IEEE J. Solid-State Circuits, SC-9* (6), 314–332 (1974).
11. J. C. Bertails, *IEEE J. Solid-State Circuits, SC-13* (6), 791–798 (1978).
12. M. Banu, J. M. Khoury, and Y. Tsividis, *IEEE J. Solid-State Circuits, SC-23* (6), 1410–1414 (1988).
13. D. Senderowicz, S. F. Dreyer, J. H. Huggins, C. F. Rehim, and C. A. Laber, *IEEE J. Solid-State Circuits, SC-17* (6), 1014–1023 (1982).
14. J. N. Babanezhad, *IEEE J. Solid-State Circuits, SC-23* (6), 1414–1417 (1988).
15. P. R. Gray and R. G. Meyer, *IEEE J. Solid-State Circuits, SC-17* (6), 969–982 (1982).
16. K. C. Hsieh, *Proc. Int. Symp. Circuits Syst.,* pp. 419–422, 1982.
17. D. A. Johns and K. Martin, *Analog Integrated Circuit Design,* Wiley, New York, 1997.
18. D. Senderowicz, NMOS operational amplifiers, in *Design of MOS VLSI Circuits for Telecommunications,* Y. Tsividis and P. Antognetti (Eds.), Prentice Hall, Upper Saddle River, N.J., 1985.
19. R. Hogervorst, Design of low-voltage low-power CMOS operational amplifier cells, Ph.D. dissertation, Delft University, 1996.
20. R. Hogervorst, J. P. Tero, R. G. H. Eschauzier, and J. H. Huijsing, *IEEE J. Solid-State Circuits, SC-29* (12), 1505–1513 (1994).
21. E. Seevinch, W. de Jager, and P. Buitendijk, *IEEE J. Solid-State Circuits, SC-23* (3), 794–801 (1988).
22. R. Castello and P. R. Gray, *IEEE J. Solid-State Circuits, SC-20* (6), 1122–1132 (1985).
23. J. N. Babanezhad and R. Gregorian, *IEEE J. Solid-State Circuits, SC-22* (6), 1080–1089 (1987).
24. B. K. Ahuja, P. R. Gray, W. M. Baxter, and G. T. Uehara, *IEEE J. Solid-State Circuits, SC-19* (6), 892–899 (1984).
25. D. M. Manticelli, *IEEE J. Solid-State Circuits, SC-21* (6), 1026–1034 (1986).
26. M. D. Pardoen and M. G. Degrauwe, *IEEE J. Solid-State Circuits, SC-25* (2), 501–504 (1990).
27. D. B. Ribner and M. A. Copeland, *IEEE J. Solid-State Circuits, SC-19* (6), 919–925 (1984).

CHAPTER 5

COMPARATORS

Comparators are the second most widely used components in electronic circuits, after operational amplifiers. A voltage comparator is a circuit that compares the instantaneous value of an input signal $v_{\text{in}}(t)$ with a reference voltage V_{ref} and produces a logic output level depending on whether the input is larger or smaller than the reference level. The most important application for a high-speed voltage comparator occurs in an analog-to-digital converter system. In fact, the conversion speed is limited by the decision-making response time of the comparator. Other systems may also require voltage comparison, such as zero-crossing detectors, peak detectors, and full-wave rectifiers. In this chapter a number of approaches to comparator design are presented. First, the single-ended auto-zeroing comparator is examined, followed by simple and multistage differential comparators, regenerative comparators, and fully differential comparators. Several design principles are introduced that can be used to minimize input offset voltage and clock-feedthrough effects.

5.1. CIRCUIT MODELING OF A COMPARATOR

One very important and widely used comparator configuration is a high-gain differential input, single-ended output amplifier [1,2]. Figure 5.1*a* shows the symbol of a differential comparator which is very similar to that of an operational amplifier. Usually, the comparator stage is followed by a latch, which is essentially a bistable multivibrator. The latch provides a large and fast output signal whose amplitude and waveform are independent of those of the input signal and is hence well suited for the logic circuits following the latch. If no latch is used, the output v_{out} should have a large swing, say, 0 to +5 V, as the input changes from −1 mV to +1 mV. Thus the required gain is around 5 V/2 mV = 2500, or 68 dB. If a latch is used,

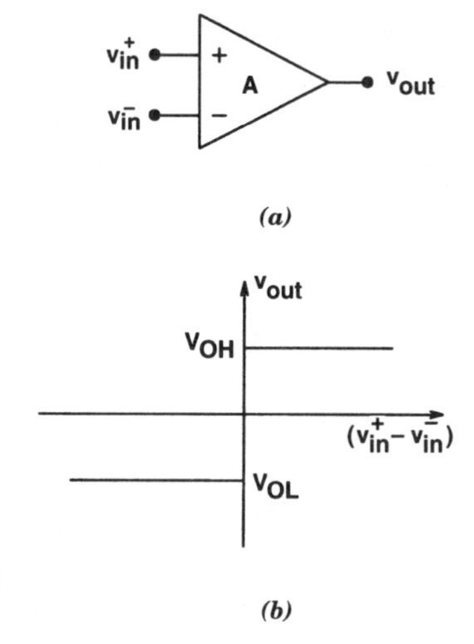

Figure 5.1. (*a*) Differential-input comparator; (*b*) transfer curve of ideal comparator.

v_{out} only needs to be higher than the combined offset and threshold voltages of the latch; this value is around 0.2 V or less. Hence a gain of 200 is now adequate. A comparator is therefore essentially a high-gain op-amp designed for open-loop operation. But unlike an op-amp it does not require frequency compensation.

The transfer curve of the ideal differential comparator is shown in Fig. 5.1*b*. In this figure the negative input of the comparator in Fig. 5.1*a* is tied to a reference voltage V_{ref}. When the positive input is greater than V_{ref} the output is high (V_{OH}), and when it is less than V_{ref}, the output is low (V_{OL}). The ideal transfer curve of Fig. 5.1b corresponds to a differential gain of infinity. In actuality the differential gain has a finite value equal to A_v. The dc transfer curve of such a comparator is shown in Fig. 5.2, where V_{iL} and V_{iH} are the input excess voltages called the *overdrive.* The overdrive is the input level that drive the comparator from some initial saturated input condition to an input level barely in excess of that required to cause the output to switch state. Another nonideal effect of the differential comparator is the input-referred dc offset voltage, V_{off}. In the absence of the offset voltage ($V_{\text{off}} = 0$), the comparator dc transfer curve will be symmetrical around V_{ref}. However, for a finite value of V_{off}, the output will begin changing only after the input difference exceeds V_{off}. The dc transfer curve of a practical differential comparator with finite

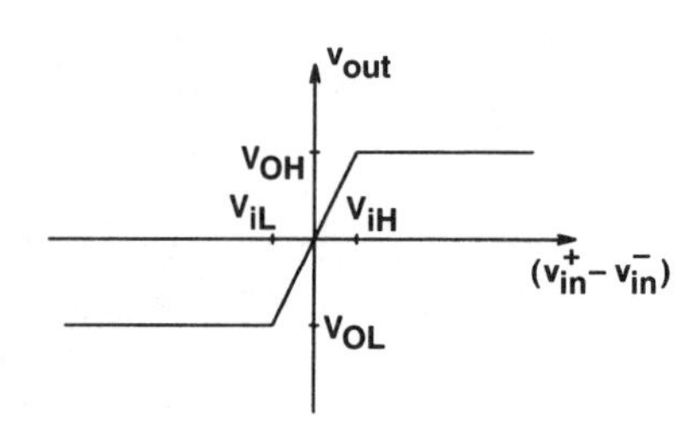

Figure 5.2. Transfer curve of comparator with finite gain.

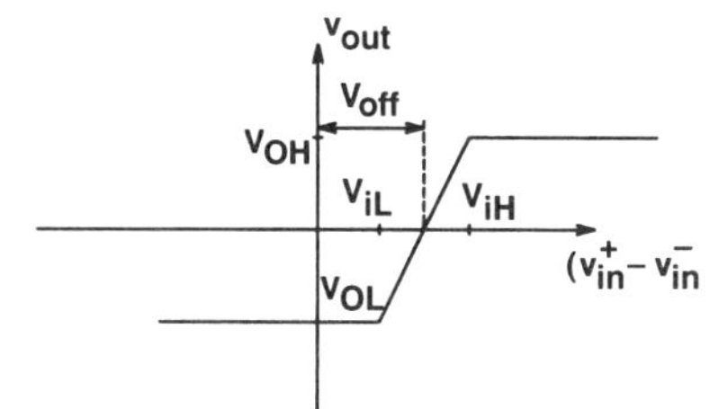

Figure 5.3. Transfer curve of comparator with finite gain and dc offset voltage.

gain of A_v and a dc offset voltage of $V_{\rm off}$ is shown in Fig. 5.3. Finally, the speed or response time is another important parameter of a comparator. In most applications it is required that following an appropriate input level change, the comparator must switch between two output levels with fast rise and fall times in the shortest amount of time. The response time is a function of the input overdrive voltage and it speeds up as the overdrive is increased.

Another class of comparators, which contain a cascade of inverter stages, is shown in Fig. 5.4. These comparators have a single-ended input/single-ended output configuration and operate in two steps: an auto-zeroing function followed by a comparison cycle. These comparators, which are used primarily in flash analog-to-digital converters, are discussed in the next section.

5.2. SINGLE-ENDED AUTO-ZEROING COMPARATORS [3]

Single-ended auto-zeroing comparators have been used extensively in high-speed flash A/D converters. Considering the system of Fig. 5.4, let the inverters be realized in CMOS technology; then a possible configuration for the input inverters is shown in Fig. 5.5*a*, with the clock waveforms and node voltage v_A illustrated in Fig. 5.5*b*. The operation is as follows: At $t = 0$, switches S_2 and S_3 are closed. S_2 connects the left-side terminal of the auto-zeroing capacitor C to ground, while S_3 shorts nodes A and B. As a result, the nodes assume a voltage that can be found from the intersection of the input–output dc characteristics of the inverter and the 45° line representing the $v_A = v_B$ condition. Figure 5.6 illustrates the situation: Fig. 5.6*a* shows the inverter, and Fig. 5.6*b* (center curve) illustrates an example of input–output characteristics (with S_3 open) for $V_{DD} = -V_{SS} = 5$ V, and for the threshold voltages, $V_{Tn} = -V_{Tp} = 1$ V. The intersection of this curve with the $v_A = v_B$ line occurs at the origin; this is in the middle of the linear range, where Q_1 and Q_2 are both in saturation and the gain of the inverter is at maximum. This favorable bias

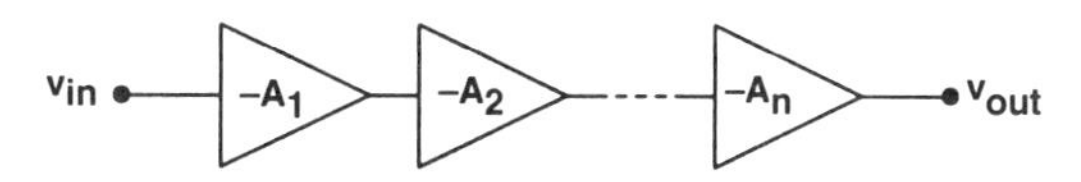

Figure 5.4. Single-ended cascade of inverter stage comparator.

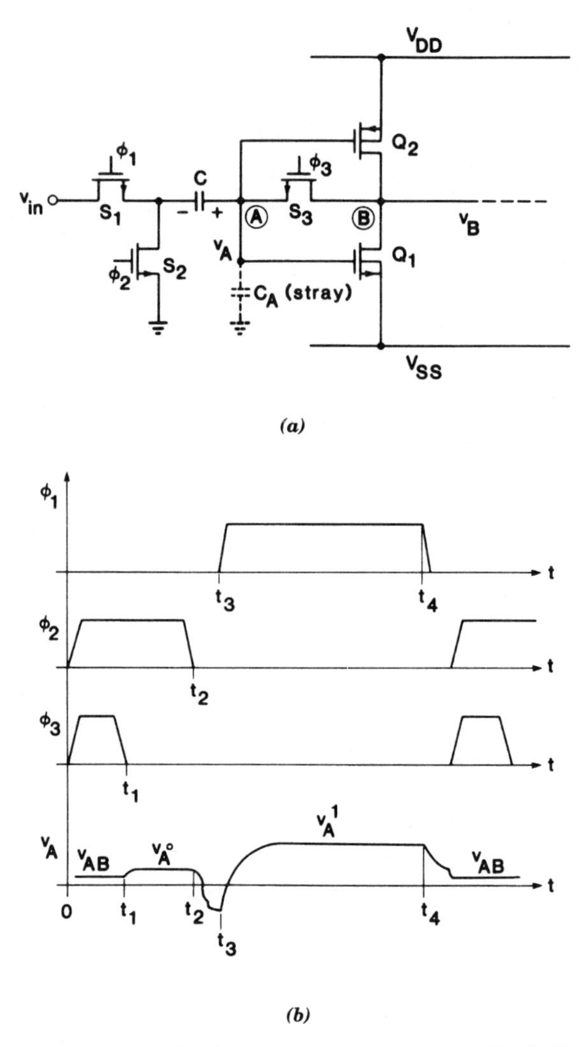

Figure 5.5. Input inverter for a CMOS cascade comparator: (*a*) circuit diagram; (*b*) waveforms.

condition is quite insensitive to variations of the threshold voltages, as illustrated by curve 1 (drawn for $V_{Tn} = 0.7$ V, $V_{Tp} = -1.2$ V) and curve 2 (for $V_{Tn} = 1.2$ V, $V_{TP} = -0.7$ V); the intersection point is in both cases near the middle of the linear range. (See Problem 5.1 for a graphical identification of the linear range for all three curves.)

Returning to the circuit of Fig. 5.5*a*, clearly during the time interval between $t = 0$ and $t = t_1$, the capacitor C charges to voltage $v_{AB} - 0 = v_{AB}$, where v_{AB} is the intersection (self-bias) voltage illustrated in Fig. 5.6*b*. (There, for nominal threshold voltages, $v_{AB} \approx 0$.) Next, at t_1, ϕ_3 goes low. This results in nodes A and B being

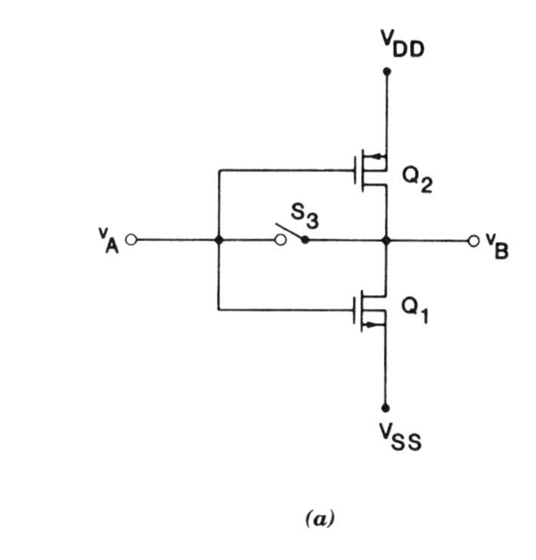

(a)

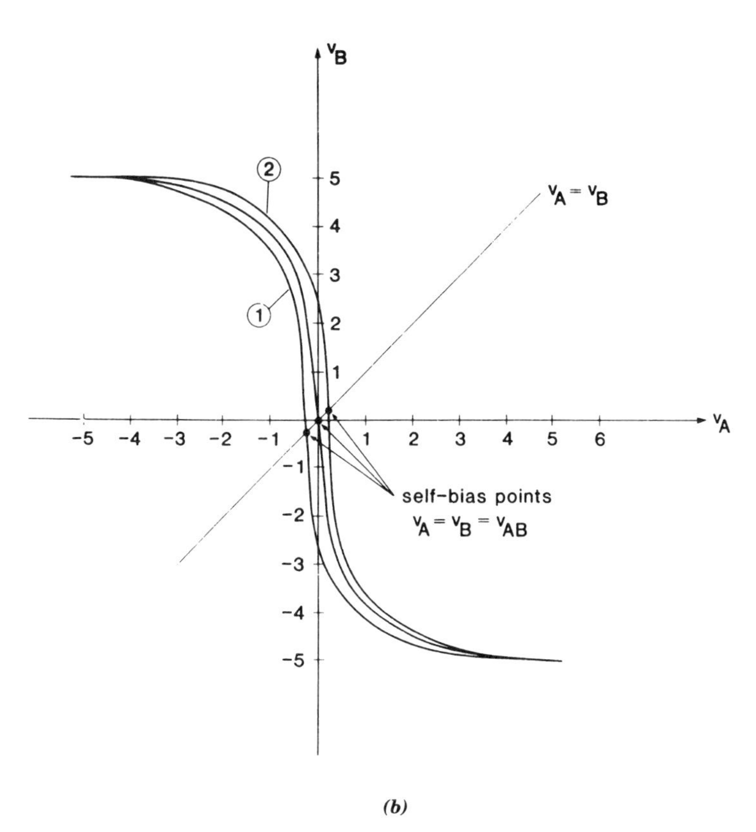

(b)

Figure 5.6. Bias conditions for the input inverter: (*a*) circuit diagram; (*b*) input–output dc characteristics for various threshold voltages (see the text).

disconnected. Also, part of the charge in the channel of S_3 enters C; in addition, through the gate-to-source overlap capacitance of S_3, additional clock-feedthrough charges enter C. The dimensions of C and S_3 must be determined such that even with the change in v_A due to these charges, the Q_1–Q_2 inverter should remain safely in its linear range. Let the resulting node voltages at A and B be denoted by v_A^0 and v_B^0, respectively.

Next (at $t = t_2$), S_2 opens. Now apart from the small stray capacitance C_A, node A is floating and hence v_A and v_B can drift to any value. However, since C is nearly open circuited at A, there will be only minimal clock feedthrough into C due to the cutoff of S_2; also, this clock-feedthrough charge is stored in C_A and will be returned to C when S_1 closes. Hence this clock feedthrough has no significant effect.

At $t = t_3$, ϕ_1 goes high and C (charged earlier to v_A^0 during the $t_1 < t \leq t_2$ interval) is connected between the input terminal and node A. Hence the node voltage v_A now becomes $v_A^1 = \alpha v_{in} + v_A^0$, where $\alpha = C/(C + C_A)$, and the voltage difference $v_A^1 - v_A^0$ has the same sign as v_{in}. The last statement holds regardless of the values of C_A and v_{AB}, the magnitude of the clock feedthrough, and any other parasitic effects. Thus, by monitoring the change $v_B^1 - v_B^0$ (which is the amplified version of $v_A^1 - v_A^0$), we can decide whether $v_{in} > 0$ or $v_{in} < 0$ holds.

A more complete diagram of a possible circuit is shown schematically in Fig. 5.7*a*, with the clock signals illustrated in Fig. 5.7*b*. The operation is as follows. As explained above, by the end of the ϕ_3 pulse, C_1 is charged to a voltage v_A^0 which is suitable for biasing the Q_1–Q_2 inverter in its linear range. If Q_3 is matched to Q_1 and Q_4 is matched to Q_2, the second inverter has the same bias point, and hence is also biased in its linear range. (Note, however, that the clock-feedthrough voltage due to the opening of S_3 is amplified by Q_1 and Q_2 and then connected to the gates of Q_3 and Q_4; hence the second stage is more vulnerable to this effect). During this time, the Q_5–Q_6 inverter is also biased in its linear range by S_4, and C_2 is precharged to the corresponding bias voltage. S_4 opens only *after* S_3 does, so that the output voltage v_C of the Q_5–Q_6 inverter is not affected by the amplified clock-feedthrough transient due to S_3. It is affected, however, by the opening of S_4. This transient can, in turn, be rendered ineffective by C_3 and S_5. When S_5 is closed, the latch is locked in a self-biased balanced state. Thus, when S_3 opens, the voltage acquired by C_3 would keep a perfectly symmetrical latch in an unstable balanced position. For the latch to function, however, the enabling (''strobe'') signal ϕ_6 must also go high.

By $t = t_4$, all precharging operations are complete. At this point, S_2 is opened and the input node A now floats. When next S_1 closes, v_D at the input of the latch will rise (if $v_{in} < 0$) or fall (if $v_{in} > 0$) from its earlier balanced value. Hence, when ϕ_6 rises, the latch will switch to the appropriate one of its two stable states.

Obviously, the system shown in Fig. 5.7 is only one example of the many different possibilities. (It is, in fact, a CMOS equivalent of the circuit described in Ref. 3.) Other circuits may contain only two inverter stages, use source followers as buffers between the various stages, use capacitive coupling also between the first and second stages, and so on. In most cases, however, biasing and auto-zeroing is accomplished using the precharging operations illustrated in Fig. 5.7.

The speed of the cascaded inverter stages is limited by the RC time constants repre-

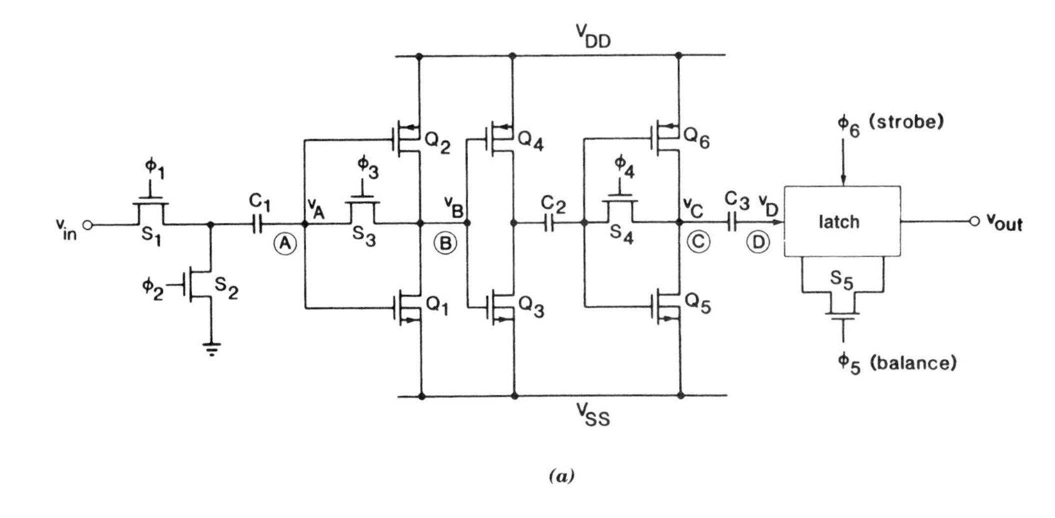

(a)

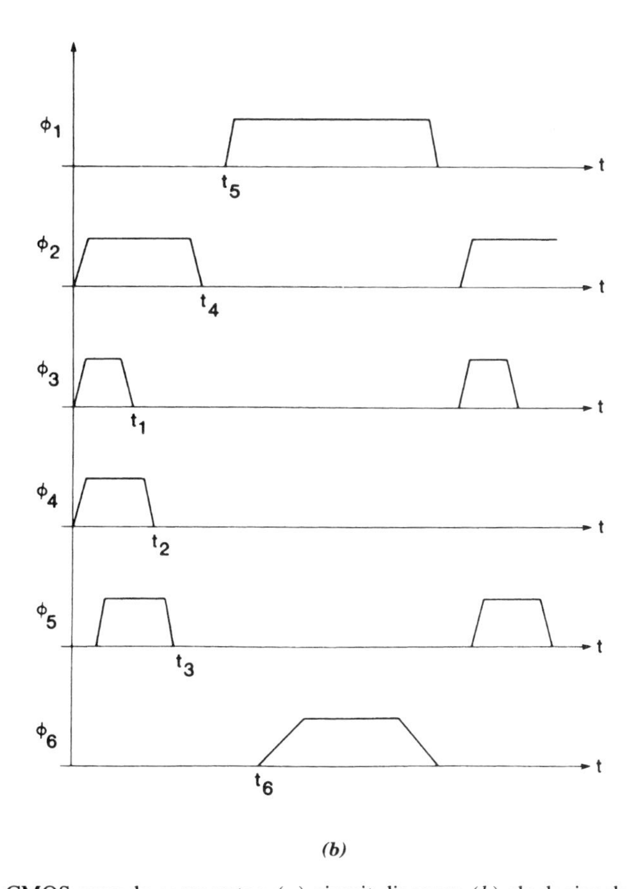

(b)

Figure 5.7. CMOS cascade comparator: (*a*) circuit diagram; (*b*) clock signals.

sented by the output resistance R_o of the ith inverter and the input capacitance C_{in} of the next [i.e., $(i + 1)$st] inverter. The former is the parallel combination of the drain resistance of the PMOS and NMOS devices in the ith stage; the latter can be approximated by the sum of the C_{gd} values in the next stage, multiplied by $(1 + |A_{i+1}|)$ due to the Miller effect. The dc gain A of the inverter is, of course, the sum of the transconductances divided by the sum of the drain conductances of the two devices in the stage. Typical values are $R_o \sim 100\ \text{k}\Omega$, $C_{in} \sim 1.5$ pF, and $A \sim 10$.

5.3. DIFFERENTIAL COMPARATORS

The simplest form of a differential comparator is the uncompensated two-stage transconductance op-amp shown in Fig. 5.8. Although removing the compensation capacitor somewhat speeds up the comparator, the response time is still slow and not adequate for many high-speed applications. As defined earlier, the response time, or propagation delay, is the delay between the time the differential input passes the comparator threshold voltage and the time the output exceeds the input logic level of the subsequent stage. The propagation delay is worst when the input signal changes from a large overdrive level to an opposite level that barely exceeds the threshold voltage. The delay time is determined primarily by the maximum current available to change the parasitic capacitances shown in Fig. 5.8. The capacitance C_{LA} consists of the junction capacitors and the gate-to-source capacitance C_{GS} of transistor Q_5. The capacitor C_{LB} is the combination of the junction capacitors and the gate-to-source capacitances of transistors Q_3 and Q_4. The response time is reduced by minimizing the parasitic capacitances C_{LA} and C_{LB} while increasing the tail current of the differential stage supplied by transistor Q_0.

One of the main applications of the CMOS comparator is the switched-capacitor comparator, which is one of the essential blocks of an A/D converter. Figure 5.9*a* shows the principle of operation of the basic switched-capacitor comparator. During

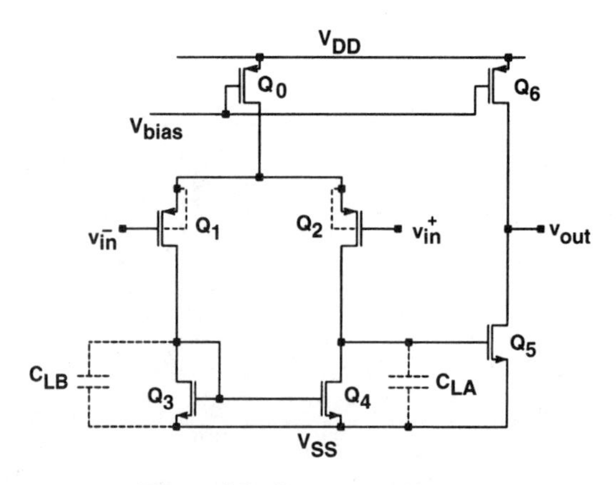

Figure 5.8. Op-amp comparator.

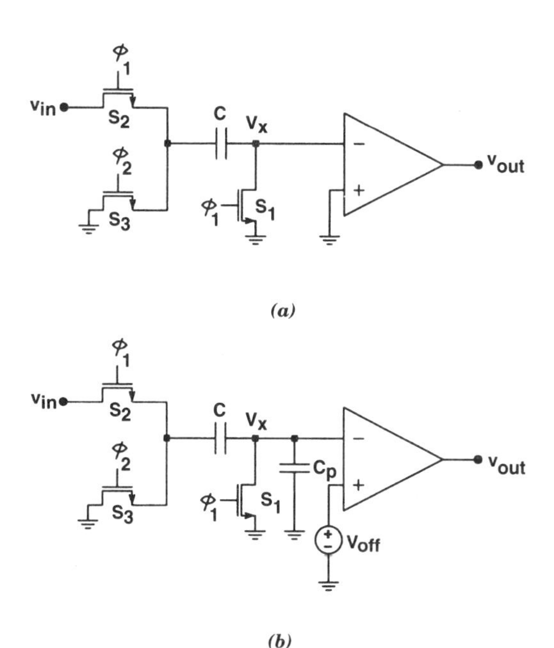

Figure 5.9. (*a*) Switched-capacitor comparator; (*b*) switched-capacitor comparator with parasitics.

the sample mode, switches S_1 and S_2 are closed while S_3 is open. During the hold and comparison mode, S_1 and S_2 are open and S_3 is closed. In the absence of any nonideal effects, the voltage levels at the end of the hold period are

$$\begin{aligned} v_x &= -v_{\text{in}}, \\ v_{\text{out}} &= -A\left(-v_{\text{in}}\right) = A_v v_{\text{in}}, \end{aligned} \tag{5.1}$$

where A_v is the open-loop voltage gain of the comparator. The circuit of Fig. 5.9*a* has several nonideal effects. One effect is the dc offset of the comparator, which similar to the CMOS op-amp, is due to the mismatch between the input and load devices of the differential stage. Another source of error is the clock-feedthrough and channel charge-pumping effect of the reset switch S_1, which contribute a residual offset charge to the holding capacitor. In Fig. 5.9*b* the comparator input offset voltage is modeled as the voltage source V_{off} that is placed in series to one of its inputs. Also shown is the total parasitic capacitance from the inverting input to ground, which is represented by capacitor C_p. Thus for Fig. 5.9*b* the differential voltage at the input of the comparator at the end of the hold period is

$$v_d = v_x - V_{\text{off}} = -\frac{C}{C + C_p} v_{\text{in}} + \frac{\Delta Q}{C + C_p} - V_{\text{off}}, \tag{5.2}$$

where ΔQ represents the total injected charge on the holding capacitor.

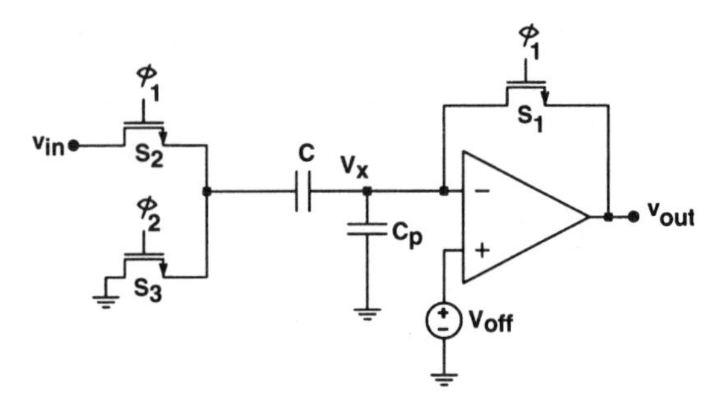

Figure 5.10. Offset-canceled switched-capacitor comparator.

An alternative circuit configuration that can provide offset cancellation is shown in Fig. 5.10. In this figure the reset switch S_1 is moved between the comparator inverting input and output. Then, during the sampling phase the capacitor C is charged between the input voltage and the offset of the comparator. In the subsequent hold and comparison phase, as before, switches S_1 and S_2 are open and S_3 is closed, connecting the left side of capacitor C to ground. The voltage at the right side of the capacitor C becomes

$$v_x = -\frac{C}{C + C_p} v_{\text{in}} + \frac{\Delta Q}{C + C_p} + V_{\text{off}}, \tag{5.3}$$

which results in the differential voltage at the input of the comparator given by

$$v_D = v_x - V_{\text{off}} = \frac{C}{C + C_p} v_{\text{in}} + \frac{\Delta Q}{C + C_p} \tag{5.4}$$

which is independent of the comparator offset voltage. Although this technique works well for removing the input offset voltage, the clock-feedthrough and channel charge-pumping effects of the reset switch are not eliminated. As Eq. (5.4) shows, the simplest way to reduce the errors due to the charge injection is to increase the size of the sampling capacitor and reduce ΔQ by reducing the size of the reset switch and by using CMOS transmission gate for first-order charge cancellation. Making the size of the holding capacitor large while reducing the size of the reset switch has the adverse effect of increasing the settling time constant of the circuit. For high-speed applications a compromise should be arrived between the sizes of the holding capacitor and the reset switch. An alternative strategy to reduce the clock-feedthrough effect is depicted in Fig. 5.11, where the reset switch is replaced with Q_1, a large device, in parallel to Q_2 a minimum-size device. At the end of the sample phase, first Q_1 turns off while Q_2 is still on. The charge injected by Q_1 is therefore absorbed by Q_2. Subsequently, Q_2 turns off and injects some charge to the holding capacitor. The amount of the charge is minimized because the switch has the smallest possible dimensions.

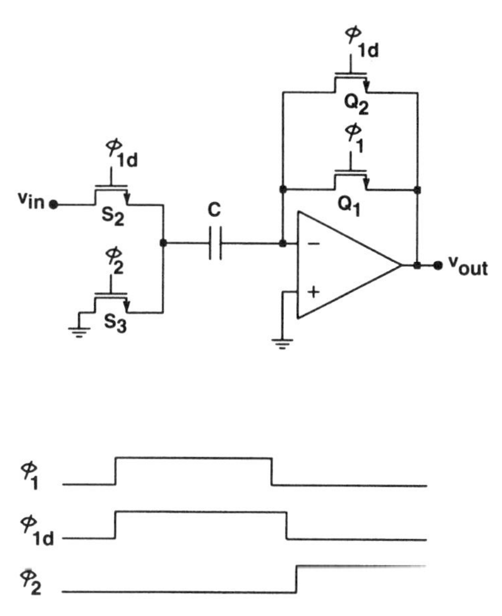

Figure 5.11. Offset-canceled switched-capacitor comparator with reduced clock-feedthrough effects.

The offset-canceling scheme of Fig. 5.10 can be used to sense the difference of two voltages. This is shown in Fig. 5.12, where during the sample phase switches S_1 and S_2 are closed and the capacitor C is charged between the offset of the comparator and the input voltage v_a. During the comparison phase switches S_1 and S_2 are opened and S_3 is closed, connecting the left side of the capacitor to the voltage source v_b. Thus, during the comparison phase the voltage at the right side of capacitor C is given by

$$v_x = v_b - v_a + \frac{\Delta Q}{C + C_p} + V_{\text{off}} \tag{5.5}$$

and the differential voltage at the input of the comparator is

$$v_d = v_x - V_{\text{off}} = v_b - v_a + \frac{\Delta Q}{C + C_p}, \tag{5.6}$$

which is, like before, independent of the comparator offset voltage and has a dc offset voltage $\Delta Q/(C + C_p)$ due to the residual feedthrough charge.

In this method, since the comparator acts as a unity-gain amplifier during the sample phase, it has to remain stable. A typical circuit is shown in Fig. 5.13. The comparator illustrated is simply a two-stage amplifier with an *RC* compensating branch $Q_c - C_c$. This branch is effective, however, only during $\phi_1 = 1$ half-period, when C is charging through S_1 and S_2, and hence the comparator functions as an

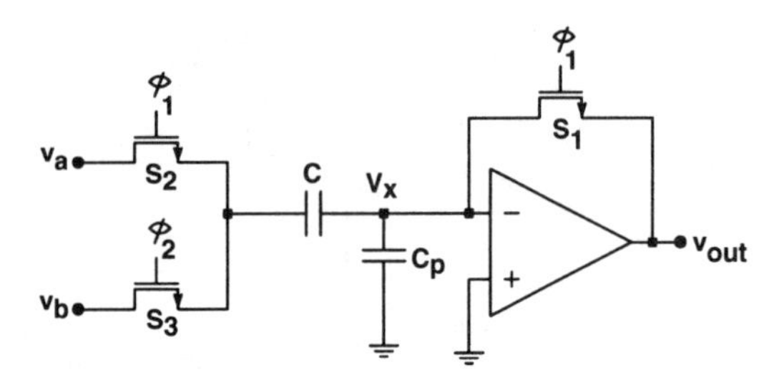

Figure 5.12. Offset-canceled differential switched-capacitor comparator.

op-amp in a unity-feedback configuration. During $\phi_2 = 1$ interval, the amplifier is in an open-loop configuration, and hence compensation (which slows down its operation) is not needed.

As mentioned before, the offset-canceling comparator still suffers from the dc offset voltage due to the residual feedthrough charge. An effective way for minimizing the errors due to the charge injection effects is the fully differential design scheme, which is discussed later.

In the offset-canceled switched capacitor comparator of Fig. 5.10, when $\phi_1 = 1$, capacitor C is charged between the input voltage and the offset of the comparator. At the instant that ϕ_1 goes low, the input signal v_{in} is sampled and held on capacitor C. In the subsequent comparison phase, ϕ_2 goes high and the left side of the capacitor is connected to ground, while the right side, in the absence of any offset voltage and clock-feedthrough effect, becomes $-v_{in}$. This circuit has the important advan-

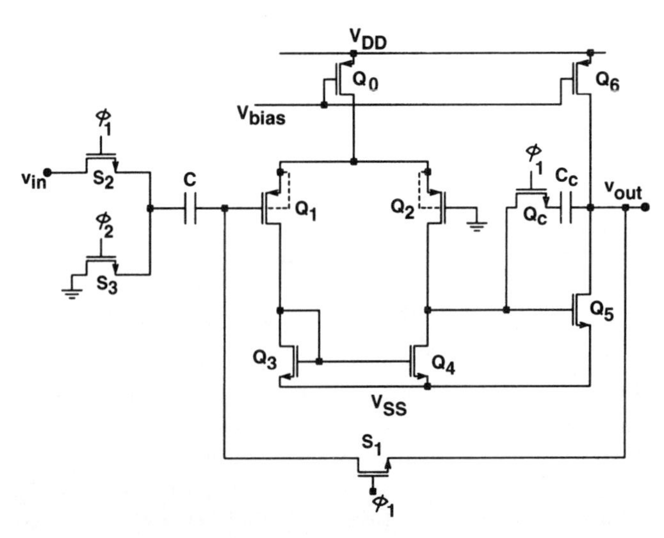

Figure 5.13. Op-amp offset-canceled switched-capacitor comparator with compensation capacitor disconnected during the comparison phase.

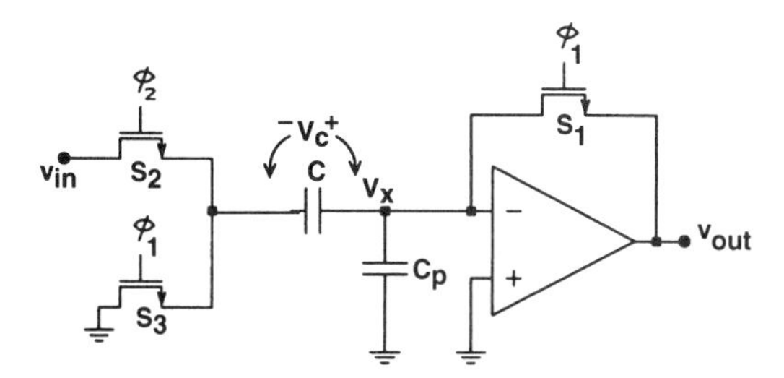

Figure 5.14. Alternative form of switched-capacitor comparator.

tage that the input signal is sampled and held during the comparison phase, and the comparator has the full period to determine the polarity of the input signal. One shortcoming of this circuit is that during every offset-canceling phase, while the right side of the capacitor is held at virtual ground, the left side is fully charged to the magnitude of the input signal. As mentioned in an earlier discussion, a large sampling capacitor should be used to minimize the residual offset voltage due to the clock-feedthrough effect. A large sampling capacitor affects the settling time and imposes stringent requirements on the charging capabilities of the input signal source and the comparator output stage. Alternatively, as shown in Fig. 5.14, the phasing of the switches connected to the left side of capacitor C can be interchanged. In this case, during the phase that ϕ_1 is high, the capacitor is charged between the offset of the comparator and ground. In the next phase, when ϕ_2 goes high, the right side of the capacitor that is connected to the inverting input of the comparator is disconnected from the output, and the left side of the capacitor is connected to the input voltage. At this time, the voltage v_x at the inverting input of the comparator is

$$v_x = v_{\text{in}} \frac{C}{C + C_p} + V_{\text{off}} \tag{5.7}$$

and the voltage across the capacitor is

$$v_c = v_x - v_{\text{in}} = V_{\text{off}} - v_{\text{in}} \frac{C_p}{C + C_p}. \tag{5.8}$$

If $C \gg C_p$, $v_x \approx v_{\text{in}} + V_{\text{off}}$, and $v_c \approx V_{\text{off}}$, the voltage across the capacitor C during ϕ_1 and ϕ_2, remains unchanged at V_{off}. This greatly relaxes the charging requirements from the input voltage source and is specially suitable for high-speed applications. On the other hand, the disadvantage of this approach is that the sample-and-hold situation no longer exists at the input stage and the comparator inverting input tracks the input signal. For proper operation, the comparator therefore has to make a decision based on the instantaneous value of the input signal.

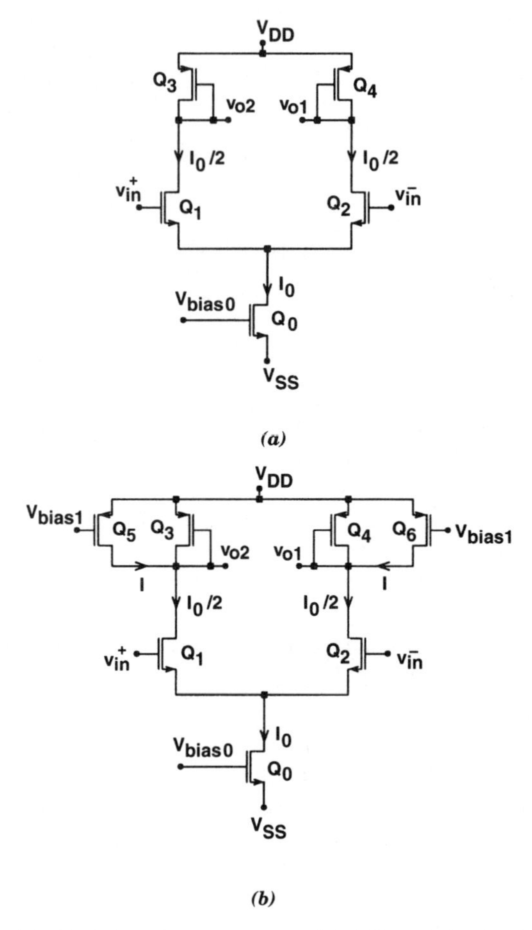

Figure 5.15. (*a*) Gain block based on source-coupled differential pair with diode-connected load devices; (*b*) gain-enhanced source-coupled differential pair.

It was mentioned earlier that the response time of the op-amp comparator shown in Fig. 5.8 improves greatly as the input overdrive voltage is increased. A high-resolution high-speed comparator can be realized by using a multistage approach, where several high-speed low-gain differential amplifiers are cascaded followed by a high-gain differential comparator or latch. Figure 5.15*a* shows a differential amplifier that can be used as the input stage of a comparator. The differential gain of the source coupled pair is given by

$$A_d = \frac{v_{o1} - v_{o2}}{v_{\text{in}}^+ - v_{\text{in}}^-} = \sqrt{\frac{\mu_n (W/L)_i}{\mu_p (W/L)_l}}, \tag{5.9}$$

where μ_n and μ_p are the mobility of the NMOS and PMOS devices. Equation (5.9) shows the dependence of the differential gain on the square root of the device

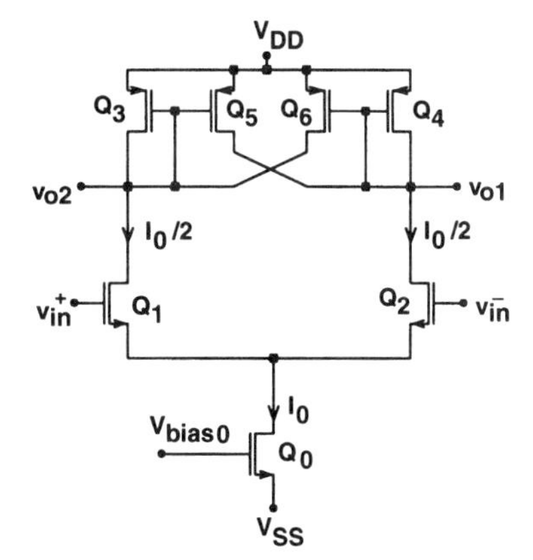

Figure 5.16. Source-coupled differential pair that uses positive feedback to provide increased gain.

dimensions. Higher-mobility NMOS transistors are used as input devices to achieve higher gain. To increase the gain further, the circuit of Fig. 5.15*b* can be used where the input transistor transconductance is increased by injecting currents I_1 and I_2 ($I_1 = I_2$) into them from the *p*-channel current source Q_5 and Q_6 [4,5]. The gain of the modified circuit is now given by

$$A_d = \sqrt{\frac{\mu_n (W/L)_1}{\mu_p (W/L)_3 (1 - 2I/I_0)}}. \tag{5.10}$$

A practical choice is $I = 0.9I_0/2$, which increases the gain by a factor of $\sqrt{10}$.

Another method that uses a controlled amount of positive feedback to effectively increase the driver devices transconductance and hence the overall gain is shown in Fig. 5.16 [1]. The gain of the positive feedback gain stage is given by

$$A_d = \sqrt{\frac{\mu_n (W/L)_1}{\mu_p (W/L)_3}} \frac{1}{1 - \alpha}, \tag{5.11}$$

where $\alpha = (W/L)_5/(W/L)_3$ is the positive feedback factor that is responsible for increasing the gain. A reasonable value for α is 0.75, which increases the gain by a factor of 4. The value of α is determined by the ratio of the load device dimensions, and although it is a reasonably well controlled parameter, a practical maximum value for α is 0.9, because beyond that any mismatches due to process variations may cause the value of α to approach unity, and per Eq. (5.15), the gain will become infinity and the stage will operate as a cross-coupled latch.

The response times of the gain stages of Figs. 5.15 and 5.16 are limited by the parasitic capacitances of the load devices. The response time of the source coupled differential stage can improve significantly if the diode-connected loads are replaced

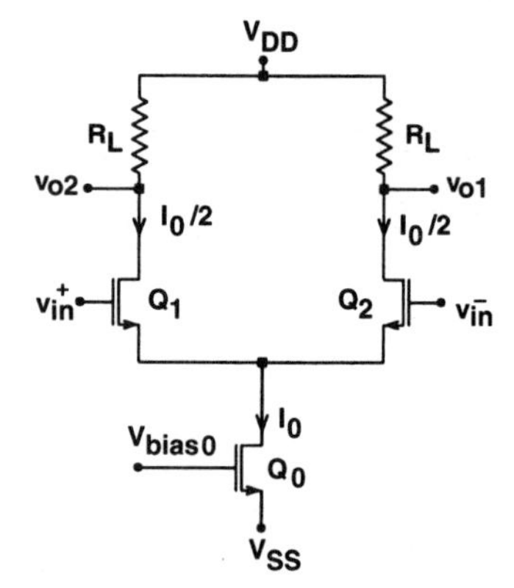

Figure 5.17. Resistive-load source-coupled differential pair.

with simple resistors, as shown in Fig. 5.17. The gain of the resistive load differential stage is given by

$$A_d = g_{mi} R_L, \tag{5.12}$$

where g_{mi} is the transconductance of the NMOS input devices and R_L is the load resistance. One drawback of this circuit is that the voltage drop across the load resistors can vary significantly due to variations in the resistance or the magnitude of the differential stage tail current. The variations of the output common-mode voltage makes it difficult to design multistage amplifiers and bias the circuit to operate under all process variations. One solution to this problem is to use a replica biasing scheme. A schematic of the resistive load differential stage with a simple V_{BE}-based bias generator is shown in Fig. 5.18. The bias current is given by

$$I_B = \frac{V_{BE}}{R_B}. \tag{5.13}$$

Assuming that $(W/L)_7 = (W/L)_6$, the voltage drop across the load resistors is given by

$$V_{RL} = \frac{(W/L)_{10}}{(W/L)_9} \frac{I_B}{2} R_L = \frac{(W/L)_{10}}{(W/L)_9} \frac{V_{BE}}{2} \frac{R_L}{R_B}. \tag{5.14}$$

This generates a very reproducible fraction of V_{BE} since it is determined by resistor and transistor ratios. In addition to providing known bias voltage, the maximum positive output swing of the differential stage is well controlled at a value of

$$\Delta V_o = 2 \frac{I_o}{2} R_L = \frac{(W/L)_{10}}{(W/L)_9} V_{BE} \frac{R_L}{R_B}. \tag{5.15}$$

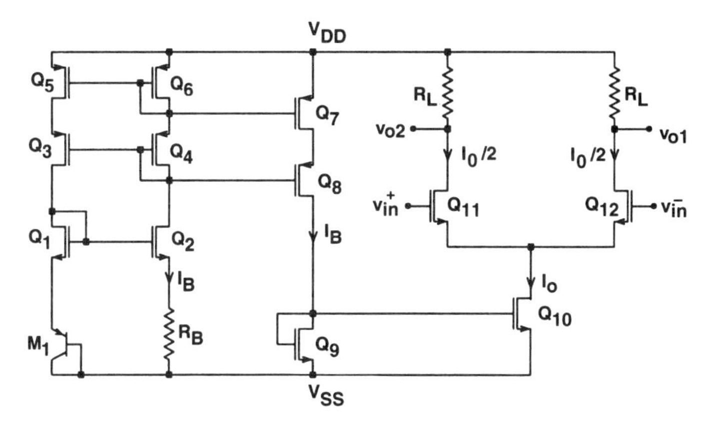

Figure 5.18. Resistive-load source-coupled differential stage with replica biasing.

Note that to first order the stage output voltage swing is independent of V_{DD}. As an example, if we assume that the bias resistance is $R_B = 3500\ \Omega$, the bias current is given by

$$I_B = \frac{0.7}{3500} = 200\ \mu\text{A}.$$

Using a value of $k'_n = 55\ \mu\text{A/V}^2$ for the NMOS transconductance factor and a threshold voltage of $V_{Tn} = 0.8$ V, then for $(W/L)_{10}/(W/L)_9 = 2$ and $(W/L)_{11}/(W/L)_{12} = 100$, the transconductance of the input devices will be

$$g_{mi} = 2\sqrt{k'_n\left(\frac{W}{L}\right)I} = 2.1 \times 10^{-3} \text{ mhos}.$$

For $R_L = 5000\ \Omega$ the differential gain is given by

$$A_d = g_{mi}R_L = 2.1 \times 10^{-3} \times 5000 = 9.8$$

and the maximum output voltage swing is, from Eq. (5.15),

$$\Delta V_o = 2 \times 0.7 \times \frac{5000}{3500} = 2 \text{ V}.$$

A three-stage direct-coupled comparator circuit comprised of two cascaded resistive load and source-coupled differential stages followed by an op-amp comparator is

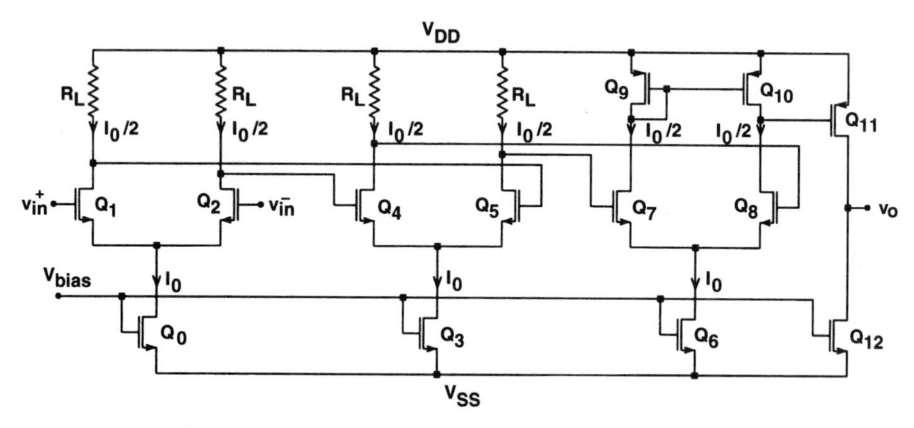

Figure 5.19. Direct-coupled three-stage comparator circuit.

shown in Fig. 5.19. The overall gain of the first two stages for identical sections is given by

$$A_d = (g_{mi}R_L)^2. \tag{5.16}$$

So if each stage has a gain of 10, the overall gain is $A_d = 100$, and a differential input of 100 μV appears as a 10-mV signal at the input of the final stage. This voltage is large enough to result in a fast response time in the last stage of the comparator. Since the stages are direct coupled, the offset is not canceled. If V_{off1}, V_{off2}, and V_{off3} represent the offset voltages of the first, second, and third stages, the input referred dc offset voltage is given by

$$V_{\text{off}} = V_{\text{off1}} + \frac{V_{\text{off1}}}{A} + \frac{V_{\text{off2}}}{A^2}. \tag{5.17}$$

Offset cancellation techniques for the multistage comparators will be introduced later.

5.4. REGENERATIVE COMPARATORS (SCHMITT TRIGGERS)

Often, comparators are used to convert a very slowly varying input signal into an output with abrupt edges, or they are used in a noisy environment to detect an input signal crossing a threshold level. If the response time of the comparator is much faster than the variation of the input signal around the threshold level, the output will chatter around the two stable levels as the input crosses the comparison voltage. Figure 5.20*a* shows the input signal and the resulting comparator output. In this situation, by employing positive (regenerative) feedback in the circuit, it will exhibit a phenomenon called *hysteresis*, which will eliminate the chattering effects. The

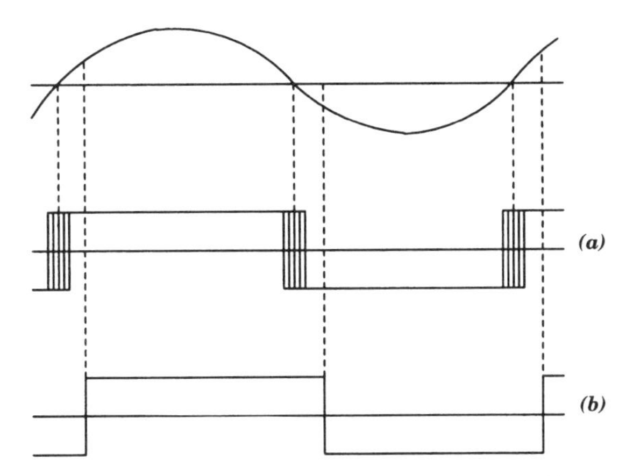

Figure 5.20. Response of a fast comparator to a slowly varying signal in a noisy environment: (a) without hysteresis; (b) with hysteresis.

regenerative comparator is commonly referred to as a *Schmitt trigger*. The response of the comparator with hysteresis to the input signal is shown in Fig. 5.20b.

A Schmitt trigger can be implemented by using positive feedback in a differential comparator, as shown in Fig. 5.21a. Assume that $v_i < v_1$, so that $v_o = V_o$; then v_1 is given by

$$v_1 = V_p = V_o \frac{R_2}{R_1 + R_2}. \tag{5.18}$$

If v_i is now increased, v_o remains constant at V_o until $v_i = v_1$. At this triggering voltage, the output regeneratively switches to $v_o = -V_o$ and remains at this value as long as $v_i > v_1$. The voltage at the noninverting terminal of the comparator for $v_i > v_1$ is now given by

$$v_1 = V_n = \frac{-R_2}{R_1 + R_2} V_o. \tag{5.19}$$

If we now decrease v_i, the output remains at $v_o = -V_o$ until $v_i = v_1$. At this voltage a regenerative transition takes place and the output returns to V_o almost instantaneously. The complete transfer function is indicated in Fig. 5.21b. Note that because of the hysteresis, the circuit triggers at a higher voltage for increasing than for decreasing signals. Note that $V_n < V_p$, and the difference between these two values is the hysteresis V_H given by

$$V_H = V_p - V_n = \frac{2R_2 V_o}{R_1 + R_2}. \tag{5.20}$$

Another method to eliminate the chattering effect around the zero crossing of the input signal is to use the comparator with dynamic hysteresis shown in Fig. 5.22.

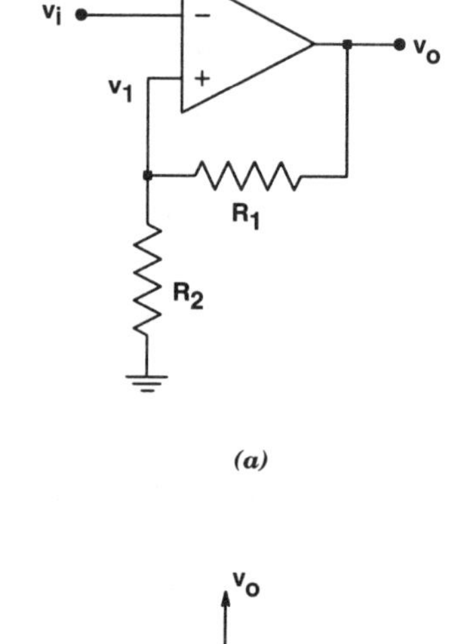

(a)

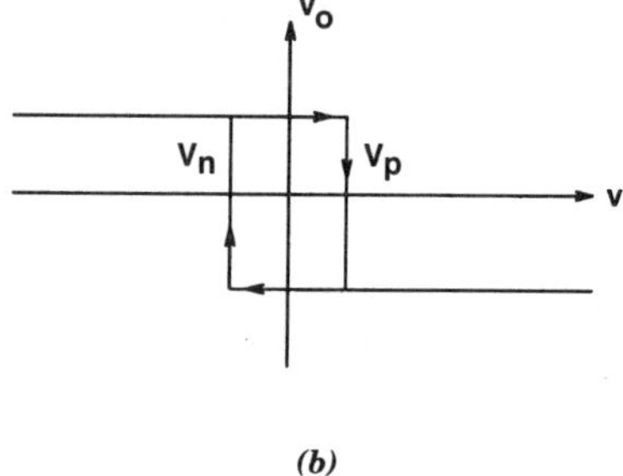

(b)

Figure 5.21. (*a*) Schmitt trigger; (*b*) composite input–output curve.

Assume that $v_{\text{in}} < v_1$, so that $v_o = +V_o$ and $v_1 = 0$. If v_{in} is now increased until $v_{\text{in}} > 0$, the output switches to $v_o = -V_o$. The negative transition of the output will capacitively be coupled to the positive input of the comparator, which also makes a negative transition. This will cause the differential input voltage between the negative and positive inputs of the comparator to become larger and speed up the output transition. Since the first transition at the output of the comparator regeneratively increases the magnitude of the differential input signal, the comparator responds to the first time the input crosses zero and ignores any subsequent zero crossings due to noise. The *RC* time constant determines the length of the time that the input signal will be ignored after its first zero crossing. If the time constant is made too large, it will limit the maximum frequency that the circuit can operate.

Many other ways are available to accomplish hysteresis in a comparator. All of them use some form of positive feedback. Earlier the source-coupled differential pair of Fig. 5.16 was introduced where positive feedback was employed to increase the gain [1]. The gain of the stage is given by Eq. (5.11), where $\alpha = (W/L)_5/(W/L)_3$ is the positive feedback factor. For $\alpha < 1$ $[(W/L)_5 < (W/L)_3]$, the circuit behaves as a gain stage. For $\alpha = 1$ the stage becomes a positive feedback latch. For $\alpha > 1$ the stage becomes a Schmitt trigger circuit with the amount of hysteresis determined by the value of α. Next, the trigger points and amount of hysteresis will be calculated for the case when $\alpha > 1$.

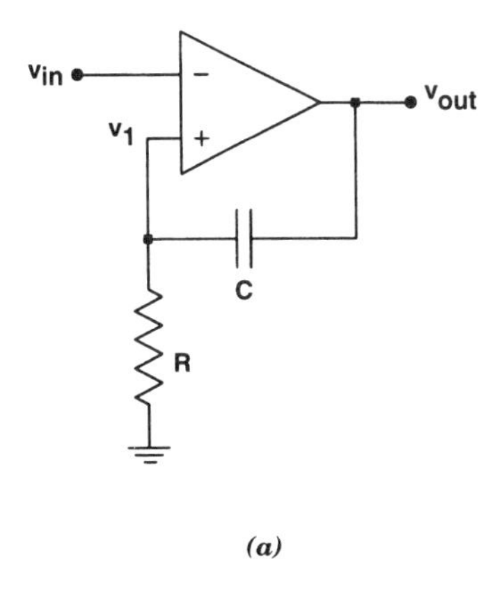

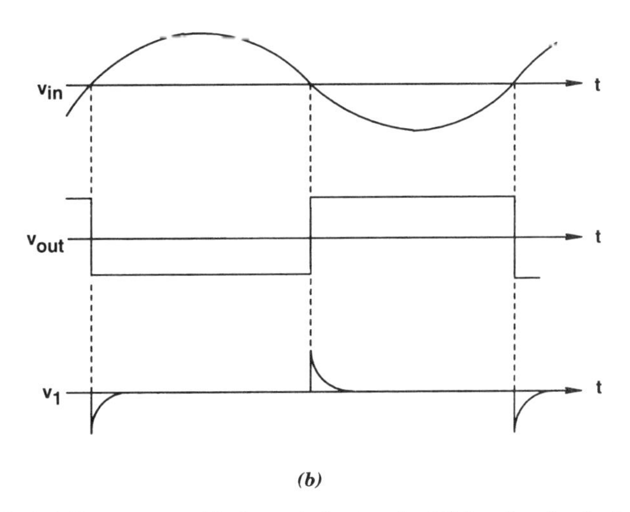

Figure 5.22. (*a*) Comparator with dynamic hysteresis; (*b*) input and output waveforms.

Consider the circuit of Fig. 5.23, where v_{in}^{+} is connected to ground (or any other reference potential) and the gate of Q_2 (v_{in}^{-}) is connected to a negative potential much less than zero. Thus Q_2 is off and Q_1 is on and all the tail current I_o flows through Q_1 and Q_3. The current through transistors Q_2, Q_4, Q_5, and Q_6 are zero and the node voltage v_{o1} is high while v_{o2} is low. Next assume that the input voltage v_{in}^{-} is gradually increased so that transistor Q_2 begins conducting and part of the tail current I_o starts flowing through it. This process continues until the current in transistor Q_2 equals the current in Q_5. Any increase of the input voltage beyond this point will cause the comparator to switch state so that Q_1 turns off and all the tail current flows through Q_2. At the switching

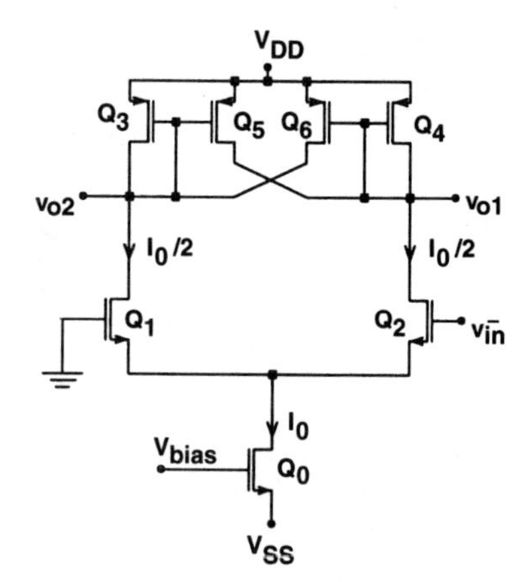

Figure 5.23. Source-coupled differential pair with positive feedback factor $\alpha > 1$ for hysteresis.

point assume that the current through transistors Q_1 and Q_2 are i_1 and i_2, respectively. Then we have

$$i_1 + i_2 = I_o \tag{5.21}$$

$$i_2 = i_5 = i_3 \frac{(W/L)_5}{(W/L)_3} = i_1 \frac{(W/L)_5}{(W/L)_3} \tag{5.22}$$

or

$$i_1 = \frac{I_o}{1 + \alpha} \tag{5.23}$$

$$i_2 = \frac{I_o \alpha}{1 + \alpha}. \tag{5.24}$$

Now the gate-to-source (v_{GS}) voltages of Q_1 and Q_2 can be calculated from their respective drain currents and are given by

$$v_{GS1} = V_{Tn} + \sqrt{\frac{i_1}{k'(W/L)_1}}, \tag{5.25}$$

$$v_{GS2} = V_{Tn} + \sqrt{\frac{i_2}{k'(W/L)_2}}. \tag{5.26}$$

In the equations above, $i_2 > i_1$, so $v_{GS2} > v_{GS1}$ and since the gate of Q_1 is tied to ground, the difference between v_{GS2} and v_{GS1} is the positive trigger level, equal to

$$V_{\text{trig}+} = \sqrt{\frac{i_2}{k'(W/L)_2}} - \sqrt{\frac{i_1}{k'(W/L)_1}}. \tag{5.27}$$

Using Eqs. (5.23) and (5.24) and since $(W/L)_1 = (W/L)_2$, Eq. (5.27) will be simplified to

$$V_{\text{trig}+} = \sqrt{\frac{I_o}{k'(W/L)_1}}\,\frac{\sqrt{\alpha} - 1}{\sqrt{1 + \alpha}}. \tag{5.28}$$

When the input voltage is increased beyond $V_{\text{trig}+}$, the comparator switches and v_{o1} turns low and v_{o2} goes high. Now transistor Q_2 is on and Q_1 is off, and all the tail current I_o flows through Q_2 and Q_4. The currents in transistors Q_1, Q_3, Q_5, and Q_6 are zero.

Next consider reducing the input voltage v_{in}^-, so that the current in Q_2 starts decreasing and Q_1 starts conducting. A similar argument can prove that the negative trigger point occurs when $i_1 = i_6$. It can be shown that the negative trigger point V_{trig}^- is given by

$$V_{\text{trig}-} = \sqrt{\frac{I_o}{k'(W/L)_1}}\,\frac{1 - \sqrt{\alpha}}{\sqrt{1 + \alpha}}. \tag{5.29}$$

The hysteresis can be calculated as

$$V_H = V_{\text{trig}+} - V_{\text{trig}-} = 2 \times \sqrt{\frac{I_o}{k'(W/L)_1}}\,\frac{\sqrt{\alpha} - 1}{\sqrt{1 + \alpha}}, \tag{5.30}$$

where $\alpha = [(W/L)_5/(W/L)_3] = [(W/L)_6/(W/L)_4]$.

The complete schematic of a comparator with hysteresis, which consists of a source-coupled differential pair with positive feedback, and a differential-to-single-ended converter is shown in Fig. 5.24. For this circuit the value of $\alpha = [(W/L)_5/(W/L)_3] = [(W/L)_6/(W/L)_4]$ is greater than 1 and the hysteresis is given by Eq. 5.30.

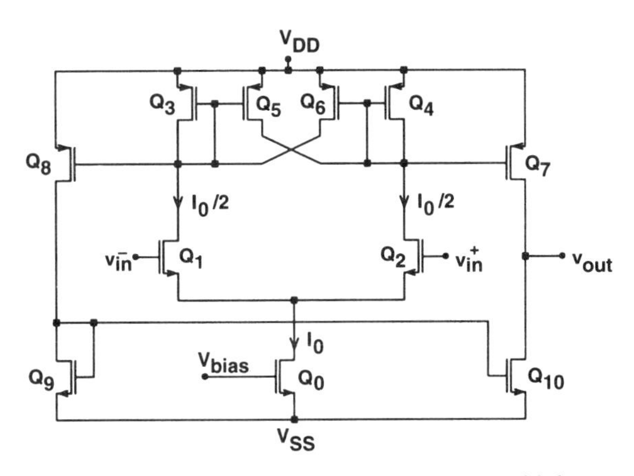

Figure 5.24. Complete schematic of a comparator with hysteresis.

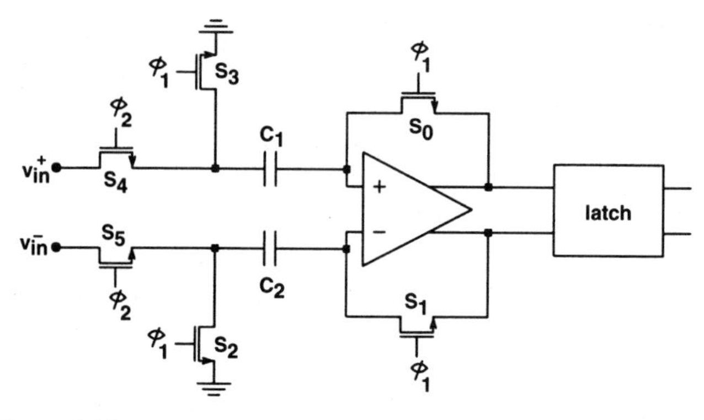

Figure 5.25. Fully differential input offset storage (IOS) comparator.

5.5. FULLY DIFFERENTIAL COMPARATORS

For high-accuracy applications an effective way for reducing the dc offset voltage due to the feedthrough charge is to use a fully differential scheme for the comparators. In such circuits, not only are clock-feedthrough effects reduced, but power supply noise and $1/f$ noise also tend to cancel. An offset-canceling fully differential comparator is shown in Fig. 5.25. In this scheme, during the offset cancellation mode, switches S_0 to S_3 are on while switches S_4 and S_5 are off. This will cause a unity-gain feedback loop to be established around the comparator and the two sampling capacitors to be charged between ground and the offset voltage of the comparator. During the tracking mode, switches S_0 to S_3 turn off, breaking the feedback loop of the comparator; S_4 and S_5 turn on and connect the capacitors to the input signal. The input differential voltage is amplified by the comparator and is sensed by the latch, which provides a logic level at its output, representing the polarity of the input differential voltage. If $V_{\text{off}A}$ and $V_{\text{off}L}$ represent the input offset voltages of the comparator and the latch, Q_0 and Q_1, represent the feedthrough charges of switches S_0 and S_1, the residual input referred offset voltage is given by

$$V_{\text{off}} = \frac{V_{\text{off}A}}{1 + A} + \frac{Q_0 - Q_1}{C} + \frac{V_{\text{off}L}}{A}. \tag{5.31}$$

From Eq. (5.31), if the charges injected by switches S_0 and S_1 match while the common-mode voltage will be slightly affected by an amount equal to $(Q_0 + Q_1)/2C$, the differential input voltage, $\Delta Q/C = (Q_0 - Q_1)/C$ will be zero. In practice, the charge injected by the two switches will never match, but the residual offset voltage due to the mismatches in the clock feedthrough will be an order of magnitude less than in the single-ended case. For this reason most advanced integrated comparators use fully differential design technique.

The offset compensation technique shown in Fig. 5.25 is known as the *input offset storage* (IOS) topology [4,5]. It is characterized by closing a unity-gain feed-

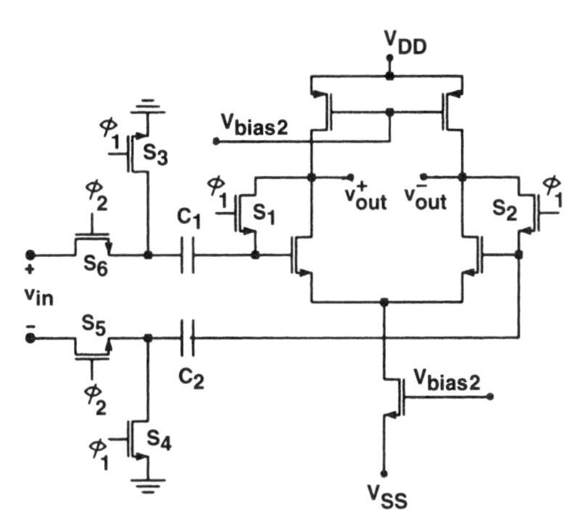

Figure 5.26. Fully differential offset canceling comparator.

back loop around the comparator and storing the resulting offset voltage on the input capacitors. In this configuration, since the input signal is capacitively coupled, it has a wide input dynamic range. Also, since during the cancellation phase, the comparator is connected as a unity-gain amplifier, it recovers from the input overdrive as well as starting the tracking phase in its active region, thereby improving its response time. The disadvantage of this topology is evident from Eq. (5.31), where it can be seen that to reduce the residual input offset voltage, the comparator should have a high gain and a large value input capacitor should be used. A comparator with high gain consumes more power and is harder to compensate for stable operation during the offset cancellation phase. A large sampling capacitor increases the settling time and hence degrades the response time of the comparator. It also loads the preceding circuit and causes a large amount of transient noise.

Figure 5.26 shows a simple fully differential stage employing the input offset storage topology [6, pp. 99–102]. In this circuit the two auto-zeroing capacitors C_1 and C_2 are precharged during the ϕ_1 interval between ground and v_s, where v_s is the self-biased input voltage of the amplifier. When ϕ_2 goes high, the input voltage v_{in} is sensed by the comparator and an amplified differential voltage $v_{\text{out}}^{+} - v_{\text{out}}^{-}$ appears at the output. There is also a clock-feedthrough signal at each input node, which appears as a common-mode signal and is hence suppressed. A high-gain (>80 dB) comparator, using a folded-cascode first stage followed by a cascode second stage with a switched compensation capacitor is described in Ref. 7.

In high-resolution applications a single-stage high-gain offset-canceling comparator such as the one shown in Fig. 5.26 will have a long response time. Therefore, high-resolution comparators use a multistage design. Each stage of the multistage design uses one of the low-gain amplifier stages shown in Figs. 5.15 to 5.17. A three-stage fully differential input offset storage (IOS) multistage comparator and the individual gain stage are shown in Fig. 5.27*a* and *b*, respectively [8]. The gain stage is similar to the gain-enhanced amplifier stage described earlier and shown in Fig. 5.15*b*. The two cascode transistors Q_7 and Q_8 have been added to reduce the Miller capacitance at the input. To recover from overdrive voltage and speed up the settling time, the reset switch Q_9 is used to equalize the differential output of the gain stage for a short period during the offset storage cycle. Each stage is capacitively coupled to the next one, and by closing the feedback loop around each stage independently, the possible instability problem of a three-stage amplifier with one feedback loop around all three stages is eliminated. The circuit operates as follows. During the offset storage mode, the feedback switches are closed, a unity-gain feedback loop is established around each gain stage, and the offset of the comparators is stored on the input capacitors. In the tracking mode the feedback around the comparators is opened and the input differential voltage is sensed and amplified by A^3, where A is the voltage gain of each amplifier stage. The output of the comparator is strobed by a latch that produces a logic level at its output. The mismatch between the channel charge injected from the feedback switches introduces an uncanceled offset at the input of each gain stage. The total input-referred dc offset voltage is dominated by the offset of the first stage. It is possible to reduce this type of error by implementing the sequential clocking scheme shown in Fig. 5.27*c*, where the gain stages are brought out of the offset cancellation mode sequentially, A_0 first and A_2 last [1,8]. When switches S_1 and S_2 are opened, A_0 leaves the offset cancellation mode while A_1 and A_2 remain in that mode. The offset voltages due to the charge injection mismatches of S_1 and S_2 on capacitors C_1 and C_2 are amplified by the first stage and stored on capacitors C_3 and C_4 before the other feedback switches are opened. This can be repeated with switches S_3 and S_4 opening before S_5 and S_6 switches to cancel its charge injection error. To ensure proper operation, the delay between the clock edges shown in Fig. 5.27*c* should be long enough to allow the input capacitors of the next stage to fully absorb the offset voltage of the previous stage.

An alternative offset cancellation technique is shown in Fig. 5.28 [4]. In this method the offset is canceled by shorting the preamplifier inputs and storing the amplifier offset on the output coupling capacitors. This topology, known as *output offset storage* (OOS), operates as follows. During offset cancellation, switches S_0 to S_3 are on, switches S_4 and S_5 are off, and the output of the gain stage is AV_{offa}, where A is the gain of the amplifier. In this phase the amplified offset of the gain stage is stored on C_1 and C_2. During the tracking mode, S_1 to S_4 are off, S_5 and S_6 are on, and the amplifier senses and amplifies the analog differential voltage. This voltage is subsequently sensed and amplified by a latch which provides a logic level at its output. In this comparator employing OOS, the residual offset is

$$V_{\text{OFF}} = \frac{\Delta Q}{AC} + \frac{V_{\text{OFFL}}}{A}, \tag{5.32}$$

where V_{OFFL} is the latch offset and $\Delta Q = Q_0 - Q_1$ is mismatch in charge injection from switches S_0 and S_1 onto capacitors C_1 and C_2. As suggested by Eq. (5.32), the offset of the amplifier is canceled completely. This is evident from Fig. 5.28, where

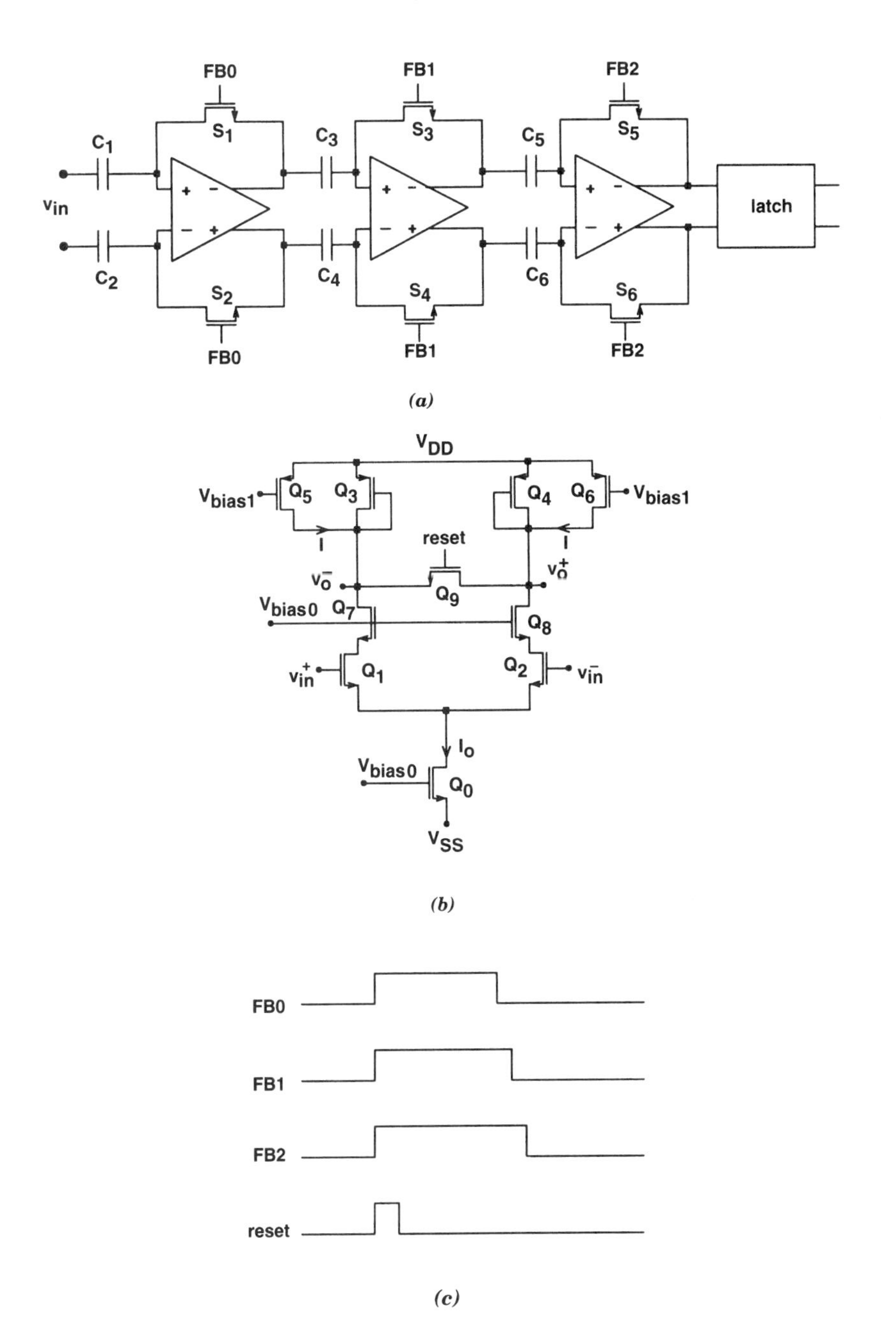

Figure 5.27. (*a*) Three-stage IOS comparator; (*b*) gain-enhanced amplifier stage; (*c*) sequential clocking scheme.

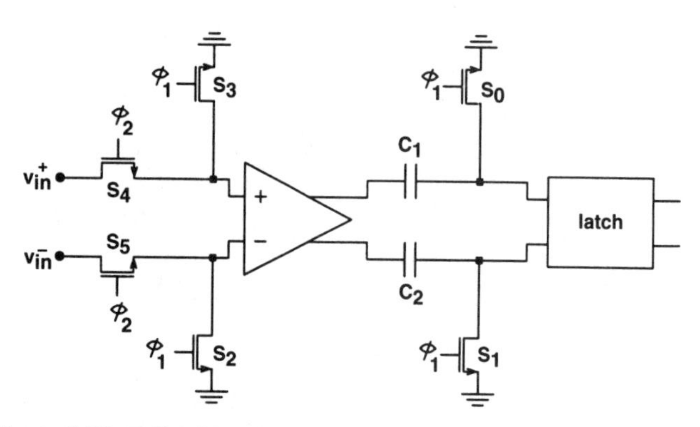

Figure 5.28. Fully differential output offset storage (OOS) comparator.

during the cancellation mode the inputs of the amplifier and the latch are both zero; hence a zero difference of the comparator input gives a zero difference at the latch input. Also, as shown in Eq. (5.32), the offset due to the charge injection of the switches is divided by the gain of the amplifier, resulting in a smaller overall input-referred offset voltage compared to that of IOS, as evident from Eq. (5.31).

In addition to having a lower residual dc offset voltage, the OOS topology generally has a smaller input capacitance, which is limited to the input capacitance of the amplifier, which can be maintained well below 100fF. The OOS topology is therefore the preferred choice in applications where a low input capacitance is required, such as a flash A/D converter, where many comparators are connected in parallel. One drawback of the OOS topology is its limited-input common-mode range, which is due to the dc coupling at its input. Another drawback is that during offset cancellation mode, the amplifier in an OOS operates in open-loop mode and the input offset is amplified by its gain. Therefore, a low-gain amplifier should be used to ensure operations in the active region under maximum input dc offset voltage.

Typically, the comparator is followed by a standard CMOS latch with a potentially large dc offset voltage. Therefore, to achieve a low input-referred offset voltage, as suggested by Eq. (5.32), a high-gain amplifier should be used. Consequently, in high-resolution applications, the use of a single-stage comparator is not feasible, and similar to the IOS topology, a multistage calibration technique will be required. Figure 5.29*a* illustrates a three-stage OOS comparator where each stage can be constructed from one of the low-gain amplifier stages shown in Figs. 5.15 to 5.17. A sequential clocking scheme such as the one shown in Fig. 5.29*b* will then reduce the dc offset due to the clock feedthrough. If $V_{\mathrm{OFF}L}$ represents the offset of the latch, the number of amplifier stages n, and the corresponding gain A, should be selected such that the input-referred offset voltage is less than 0.5LSB, or

$$\frac{V_{\mathrm{OFF}L}}{(A)^n} < 0.5\mathrm{LSB}. \tag{5.33}$$

After the value of the required gain $(A)^n$ is determined, the number of the stages is selected to provide the smallest delay [8].

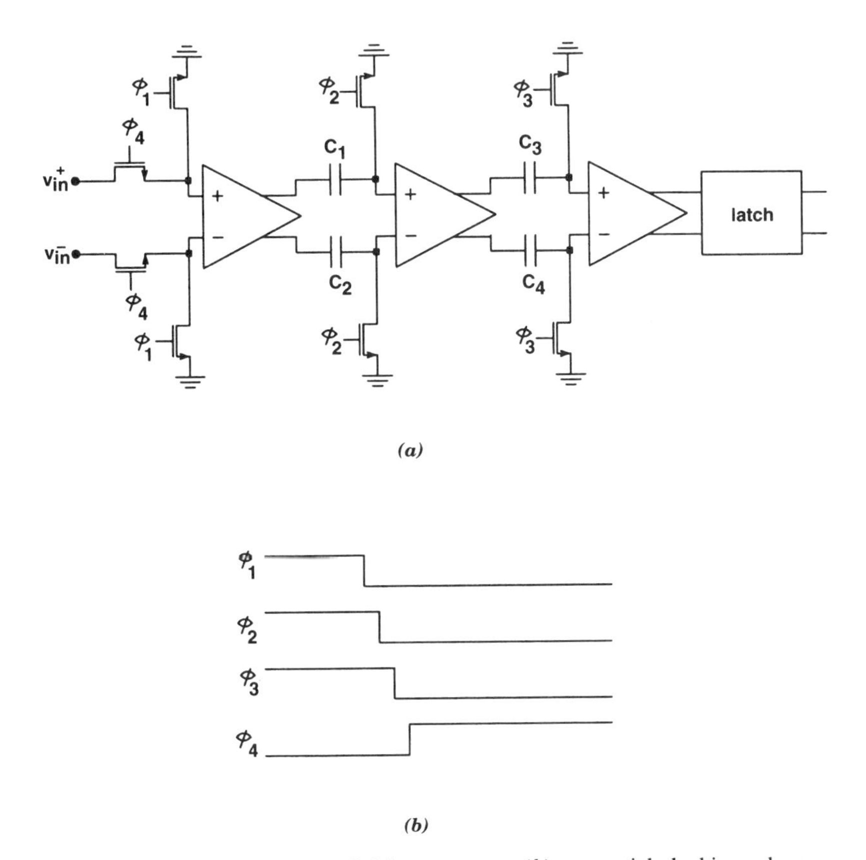

Figure 5.29. (*a*) Three-stage OOS comparator; (*b*) sequential clocking scheme.

The multistage calibration technique is an effective method to reduce the contribution of the latch offset to the overall residual input-referred dc offset voltage. An alternative method to improve the performance of a fully differential comparator is to apply the offset cancellation to both the gain stage and the latch [4]. By canceling the offset of the latch, it will not be necessary to use multistage low-gain amplifiers, and high performance can be achieved by using a single low-gain amplifier that is optimized for both speed and power dissipation.

The simplified block diagram of a CMOS comparator that applies offset cancellation to both the amplifier and the latch is shown in Fig. 5.30. It consists of two transconductance amplifiers, G_{m1} and G_{m2}, that share the same output nodes, load resistors, and offset storage capacitors. In this circuit B_1 and B_2 are buffers that isolate the common output nodes from the feedback capacitors. The operation of the circuit is as follows. During the offset cancellation mode, S_1 to S_6 are on, S_7 to S_{10} are off, the inputs of G_{m1} and G_{m2} are grounded, and their offsets are amplified and stored on capacitors C_1 and C_2. During the comparison mode, S_1 to S_6 turn off, S_7 to S_{10} turn on, the capacitors are connected in the feedback loop of G_{m2}, the

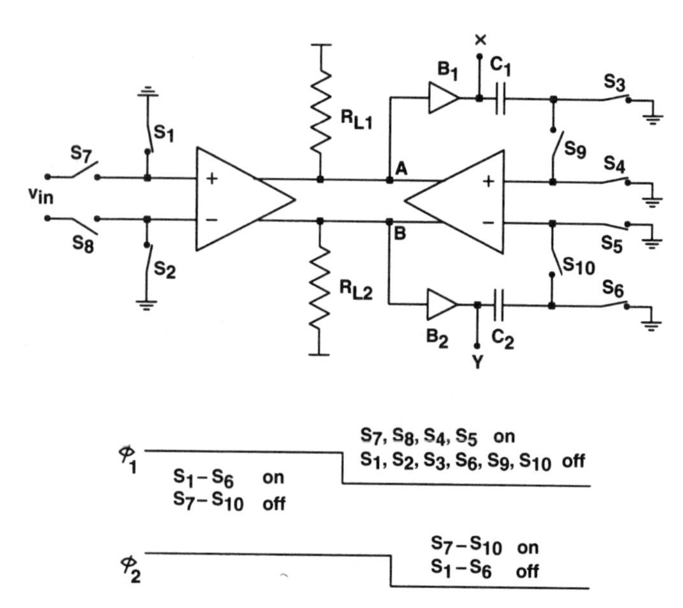

Figure 5.30. CMOS comparator block diagram and timing.

inputs of G_{m1} are released from ground, and the input voltage is sensed. The input voltage is amplified by G_{m1} to establish an imbalance in output nodes A and B which is coupled to the inputs of G_{m2} through C_1 and C_2, initiating regeneration around G_{m2}. This calibration method can be viewed as an output offset storage applied to both the amplifier G_{m1} and the latch G_{m2}, resulting in complete cancellation of their offsets.

A CMOS comparator based on the topology of Fig. 5.30 is shown in Fig. 5.31. In this circuit the source-coupled pairs Q_1–Q_2 and Q_3–Q_4 constitute amplifiers G_{m1} and G_{m2}, respectively. The loads are formed from the diode-connected transistors Q_5 and Q_6 and gain enhancement devices Q_7 and Q_8. The output common-mode voltage is set by transistors Q_5 and Q_6. The source followers Q_9 and Q_{10} serve as buffers B_1 and B_2 in Fig. 5.30. As described earlier and shown in Fig. 5.15*a*, the additional current sources, Q_7 and Q_8, increase the gain and decrease the voltage drop across Q_5 and Q_6. Normally, the gates of Q_7 and Q_8 would be connected to a fixed bias voltage. However, by connecting the gates to the inputs of the differential stage, the push-pull operation of Q_3 with Q_7 and Q_4 with Q_8 improves the charge and discharge response times of nodes A and B in the following way: When, for example, node F goes low and node E goes high, the current in Q_8 is increased, thereby pulling node B more quickly to a higher voltage. The current in Q_7 is reduced, thus allowing Q_3 to discharge node A to a lower voltage much faster.

In this circuit, during the calibration mode, S_1 to S_6 are on, S_7 to S_{10} are off, and the combined offset voltages of G_{m1} and G_{m2} are stored on capacitors C_1 and C_2. During the comparison mode the comparators are connected in the feedback loops of G_{m2} and the input voltage is sensed by G_{m1}. Any differential input voltage will

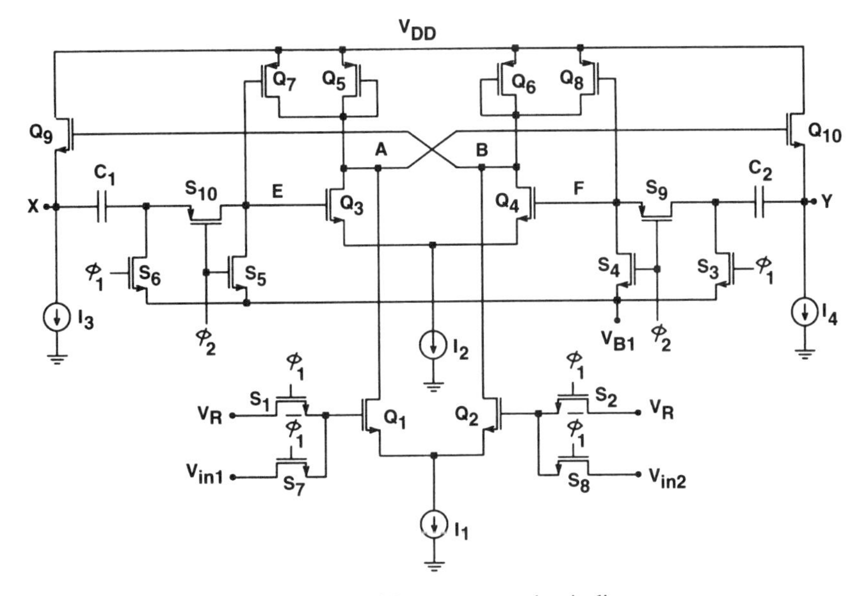

Figure 5.31. CMOS comparator circuit diagram.

off-balance the currents of the differential pair Q_1–Q_2, which reflects into a differential voltage at nodes A and B. This voltage is regeneratively coupled to the inputs of G_{m2} through capacitors C_1 and C_2, causing the outputs to switch. Since the comparator of Fig. 5.31 includes calibration of both the amplifier and the latch, its residual offset is due primarily to mismatches in the charge injection of switches S_3 to S_6, S_9, and S_{10}.

The differential output voltage swing of the comparator is normally on the order of 1 to 2 V. The comparator can be followed by a latch or a nonregenerative amplifier such as the one shown in Fig. 5.32 to develop full CMOS levels from the differential output. The bias circuit consisting of transistor Q_{15}–Q_{17} and current source I_1 replicates the X and Y common-mode voltage at the source of Q_{17} and generate pull-up currents in Q_{13} and Q_{14} that, during reset, are equal to pull-down currents in Q_{11} and Q_{12} if their gates are driven from X and Y.

5.6. LATCHES

As mentioned earlier, to provide the gain needed to generate logic levels at the output, as well as synchronize the operation of the comparator with other parts of a system, the amplifier of the comparator is usually followed by a latch. The latter can simply be a cross-coupled bistable multivibrator. Some possible latch circuits are shown in Figs. 5.33 to 5.35. The circuit of Fig. 5.33 is a dynamic CMOS latch [9] used to amplify small differences to CMOS levels. In this circuit, when ϕ is

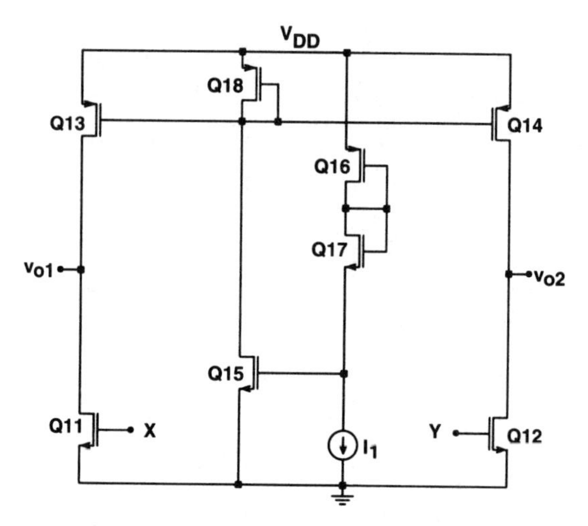

Figure 5.32. CMOS comparator output stage.

low, Q_5 is off, S_1 and S_2 are on, and the input capacitances are precharged to v_{in1} and v_{in2}. Subsequently, when ϕ goes high it turns off S_1 and S_2 to isolate nodes X and Y from the input terminals and turns on Q_5 to initiate regeneration, and the latch assumes one of its stable states, depending on the sign of $v_{in1} - v_{in2}$.

The latch of Fig. 5.34 also includes self-biasing and auto-zeroing circuitry [6, pp. 99–102]. The operation is as follows. When ϕ_2 goes high, S_4 and S_5 short circuit the gates and drains of Q_2 and Q_3, respectively. This action biases the two inverters Q_2–Q_5 and Q_3–Q_6 (which form the multivibrator) in their linear regions. It also precharges the capacitors C_3 and C_4 such that any asymmetry between the two inverters is compensated for by the slightly different bias voltages provided by C_3

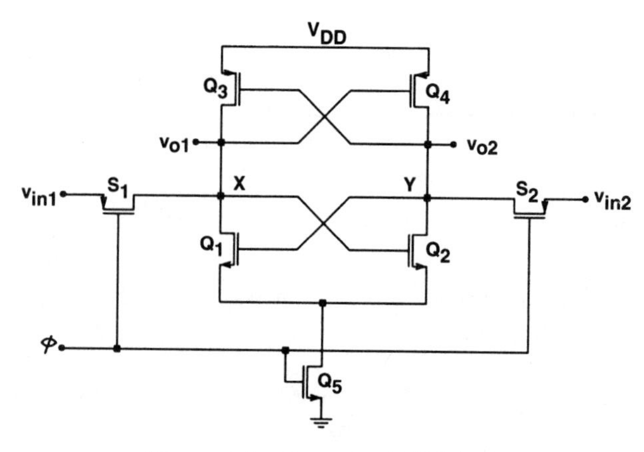

Figure 5.33. Dynamic CMOS latch.

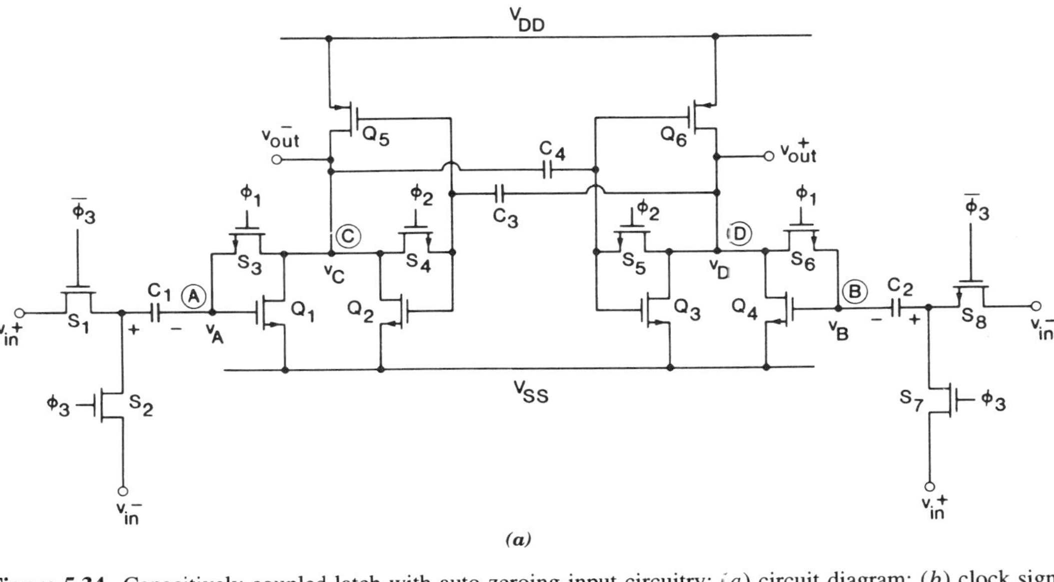

(a)

Figure 5.34. Capacitively coupled latch with auto-zeroing input circuitry: (*a*) circuit diagram; (*b*) clock signals.

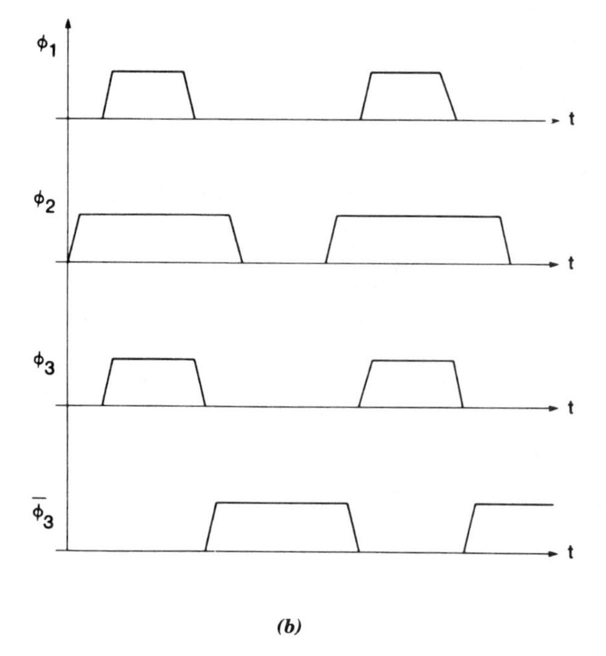

(b)

Figure 5.34. *(continued)*

and C_4. During this time the multivibrator has a loop gain less than 1, and hence it does not switch into either one of its stable states. Next, when ϕ_1 and ϕ_3 go high, S_2 and S_3 precharge C_1 to $v_{in}^- - v_A$, while S_6 and S_7 precharge C_2 to $v_{in}^+ - v_B$. Next, ϕ_1 goes low and the input devices Q_1 and Q_4 are released from their self-biased states, leaving C_1 and C_2 floating. Then ϕ_3 goes low and $\bar{\phi}_3$ high; v_A now changes by $v_{in}^+ - v_{in}^-$, while v_B changes by $v_{in}^- - v_{in}^+$ from the self-biased values. This also causes corresponding changes in v_C and v_D inside the multivibrator. When ϕ_2 finally goes low, unleashing the multivibrator, the voltage differences between v_A and v_B as well as v_C and v_D cause it to go to the appropriate stable state; v_C will be high and v_D low if $v_{in}^+ < v_{in}^-$ holds, or v_C low and v_D high if $v_{in}^+ > v_{in}^-$ is valid.

The sequence in which the various clock phases rise and fall is important for proper operation of this latch circuit (as it is for almost any other). The reader is urged to analyze the operation if (say) ϕ_2 goes low before ϕ_3 goes high, and so on, to convince himself or herself of the validity of this statement.

Yet another latch [10] is shown in Fig. 5.35. In this circuit, transistors Q_1, Q_2, Q_3, Q_4, and Q_7 act as a differential preamplifier when S_5 is closed. On the other hand, when S_6 is closed, Q_3, Q_4, Q_5, Q_6, and Q_7 form a bistable multivibrator. When ϕ_1, ϕ_2, and ϕ_3 are high, the preamplifier is self-biased and C_1 and C_2 are precharged to $v_{in}^- - v_A$ and $v_{in}^+ - v_B$, respectively. Then ϕ_1 goes low, allowing the amplifier to function and leaving C_1 and C_2 floating at nodes A and B. Next, the bottom terminals of C_1 and C_2 are switched to v_{in}^+ and v_{in}^-, respectively; this causes an

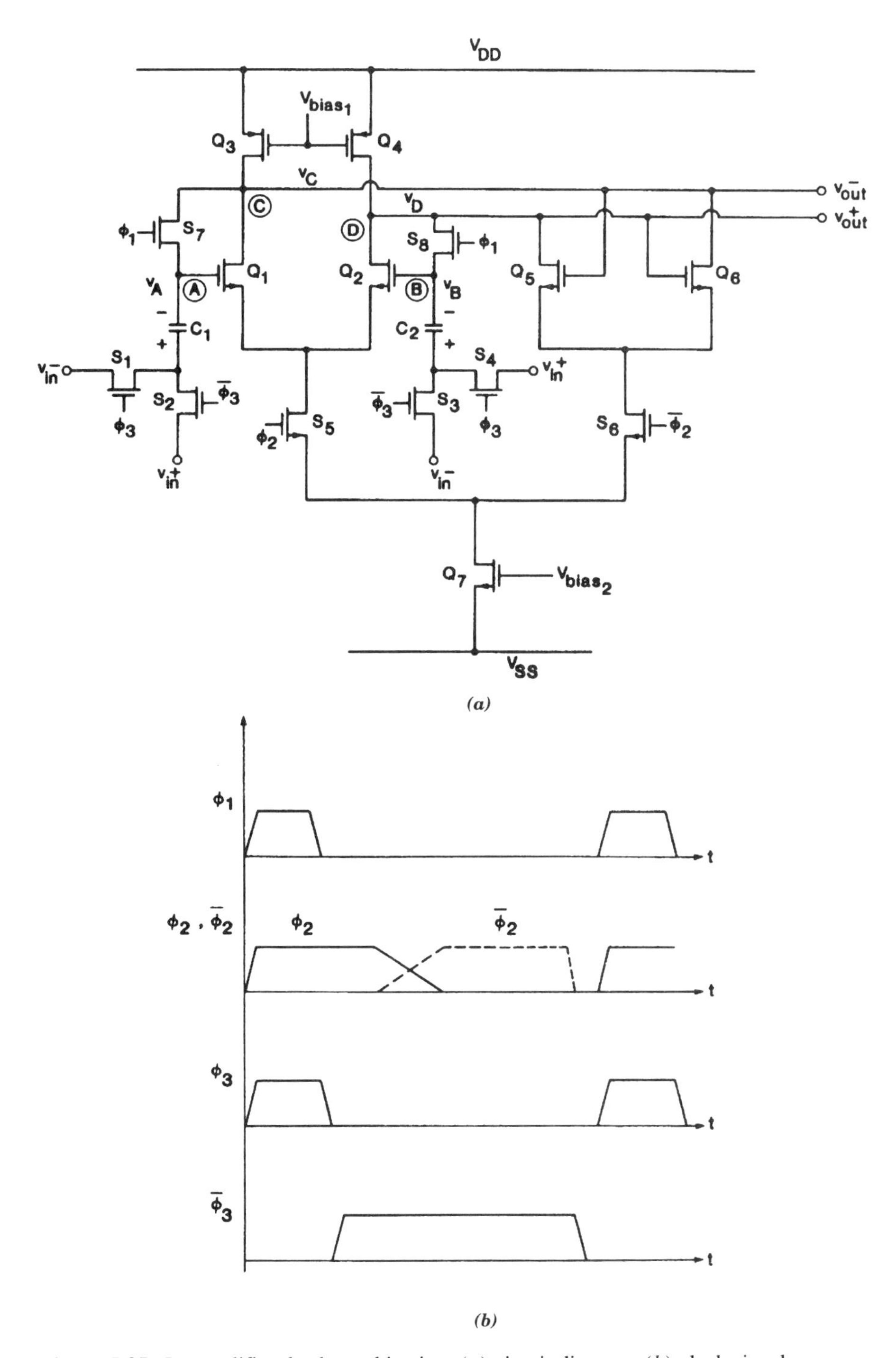

Figure 5.35. Preamplifier–latch combination: (*a*) circuit diagram; (*b*) clock signals.

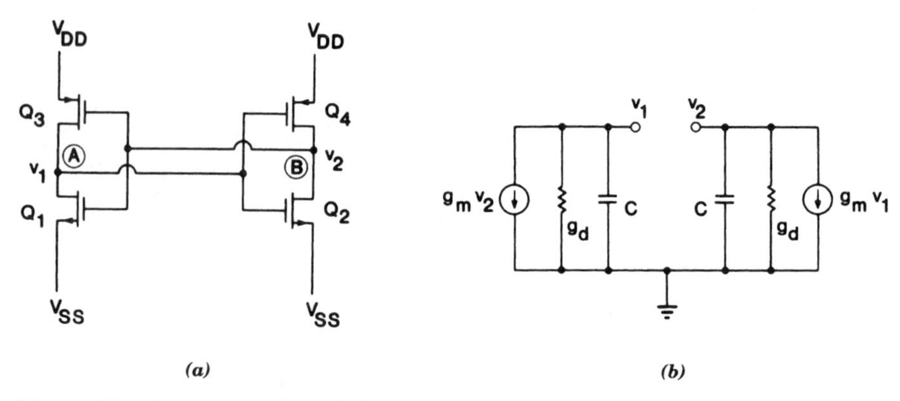

Figure 5.36. Direct-coupled multivibrator: (*a*) circuit diagram; (*b*) small-signal equivalent circuit.

amplified voltage difference between nodes C and D. At this point, S_5 slowly opens and S_6 slowly closes. This causes the multivibrator to come to life and to assume one of its stable states. The state chosen is determined by the sign of $v_C - v_D$.

Even if the amplifier and the latch are built from the same types of inverters, the rise and fall times of the amplifier will be much longer than those of the latch. To understand this phenomenon, consider the simple multivibrator of Fig. 5.36*a*. Its small-signal equivalent circuit with all devices in the saturation region is shown in Fig. 5.36*b*, where $g_m = g_{m1} + g_{m3} = g_{m2} + g_{m4}$ and $g_d = g_{d1} + g_{d3} = g_{d2} + g_{d4}$; also, C is the capacitance loading nodes A and B. It can easily be shown (Problem 5.12) that the natural modes (poles) of the circuit of Fig. 5.36 are $s_{1,2} = \pm g_m/C$. Hence its transients are exponential functions with a time constant $\tau = C/g_m$. In the absence of positive feedback, if the inverters Q_1–Q_3 and Q_2–Q_4 are simply cascaded as in an amplifier, the time constant is $\tau' = C/g_d$. The ratio of the time constants is $\tau/\tau' = g_d/g_m = 1/A$, where A is the gain of the inverter [11]. Since typically, $A = 10$, the latch can be about an order of magnitude faster than the amplifier driving it. It is possible to take advantage of the speed of the latch by using two amplifiers to feed a single latch (Fig. 5.37*a*). In this system [12], the two amplifiers have the same configuration as the two input stages in the circuit of Fig. 5.7*a*. They alternate in auto-zeroing and amplifying, but the amplifying periods have a duty cycle *longer* than 50% (Fig. 5.37*b*), so that the input of the latch can receive a *continuous-time* input signal. To assure this, the intervals during which the switches S_A and S_B connect the amplifiers to the latch overlap. Thus the latch clock frequency (which is the effective overall clock frequency of the comparator) can be different from the amplifier clock rates. This system can thus operate about 10 times faster than the usual single-amplifier version, since the limiting factor is now the speed of the latch, not that of the amplifier.

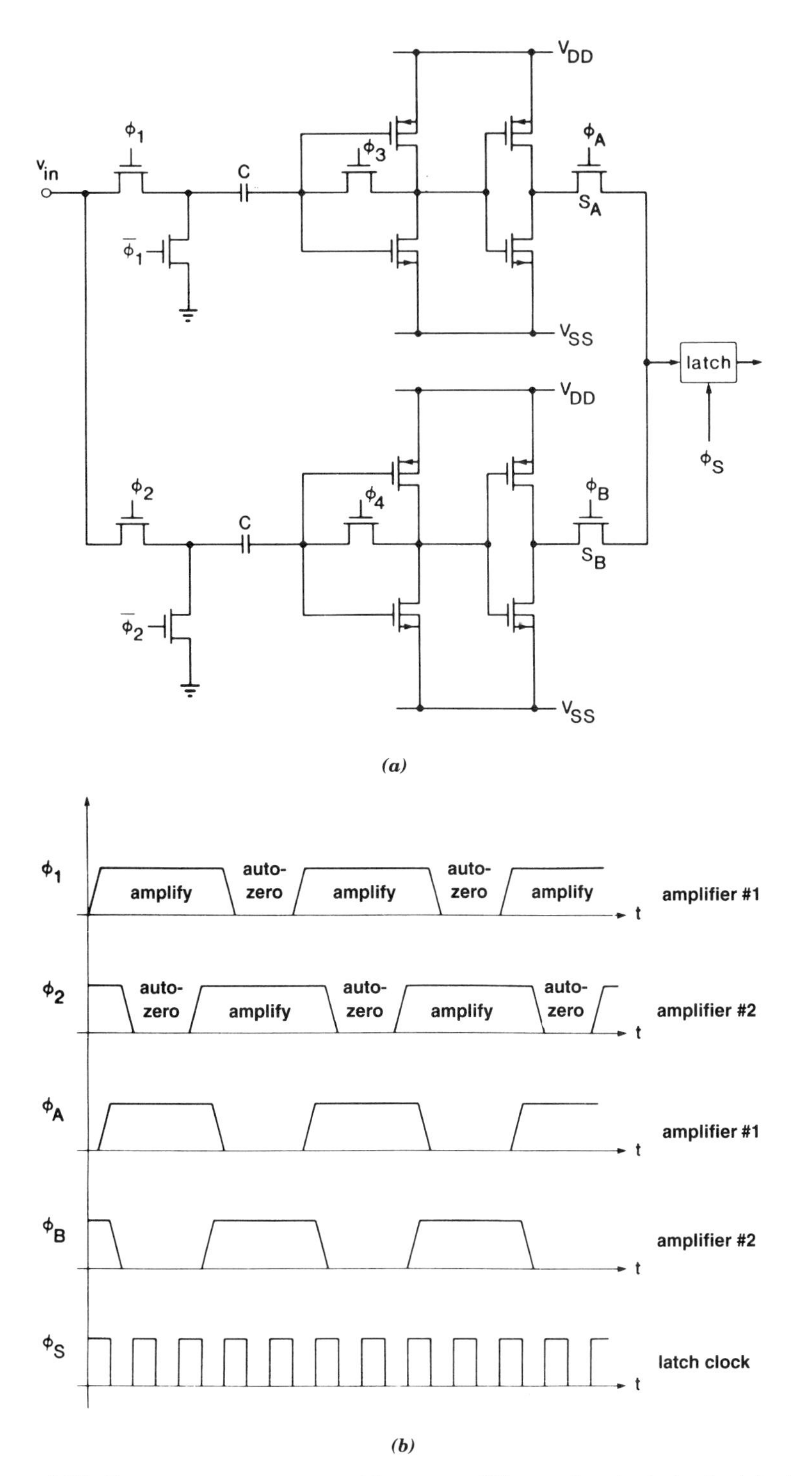

Figure 5.37. Fast comparator system with two amplifiers and a single latch: (*a*) circuit diagram; (*b*) clock signals.

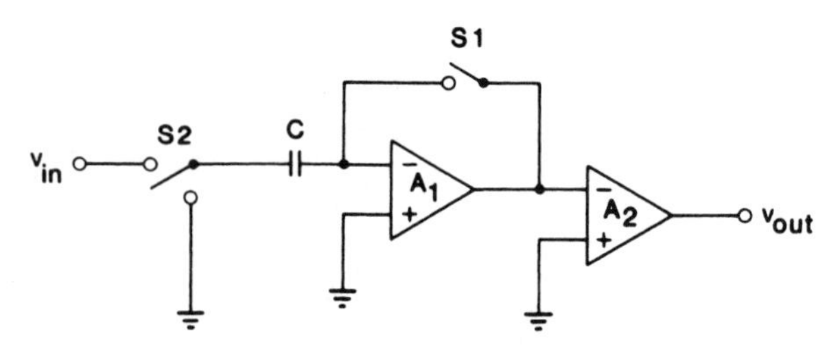

Figure 5.38. Reduction of input offset voltage by capacitive storage for a multistage comparator (for Problem 5.5).

PROBLEMS

5.1. For the inverter of Fig. 5.6*a* with the input–output characteristics shown in Fig. 5.6*b*, prove that the limits of the linear range (where Q_1 and Q_2 are both in saturation) are the intersections of the characteristics with the 45° lines $v_B = v_A + |V_{Tp}|$ and $v_B = v_A - V_{Tn}$. Draw these lines for the curves of Fig. 5.6*b*, and identify the linear ranges.

5.2. The circuit of Fig. 5.7*a* is to be fabricated using a CMOS process with the following parameters: $V_{Tn} = -V_{Tp} = 1$ V, $\mu_n = 3\mu_p = 670$ cm²/V·s, $t_{ox} = 800$ Å, $\lambda_n = 0.012$ V^{-1}, and $\lambda_p = 0.02$ V^{-1}. Design the input inverter such that for $v_A = 0$ V, $v_B = 0$ V and $i_{D1} = i_{D2} = 50$ μA. How much is the gain of the stage? (*Hint:* Use the formulas of Tables 2.2 and 2.3.)

5.3. In the circuit of Fig. 5.7*a*, the clock-feedthrough capacitance between the gate of S_3 and node A is 15fF. The clock voltage is 10 V peak to peak. How large must C_1 be if the first two inverters (Q_1–Q_2 and Q_3–Q_4) are to operate with all devices in saturation, despite the clock-feedthrough voltage at node A? Assume the *W* and *L* values obtained in Problem 5.2 for the two input inverters.

5.4. Show that the differential voltage at the input of the comparator of Fig. 5.9*b* at the end of the hold period is given by Eq. (5.2).

5.5. Calculate the effective input offset voltage of the auto-zeroed comparator of Fig. 5.38 from the individual gains and offset voltages of its two stages A_1 and A_2. Assume that the clock-feedthrough charge is ΔQ and the input capacitance of the first stage is C_p.

5.6. Prove Eq. (5.10) for the comparator of Fig. 5.15*b*.

5.7. For the source-coupled differential stage of Fig. 5.18, determine the values of R_B and R_L, the aspect ratios of Q_6–Q_7 and Q_9–Q_{10} and the dimensions of Q_{11} and Q_{12} such that the differential gain becomes 16. (Assume that $k'_n = 55$ mA/V² and $V_{Tn} = 0.8$ V.)

5.8. Prove Eq. (5.17) for the three-stage comparator circuit of Fig. 5.19.

5.9. Select values for R_1 and R_2 so that the comparator of Fig. 5.21*a* has ± 30 mV hysteresis. (Assume that $V_{DD} = 5$ V and $V_{SS} = -5$ V.)

5.10. For the circuit of Fig. 5.24, if $W_1 = W_2 = 50$ μm, $L_1 = L_2 = 2$ μm, and the tail current $I_0 = 100$ μA, calculate α from Eqs. (5.28) and (5.29) such that $V_{\text{trig}}^{+} = V_{\text{trig}}^{-} = 30$ mV.

5.11. Prove Eq. (5.31) for the circuit of Fig. 5.25.

5.12. For the multivibrator of Fig. 5.36*a:* **(a)** derive the small-signal equivalent circuit of Fig. 5.36*b*; **(b)** using Laplace transform analysis, find the natural modes of the circuit; and **(c)** find the natural modes for the case when the two inverters (Q_1–Q_3 and Q_2–Q_4) are cascaded without closed-loop feedback. **(d)** What conclusions can be drawn from the relative magnitudes of the natural modes of the two circuits?

REFERENCES

1. D. J. Allstot, *IEEE J. Solid-State Circuits, SC-17*(6), 1080–1087 (1982).
2. R. Gregorian and J. G. Gord, *IEEE J. Solid-State Circuits, SC-17*(6), 698–700 (1983).
3. Y. S. Lee, L. M. Terman, and L. G. Heller, *IEEE J. Solid-State Circuits, SC-13*(2), 294–297 (1978).
4. B. Razavi and B. A. Wooley, *IEEE J. Solid-State Circuits, SC-27*(6), 1916–1926 (1992).
5. B. Razavi, *Principles of Data Conversion System Design,* IEEE Press, New York, 1995.
6. K. W. Martin, *Project Reports,* Microelectronics Innovation and Computer Research Opportunities (MICRO) Program, University of California, Berkeley, Calif., 1983.
7. H. S. Lee, D. A. Hodges, and P. R. Gray, *IEEE Int. Solid-State Circuits Conf.*, pp. 64–65, 1984.
8. J. Doernberg, P. R. Gray, and D. A. Hodges, *IEEE J. Solid-State Circuits, SC-24*(2), 241–249 (1989).
9. S. Chin, M. K. Mayes, and R. Filippi, *ISSCC Dig. Tech. Pap.*, pp. 16–17, Feb. 1989.
10. K. Martin, personal communication.
11. W. C. Black, Jr., personal communication.
12. Y. Fujita, E. Masuda, S. Sakamoto, T. Sakane, and Y. Sato, *IEEE Int. Solid-State Circuits Conf.*, pp. 56–57, Feb. 1984.

CHAPTER 6

DIGITAL-TO-ANALOG CONVERTERS

The analog-to-digital (A/D) and digital-to-analog (D/A) converters are the main link between the analog signals and the digital world of signal processing. Data converters are generally divided into the two broad categories of Nyquist-rate and oversampling converters. *Nyquist-rate converters* are converters that operate at 1.5 to 5 times the Nyquist rate (i.e., a sample rate of 3 to 10 times the signal's bandwidth), and each input signal is uniquely represented by an output signal. Conversely, *oversampling* or *delta-sigma converters* operate at sampling rates that are much higher than the input signal's Nyquist rate and increase the output signal-to-noise ratio by subsequent filtering that removes the out-of-band quantization noise. The ratio of the sampling rate to the Nyquist rate is called the *oversampling ratio*. For most practical delta-sigma converters, the oversampling ratio is typically between 16 and 256.

A Nyquist-rate digital-to-analog converter (DAC) is a device that converts a digital input signal (or code) to an analog output voltage (or current) that is proportional to the digital signal. In this and the following chapters a variety of methods are presented for realizing Nyquist-rate converters. Oversampling converters are not discussed in this book; the interested reader is referred to Ref. 20 for an in-depth coverage of this subject. The chapter begins with a general introduction and characterization of the converters, followed by a discussion of voltage, charge, current, and hybrid-mode D/A converters.

6.1. DIGITAL-TO-ANALOG CONVERSION: BASIC PRINCIPLES

The block diagram of a D/A converter is shown in Fig. 6.1. The inputs are a reference voltage V_{ref} and an N-bit digital word $b_1\, b_2\, b_3 \cdots b_N$. Under ideal conditions, in the

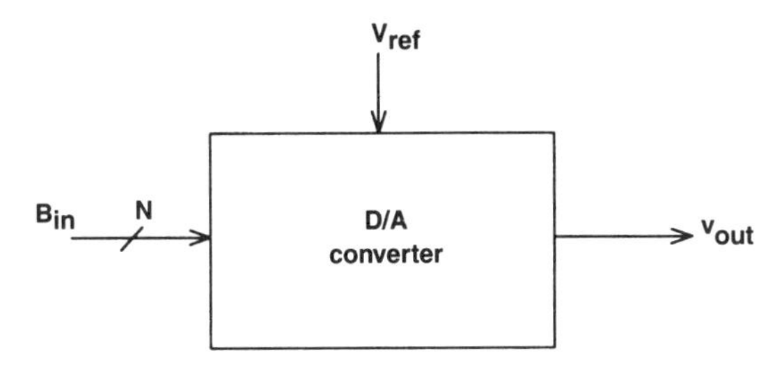

Figure 6.1. Block diagram of a digital-to-analog converter.

absence of noise and any imperfections, the D/A converter voltage output can be expressed as

$$v_{\text{out}} = (b_1 \cdot 2^{-1} + b_2 \cdot 2^{-2} + \cdots + b_n \cdot 2^{-N})V_{\text{ref}}, \tag{6.1}$$

where N is the number of the bits of the input digital word. For an N-bit D/A converter the resolution is 2^N and is equal to the number of discrete analog output levels corresponding to the various input digital codes. If V_{ref} represents the input reference voltage, the smallest analog output corresponding to one *least significant bit* (LSB) is

$$V_{\text{LSB}} = V_{\text{ref}} \cdot 2^{-N}. \tag{6.2}$$

For bipolar analog outputs, the digital input code retains the sign information in one extra bit—the *sign bit*—in the *most significant bit* (MSB) position. The most commonly used binary codes in bipolar conversion are sign magnitude, one's complement, offset binary, and two's complement. Table 6.1 shows each of the bipolar codes for a 4-bit (3-bit plus sign) digital word. The word length N determines the range of the numbers associated with each of the four binary representation systems. In all four notations, the largest positive number is given by $1 - 2^{-(N-1)}$ in decimal. For sign-magnitude and one's complement numbers, the lower bound is $-[1 - 2^{-(N-1)}]$, while in two's complement and offset binary the most negative number is -1. All four number systems have unique representations for all numbers except for the zero in sign-magnitude or one's complement notations. Positive and negative zeros have different representations in these two number systems. For two's complement and offset binary notations there is a unique zero representation.

The most useful way of indicating the relationship between analog and digital quantities involved in a conversion is the input–output transfer characteristic. Figure 6.2 shows the transfer characteristic for an ideal 3-bit unipolar D/A converter that is made up of 2^3 distinct output levels. In practical D/A converters the ideal transfer characteristic of Fig. 6.2 cannot be achieved. The types of errors or deviations from this ideal condition are graphically illustrated in Fig. 6.3. The *offset error* is illustrated in Fig. 6.3*a* and is defined as the deviation of the actual output from the ideal output when the

TABLE 6.1. Commonly Used Bipolar Codes

Number	Decimal Fraction	Sign Magnitude	One's Complement	Offset Binary	Two's Complement
+7	$\frac{7}{8}$	0111	0111	1111	0111
+6	$\frac{6}{8}$	0110	0110	1110	0110
+5	$\frac{5}{8}$	0101	0101	1101	0101
+4	$\frac{4}{8}$	0100	0100	1100	0100
+3	$\frac{3}{8}$	0011	0011	1011	0011
+2	$\frac{2}{8}$	0010	0010	1010	0010
+1	$\frac{1}{8}$	0001	0001	1001	0001
0	0_+	0000	0000	1000	0000
0	0_-	1000	1111	1000	0000
−1	$-\frac{1}{8}$	1001	1110	0111	1111
−2	$-\frac{2}{8}$	1010	1101	0110	1110
−3	$-\frac{3}{8}$	1011	1100	0101	1101
−4	$-\frac{4}{8}$	1100	1011	0100	1100
−5	$-\frac{5}{8}$	1101	1010	0011	1011
−6	$-\frac{6}{8}$	1110	1001	0010	1010
−7	$-\frac{7}{8}$	1111	1000	0001	1001
−8	$-\frac{8}{8}$			0000	1000

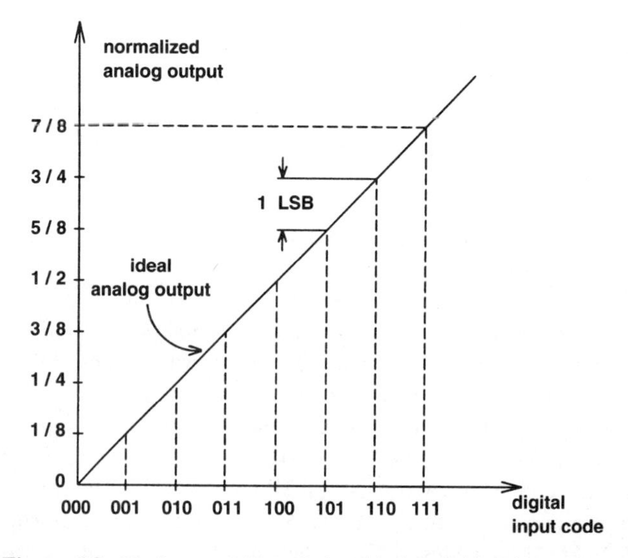

Figure 6.2. Ideal conversion relationship in a 3-bit D/A converter.

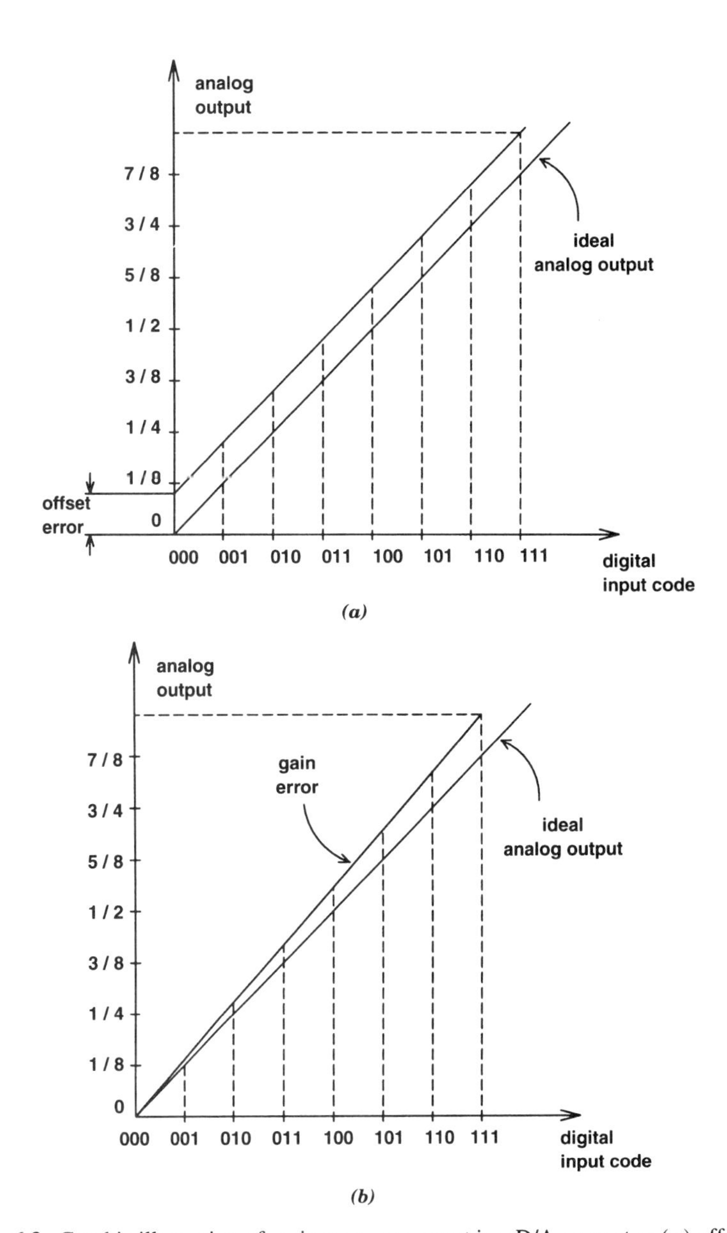

Figure 6.3. Graphic illustration of various errors present in a D/A converter: (*a*) offset error; (*b*) gain error. *(Figure continued)*

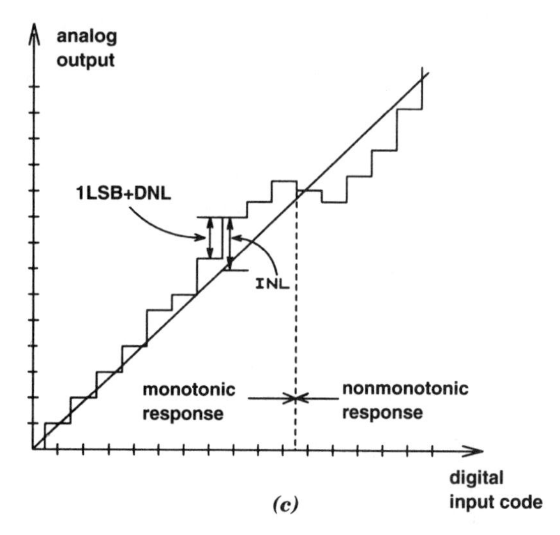

Figure 6.3. (*c*) differential nonlinearity, integral nonlinearity, and nonmonotonic response.

ideal output should be zero. The *gain error* is the change in the slope of the transfer characteristic and is shown in Fig. 6.3*b*. The gain error is due to the inaccuracy of the scale factor or the reference voltage. The *linearity* is a measure of the nonlinearity error at the output, after the offset and gain errors have been removed. There are two types of nonlinearity errors. *Integral nonlinearity* (INL) is defined as the worst-case deviation of the transfer characteristic from an ideal straight line between zero and full scale (i.e., the endpoints of the transfer characteristic). The *differential nonlinearity* (DNL) is the maximum deviation of each output step size of 1 LSB. It is a measure of the nonuniform step sizes between adjacent transitions and is normally specified as a fraction of LSB. The integral and differential nonlinearities are shown in Fig. 6.3*c*. Unlike the offset and gain errors, the nonlinearity errors cannot be corrected by simple trimming and they can only be minimized by improving the matching of the precision components of the D/A converters (i.e., resistors or capacitors). Finally, *monotonicity* in a D/A converter implies that the analog output always increases as the digital input code increases. Nonmonotonicity is due to excessive differential nonlinearity. Guaranteed monotonicity implies that the maximum differential nonlinearity is less than one LSB. Nonmonotonicity is illustrated in Fig. 6.3*c*, where the analog output decreases at some points in its dynamic range while the input code is increasing.

6.2. VOLTAGE-MODE D/A CONVERTER STAGES

An important, yet simple class of D/A converters is based on the accurate scaling of a reference voltage V_{ref}. Voltage scaling can be achieved by connecting a series of N equal segments of resistors between V_{ref} and ground. For an N-bit converter the resistor string consists of 2^N segments. The string of resistors behave as a voltage divider and the voltage across each segment is one LSB, given by

$$V_N = \frac{V_{\text{ref}}}{2^N}. \tag{6.3}$$

An N-bit D/A converter can be realized by using a string of 2^N resistors and a switching matrix implement with MOS switches [1]. The D/A conversion technique is illustrated with a conceptual 3-bit version of the unipolar converter shown in Fig. 6.4. Here the switch matrix is connected in a treelike manner, which eliminates the need for a digital decoder. For an N-bit converter, $2^{N+1} - 2$ switches are needed. As illustrated in Fig. 6.4, the voltage selected propagates through N levels of switches before getting

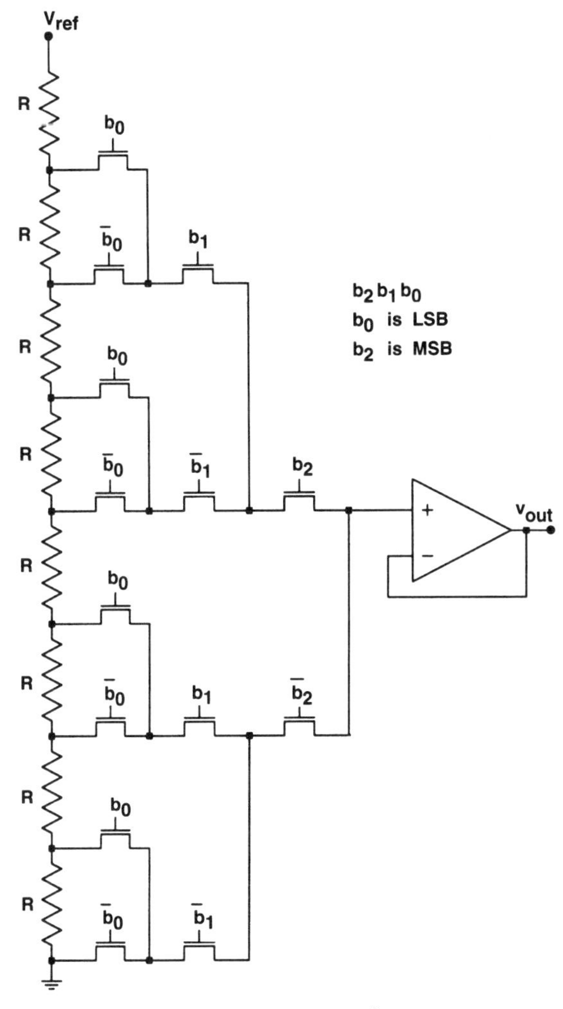

Figure 6.4. Three-bit unipolar resistive DAC with 2^3 resistors and transmission gate tree decoder.

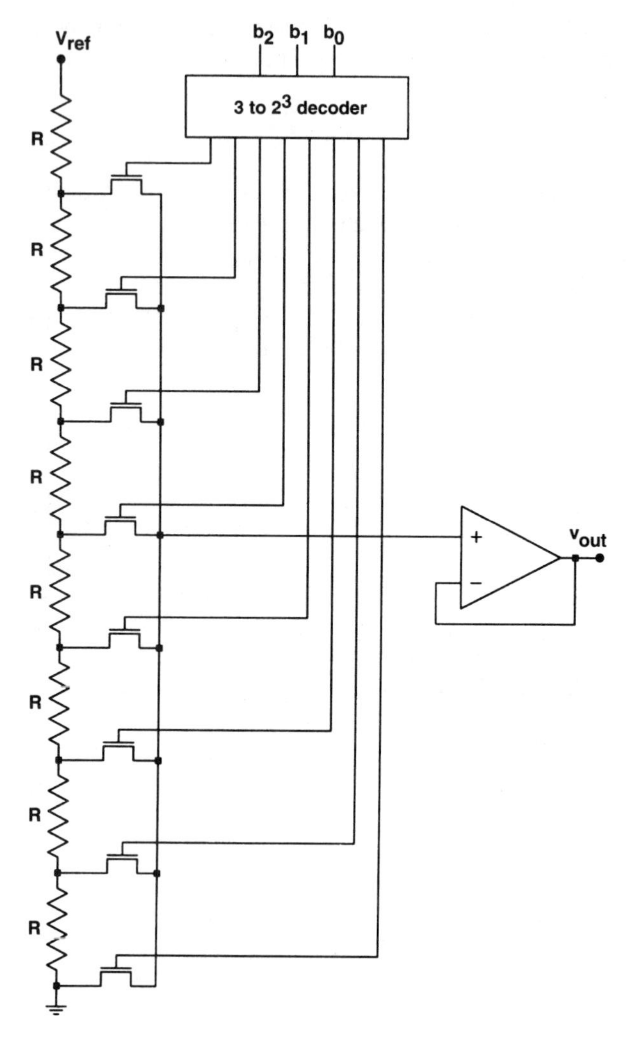

Figure 6.5. Three-bit unipolar resistive DAC with 2^3 resistors and digital decoder.

to the buffer amplifier. The buffer is necessary to provide a low-impedance output to the external load. Assuming that the buffer's dc offset voltage does not vary with its input common-mode voltage, the D/A technique has guaranteed monotonicity and can be used for converters up to 10-bit resolution. As the number of the bits increases, the delay through the switch network imposes a major limitation on the speed. The output impedance of the resistor string also varies, as a function of the closed switch position in the network. Also, the delay through the resistor string may become a major source of delay.

For high-speed applications the tree decoder is replaced with a digital decoder. One such 3-bit DAC is shown in Fig. 6.5. The logic circuit is an N-to-2^N decoder that can take a large area. The common node of all switches is directly connected to the buffer amplifier. The 2^N junctions of the transistors have a large area and result in a large capacitive load. The voltage selected propagates through one switch; hence, despite the larger capacitive load, the DAC output can achieve higher-speed operation.

A more efficient implementation of a 5-bit resistor DAC is shown in Fig. 6.6. This method, known as the *intermeshed ladder architecture,* uses a two-level row–column decoding scheme similar to one used in digital memory [2–4]. The N-to-2^N decoding is achieved by splitting N to $N_1 + N_2 = N$ and realizing it as the combination of

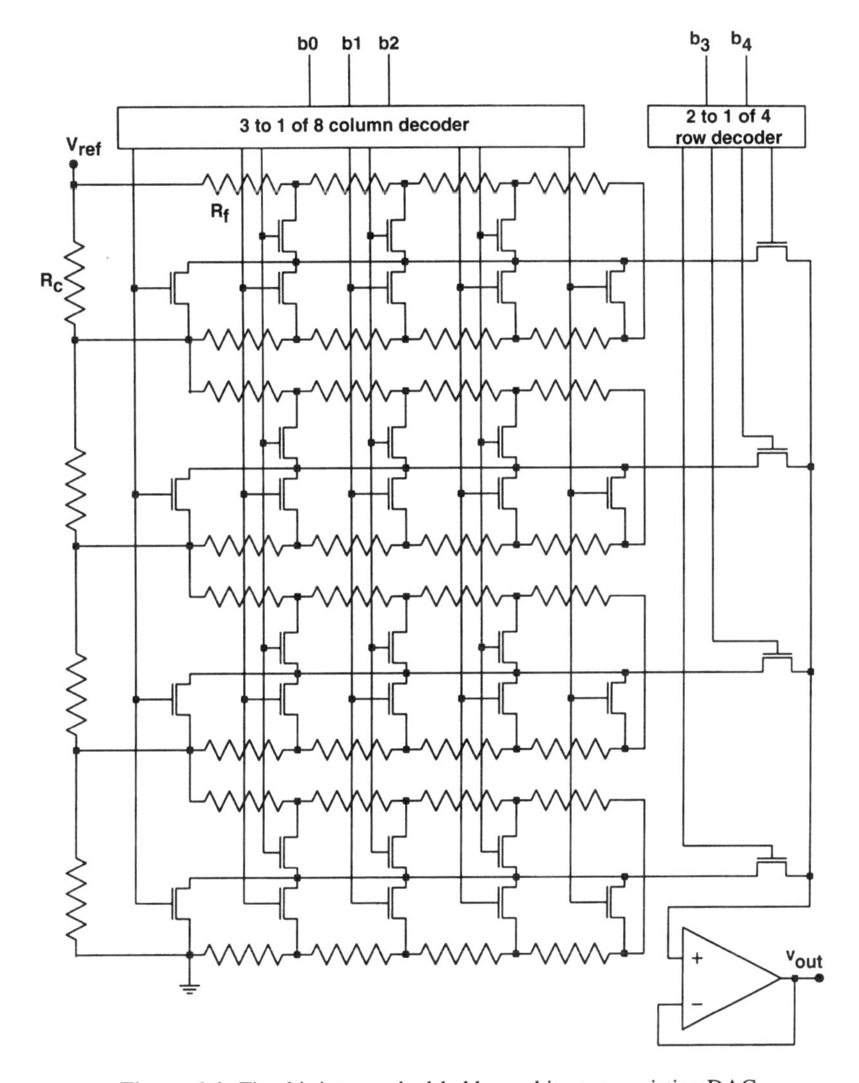

Figure 6.6. Five-bit intermeshed-ladder architecture resistive DAC.

an N_1-to-2^{N_1} row and N_2-to-2^{N_2} column decoders. For a given digital code, one of the 2^{N_2} columns is selected, all transistor switches in that column are turned on, and the 2^{N_1} resistor nodes are connected to 2^{N_1} rows. The row decoder selects one of the 2^{N_1} rows and connects it to the input of the buffer amplifier.

This approach uses 2^{N_1} subsegments, each consisting of 2^{N_2} resistor segments. For large N the number of the resistors, 2^N, grows exponentially and the impedance of the resistor array becomes very large. The output impedance also varies as a function of the position of the closed switch in the network. To reduce the impedance of the array and hence its settling time, a coarse array is placed in parallel to the fine one. The coarse array consists of 2^{N_1} (N_1 is the number of the rows) resistors and is denoted by R_c in Fig. 6.6. The coarse array in this way determines 2^{N_1} accurate voltages and determines the value corresponding to the N_1 MSBs. In this arrangement the 2^{N_1} resistor subsegments each consist of 2^{N_2} segments, and the endpoints of the subsegments are connected to the 2^{N_1} coarse resistors. If the coarse resistors are represented by R_c and the resistors in the subsegment by R_f, the total resistance of the array is given by

$$R_{\text{array}} = 2^{N_1} (R_c \parallel 2^{N_2} R_f). \tag{6.4}$$

As a result of this modification, the worst-case output resistance of the array seen by the buffer is reduced to $R_{\text{array}}/4$.

For high-resolution applications the resistor string DAC suffers from several drawbacks: The number of the resistors and switches grows exponentially and it exhibits a long delay at the output. Hence the resistive DACs of Figs. 6.4 to 6.6 are not practical when the number of the bits grows beyond 10. To take advantage of the inherent monotonicity of the voltage-division DAC while keeping the number of resistors to a manageable level, it is possible to use a two-stage DAC such as the one shown in Fig. 6.7 [5]. As the figure shows, the 6-bit DAC consists of two resistor strings each having eight segments. The coarse resistor string is connected between V_{ref} and ground. Two operational amplifiers connected as voltage followers buffer consecutive voltages of the coarse DAC. The fine resistor string is connected between the two outputs of the followers. The monotonocity of the two-stage DAC cannot be guaranteed, due to the offset voltage of the unity-gain buffers. A special sequence can be used to operate the coarse array switches to make the operation of the DAC independent of the buffer offset voltages. To understand this, consider a portion of the coarse array shown in Fig. 6.8. This figure corresponds to the ith code of the coarse bits, where buffer 1 is connected to segment voltage V_i and buffer 2 to segment voltage V_{i-1}. The output of the buffers will be $(V_1)_i = V_i - V_{\text{off1}}$, and $(V_2)_i = V_{i-1} - V_{\text{off2}}$, where V_{off1} and V_{off2} are the corresponding offset voltages of the first and second op-amps. For the next sequential code, if node A moves to V_{i+1} and node B to V_i, the buffer outputs will be $(V_1)_{i+1} = V_{i+1} - V_{\text{off1}}$ and $(V_2)_{i+1} = V_i - V_{\text{off2}}$. For monotonic operation it is necessary for $(V_2)_{i+1} = (V_1)_i$; otherwise, the consecutive coarse output voltages will not be continuous. This is possible only if $V_{\text{off1}} = V_{\text{off2}}$ or $V_{\text{off1}} = V_{\text{off2}} = 0$, which cannot be guaranteed in practice. Alternatively, for the sequential code, node A can be kept at V_i and node B switched

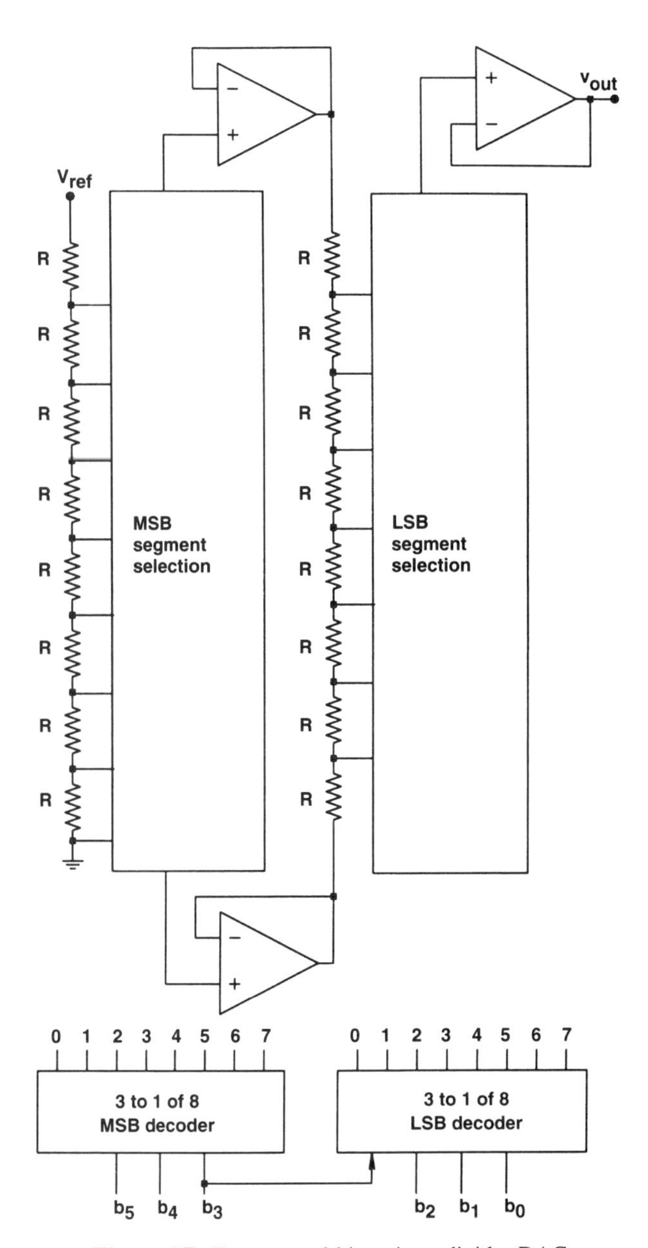

Figure 6.7. Two-stage 6-bit resistor divider DAC.

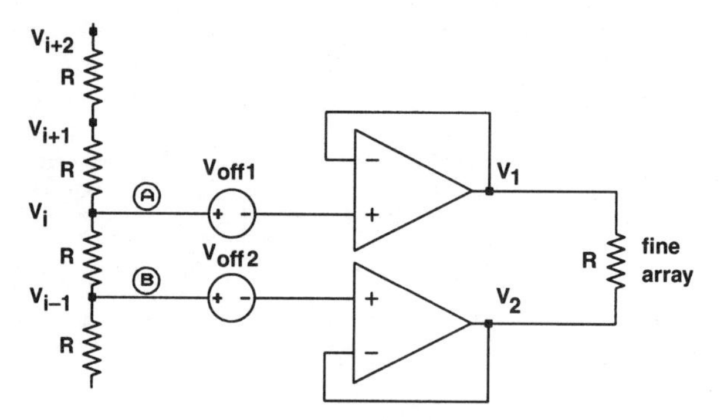

Figure 6.8. Portion of the coarse resistor segment.

to V_{i+1}. This choice guarantees a continuous output from the coarse array and hence monotonic operation for the DAC. However, since the top and bottom of the fine resistor string switches for consecutive codes, the decoding and switching of the fine DAC should be modified accordingly. Figure 6.9 shows the complete circuit diagram of the 6-bit DAC, which includes the switching details of the coarse and fine DACs. As the figure shows, for an N-bit DAC, the converter functions by applying voltage V_{ref} to the top of the coarse resistor array and dividing it to $2^{N/2}$ nominally equal voltage segments. For a given code, buffer A_1 transfers the voltage at the ith tap to the top of the fine string, while A_2 applies the voltage at tap $i - 1$ to the bottom of fine string. The A_3 output results from linearly interpolating the voltage drop between taps i and $i - 1$, weighted by the $N/2$ lower bits of the N-bit input digital word. For the next coarse digital code, the polarity of the voltage across the fine resistor string is reversed. This reversal occurs at every other adjacent resistor segment and is corrected in the second stage by alternating between two switch arrays. The least significant bit of the digital input code of the coarse DAC makes this selection. This bit selects between the odd and even segments, which allows the analog output of the coarse divider to have a continuous output voltage independent of the offset of the two voltage followers. By using this method it is possible to obtain a 16-bit DAC which has a guaranteed monotonicity without the need for a straight 2^{16}-segment resistor divider.

For an N-bit DAC, the circuit of Fig. 6.9 uses two $2^{N/2}$-segment resistor strings, two $N/2$-to-$2^{N/2}$ decoders, and four sets of switching elements. Alternatively, the two-stage DAC can be implemented using a single $2^{N/2}$ segment resistor string, two sets of switching elements, and two $N/2$-to-$2^{N/2}$ decoders. A 6-bit version of such a DAC is shown in Fig. 6.10. Here the two inputs of the A_1 and A_2 buffers are independently switched to the taps of the eight-segment resistor string. The three MSBs of the input digital word are decoded to eight lines that control the connection of A_1 to the taps of the resistor string. The next three LSBs control the connection of A_2 to the resistor string in a similar fashion. A nine-segment resistor string is placed between the outputs of A_1 and A_2. Tap 8, from the bottom of the string, is buffered by A_3 and brought out as the output of the DAC. From Fig. 6.10, V_1 and

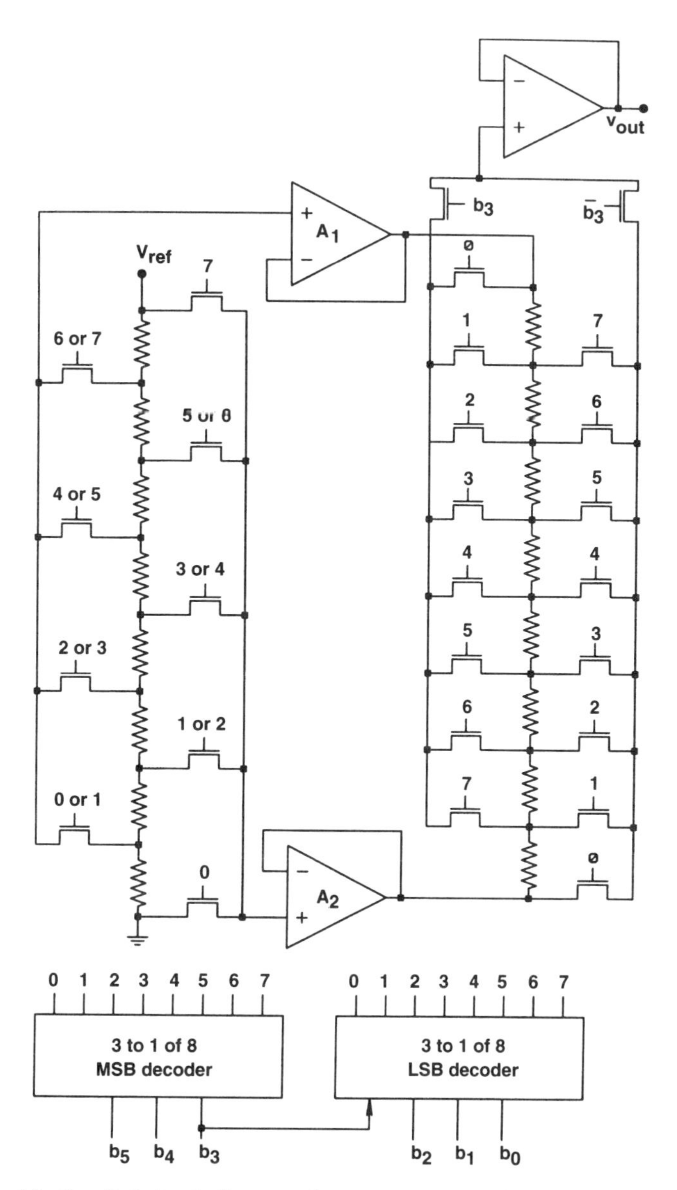

Figure 6.9. Detailed circuit diagram of two-stage monotonic voltage-divider DAC.

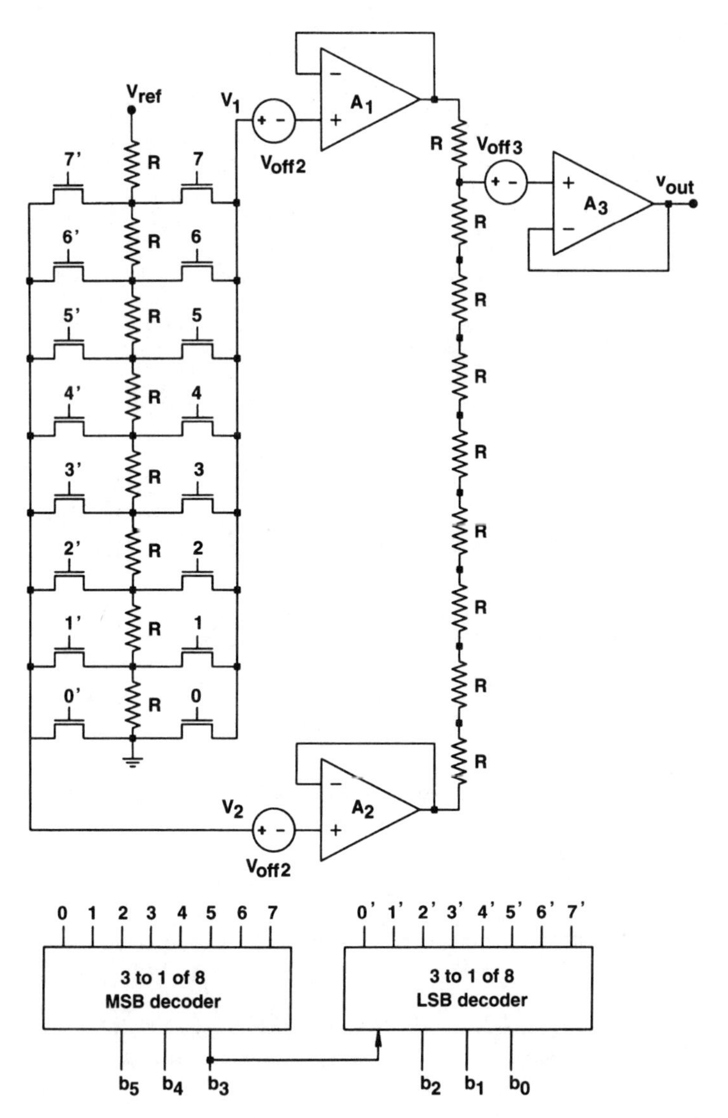

Figure 6.10. Alternative form of 6-bit unipolar two-stage resistive-divider DAC.

V_2 are the outputs of two independently controlled 3-bit DACs. One converts the three MSBs and the other the three LSBs of the input digital word to analog voltages. Ignoring the dc offset voltages of the three op-amps, v_{out} can be calculated as

$$v_{\text{out}} = V_2 + \frac{8}{9}(V_1 - V_2)$$

or

$$v_{\text{out}} = \frac{8}{9}V_1 + \frac{1}{9}V_2 = \frac{8}{9}V_1 + \frac{8}{9} \times \frac{1}{8}V_2,$$

so that

$$v_{\text{out}} = \frac{8}{9}\left(V_1 + \frac{1}{8}V_2\right). \tag{6.5}$$

Ignoring the factor $\frac{8}{9}$, the relationship inside the parentheses represents the output of a 6-bit segmented DAC. If the effect of the op-amp offsets is included in the calculations, Eq. (6.5) will be modified in the following way:

$$v_{\text{out}} = \frac{8}{9}\left(V_1 + \frac{1}{8}V_2\right) - \frac{8}{9}\left(V_{\text{off1}} + \frac{V_{\text{off2}}}{8}\right) + V_{\text{off3}}. \tag{6.6}$$

As Eq. (6.6) shows, the dc offset of the op-amps appears as a constant dc voltage at the output without interfering with the DAC operation. Therefore, if the op-amp offset voltages do not change with the common-mode voltage, this structure is inherently monotonic and the complex switching scheme of the DAC shown in Fig. 6.9 will not be necessary. The only disadvantage of this scheme is the $\frac{8}{9}$ attenuation factor for the 6-bit DAC. For an N-bit DAC, if we split the input code into two $N/2$ bits (N is even) segments, the output relationship will be given by

$$v_{\text{out}} = \frac{2^{N/2}}{2^{N/2} + 1}\left(V_1 + \frac{1}{2^{N/2}}V_2\right). \tag{6.7}$$

As an example for a 16-bit DAC, $N = 16$ and we have

$$v_{\text{out}} = \frac{256}{257}\left(V_1 + \frac{1}{256}V_2\right). \tag{6.8}$$

The first stage will be an 8-bit DAC followed by two buffer amplifiers with a 257-segment resistor string connected between their outputs. The final 16-bit DAC output will be taken from tap 256 of the second-stage resistor string. The 256/257 attenuation can be ignored or compensated by modifying the reference voltage, V_{ref}, which is connected to the top of the first-stage resistor string.

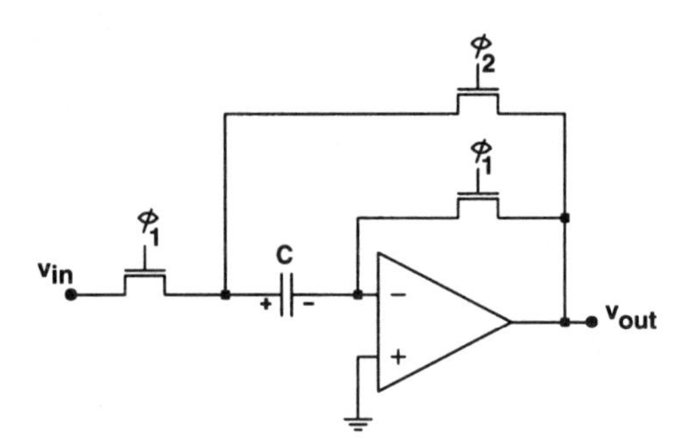

Figure 6.11. Offset-compensated switched-capacitor sample-and-hold circuit.

The DACs described so far in this section all use unity-gain buffers to isolate the resistor string from the external load. The unity-gain buffer has several disadvantages. For high-resolution DACs, the op-amp should have a large common-mode-rejection ratio (CMRR) to maintain the accuracy over the entire input common-mode range. Also, for low-voltage operation, complex op-amps with rail-to-rail input stage should be used to facilitate operation over the wide input range. For bipolar outputs, the bottom of the resistor string will be connected to a negative reference. The absolute values of the positive and negative reference voltages should match closely. Any mismatch will introduce an offset and linearity error.

An alternative to the unity-gain buffer is the offset-compensated switched-capacitor sample-and-hold stage shown in Fig. 6.11, where ϕ_1 and ϕ_2 are nonoverlapping two-phase clocks [6]. When ϕ_1 is high, capacitor C will be charged between the output of the DAC and the op-amp offset voltage and acquires the voltage $v_{in} - V_{off}$, where $v_{in} = V_{dac}$. When ϕ_2 goes high, the output becomes $V_{DAC} - V_{off} + V_{off} = V_{DAC}$, which is independent of the op-amp offset voltage.

It is also possible to use the switched-capacitor gain stage shown in Fig. 6.12*a* as a buffer [7,8]. In this circuit, when $\phi_1 =$ "1," the op-amp has its inverting input terminal shorted to its output node and hence performs as a unity-gain voltage follower with output voltage V_{off}. Hence capacitor αC charges to $V_{off} - v_{in}$ while C changes to V_{off}. When next ϕ_2 goes high, αC recharges to V_{off} and C to $V_{off} - v_{out}$. If the time when this happens is $t = NT$, by charge conservation at node A,

$$\alpha C \{V_{off} - [V_{off} - v_{in}(nT - T/2)]\} + C\{[V_{off} - v_{out}(nT)] - V_{off}\} = 0. \tag{6.9}$$

In this equation, V_{off} terms cancel out and $v_{out}(nT) = \alpha v_{in}(nT - T/2)$ results. Thus a positive gain of α and a delay $T/2$ are provided by the stage, and the output offset voltage is $V_{os} = 0$. Note also that the circuits of Figs. 6.11 and 6.12 are fully stray insensitive.

By interchanging the clock phases at the input terminals, an inverting voltage amplifier can also be obtained (Fig. 6.13). By an analysis similar to that performed

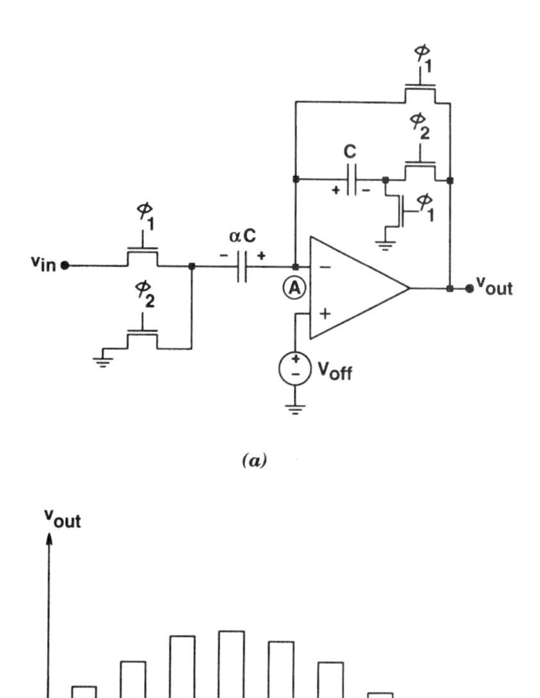

Figure 6.12. Offset-compensated noninverting voltage amplifier: (*a*) circuit; (*b*) output waveform.

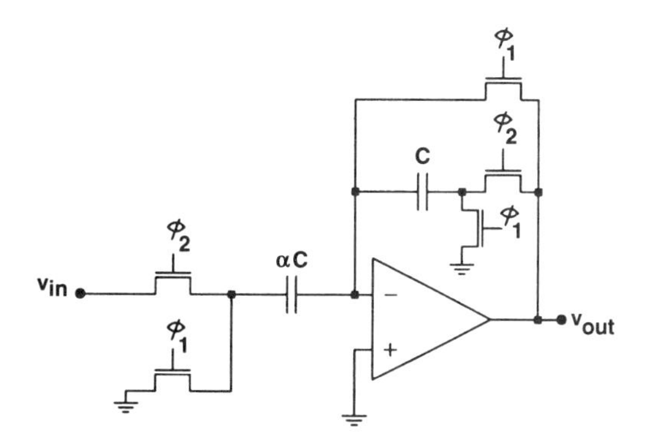

Figure 6.13. Offset-compensated inverting voltage amplifier.

for the circuit of Fig. 6.12, it can be shown (Problem 6.3) that this circuit is a delay-free inverting amplifier with gain α. As before, V_{off} is canceled by the switching arrangement and does not enter v_{out} if the op-amp gain is infinite. (See Problem 6.3 for the finite-gain case.)

As mentioned earlier, in the circuits of Figs. 6.11, 6.12, and 6.13, the output voltage is V_{off} whenever ϕ_1 is high, hence the output is valid only when ϕ_2 is high. As an example, for the circuit of Fig. 6.12*a,* the output waveform is as illustrated schematically in Fig. 6.12*b*. Clearly, the op-amp must have a high slew rate and fast settling time, especially if the clock rate is high. At the cost of a few additional components [9], this disadvantage of offset compensation can be eliminated (Problem 6.4).

Low-frequency noise signals, which do not change substantially during clock period T, are similarly canceled by offset compensation. Thus the troublesome $1/f$ noise discussed in Section 2.7 is greatly reduced. Figure 6.14*a* and *b* illustrate the output noise spectra with and without offset compensation, respectively. Note that cancellation also occurs at $2f_c$, $4f_c$, . . . , which are equivalent to dc for the sampled noise.

If the digital input signal is bipolar, that is, if it has either a positive or a negative sign as indicated by a sign bit b_0, the gain stage with polarity control shown in Fig. 6.15 can be used. When $b_0 = 0$, indicating that the digital signal is positive, the circuit functions like the noninverting unity-gain sample-and-hold stage of Fig. 6.12. If $b_0 = 1$, the digital signal is negative; then the roles of ϕ_1 and ϕ_2 interchange in

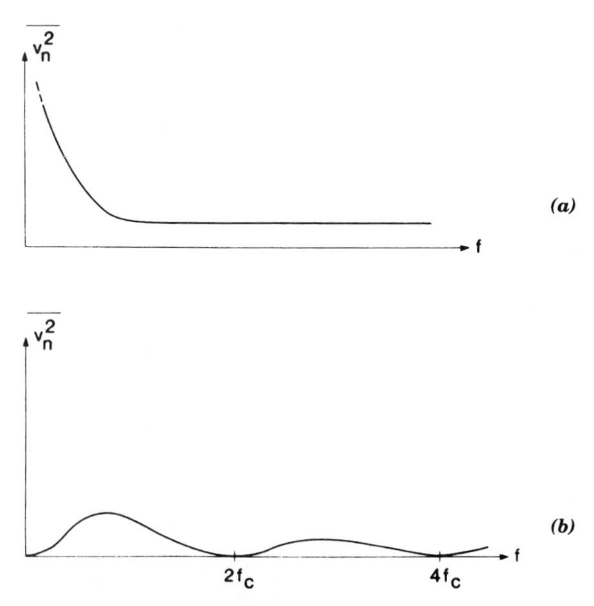

Figure 6.14. Noise power for a switched-capacitor voltage amplifier: (*a*) without offset compensation; (*b*) with offset compensation.

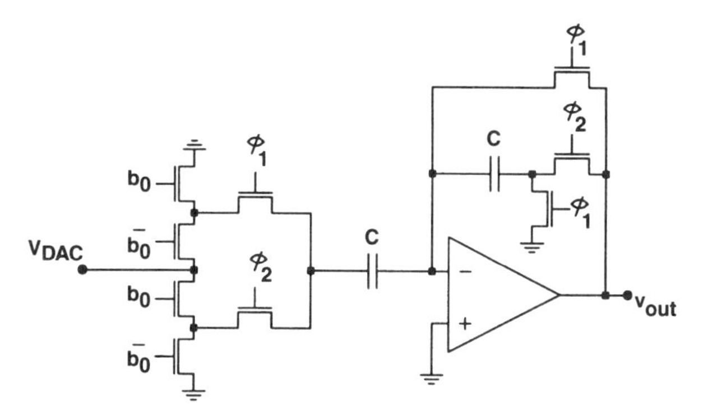

Figure 6.15. Offset-compensated switched-capacitor gain stage with polarity control.

the input branch and the circuit functions like the inverting unity-gain stage of Fig. 6.13. The advantages of using the switched-capacitor stages of Figs. 6.11 to 6.15 are that the op-amp has no common-mode input signal, and bipolar outputs can be generated from a unipolar DAC with a single positive (or negative) reference voltage.

6.3. CHARGE-MODE D/A CONVERTER STAGES

An important advantage of switched-capacitor circuits is that they can be made digitally variable and thus also programmable. This is accomplished by replacing some capacitors in the circuit by *programmable capacitor arrays* (PCAs). Such a binary-programmed array [10] is shown in Fig. 6.16. In the figure the triangular symbols denote inverters, and $b_0, b_1, \ldots, b_7$, are binary-coded (high or low, 1 or 0) digital signals. Thus if (say) b_7 is *high,* the left-side switching transistor associated with capacitor C is on and the right-side transistor is off. Hence C is connected between nodes X and X′. If b_7 is low, the right-side transistor is on, and it connects the right-side terminal of C to ground rather than to X′. Therefore, C never floats, and the total capacitance loading at node X is constant. The value of the capacitance between X and X′ in the 8-bit PCA of Fig. 6.16 is thus clearly

$$C_T = \sum_{i=0}^{7} \frac{C}{2^{7-i}} b_i = 2^{-7} C \sum_{i=0}^{7} 2^i b_i, \tag{6.10}$$

while the total capacitance loading node X is $2C(1 - 2^{-8})$ (Problem 6.5).

Care must be taken in the design of the PCA to minimize noise injection from the substrate into the circuit. Thus the bottom plate of the capacitor (which is in the substrate or right above it) should never be connected to the inverting input terminal of an op-amp; otherwise, the noise from the power supply which biases the substrate will be coupled to the op-amp's input and amplified by the op-amp.

An obvious application of PCAs is the realization of charge-mode D/A converters.

It can be obtained by replacing the input capacitance αC in the offset-free voltage amplifier of Fig. 6.12*a* by a PCA. An example of an *N*-bit charge mode DAC is shown in Fig. 6.17, where V_{ref} is a temperature-stabilized constant reference voltage. If b_1 represents the most significant bit (MSB) and b_N the least significant bit (LSB), the output voltage at the end of clock period ϕ_2 is given by

$$v_{\text{out}} = V_{\text{ref}} \sum_{i=1}^{N} b_i \cdot 2^{-i}. \tag{6.11}$$

Thus the output is the product of the reference voltage V_{ref} and the binary-coded digital signal $(b_1, b_2, b_3, \ldots, b_N)$.

Note that the orientation of all capacitors is such that their *top* plates (indicated by light lines) are connected to the op-amp input terminal. This reduces substrate noise voltage injection. Also, due to the presence of the switching devices driven by $\bar{b}_1$, $\bar{b}_2$, and so on, the total capacitance connected to the op-amp input is constant, which makes its compensation an easier task.

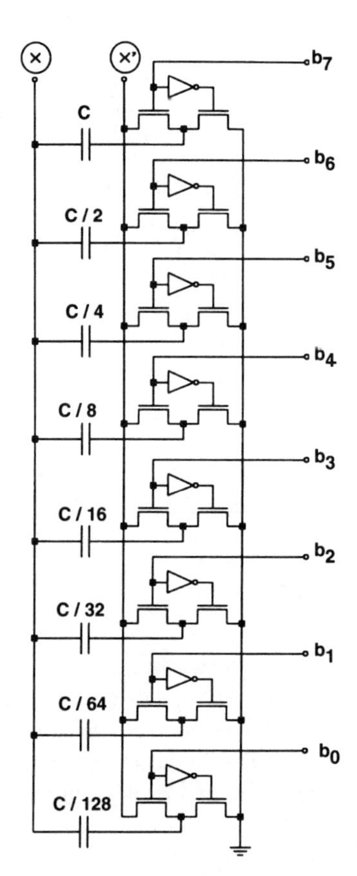

Figure 6.16. Binary-programmable capacitor array (PCA).

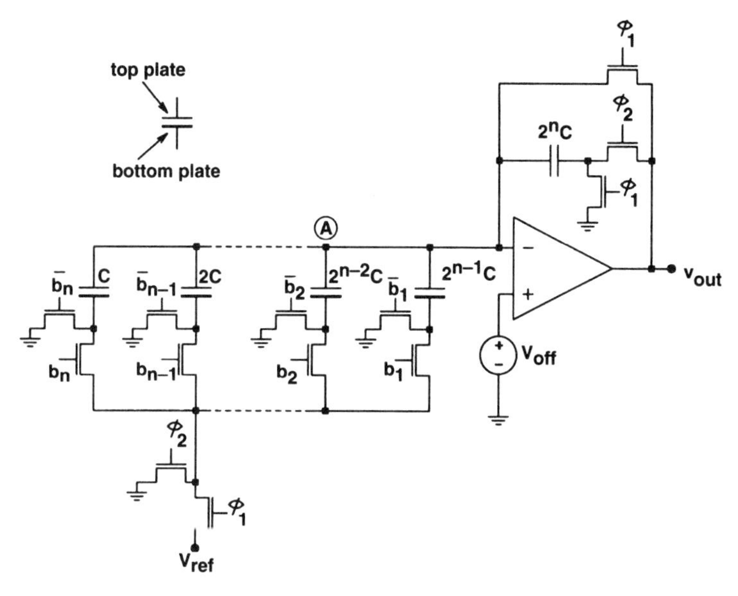

Figure 6.17. An *n*-bit charge-mode digital-to-analog converter.

If the digital input signal is *bipolar,* that is, if it has either a positive or a negative sign as indicated by a sign bit b_0, the DAC shown in Fig. 6.18 can be used. If $b_0 = 0$, indicating that the digital signal is positive, the circuit functions in exactly the same way as the DAC of Fig. 6.17. If, however, $b_0 = 1$, so that the digital signal is negative (as can easily be deduced), in the input branch ϕ_1 and ϕ_2 exchange roles. Now the circuit functions as the inverting voltage amplifier of Fig. 6.13. Thus the input–output relation is

$$v_{\text{out}} = -V_{\text{ref}} \sum_{i=1}^{N} b_i \cdot 2^{-i}, \tag{6.12}$$

as required by the negative digital signal.

For an N-bit DAC, the capacitor ratio is 2^N and the total capacitance $C_{\text{total}} = (2^{N+1} - 1)C$. For $N = 8$, $2^8 = 256$ and $C_{\text{total}} = 511C$. The ratio and the total capacitance increase rapidly with increasing N and the matching accuracy deteriorates. The offset-free scheme of Fig. 6.18 can be used in a cascade design to reduce the capacitor ratio [11]. The circuit diagram for a bipolar 8-bit D/A converter is shown in Fig. 6.19. The output of the first stage is fed to the second stage with the same weighting as the least significant bit. The output voltage at the end of clock period ϕ_2 is given by

$$v_{\text{out}} = kV_{\text{ref}} \sum_{i=1}^{4} (b_i + 2^{-4} b_{i+4}) \cdot 2^{-i}, \tag{6.13}$$

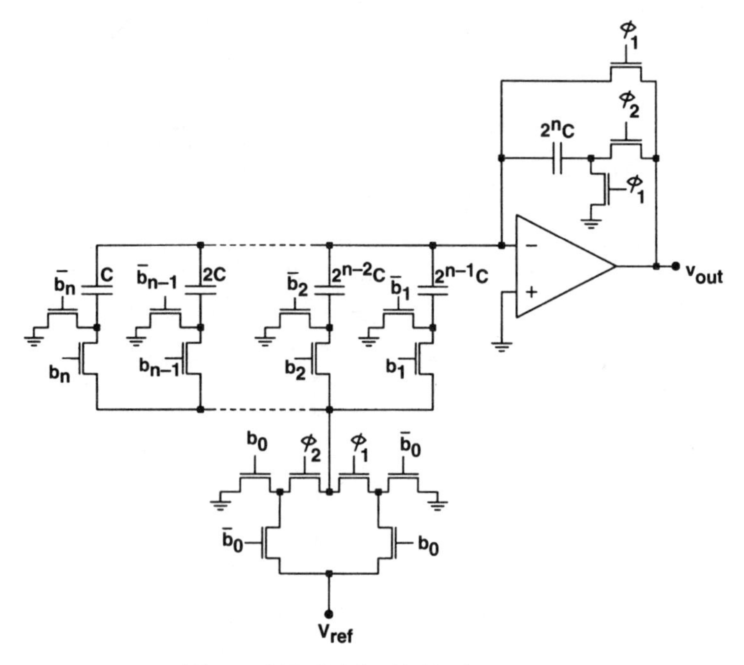

Figure 6.18. DAC with bipolar output.

which reduces to

$$v_{\text{out}} = kV_{\text{ref}} \sum_{i=1}^{8} b_i \cdot 2^{-i}, \tag{6.14}$$

where k is determined by the sign of b_1. The capacitor ratio is reduced to $2^4 = 16$ and the total capacitance is $C_{\text{total}} = 63C$. As Fig. 6.19 shows, the capacitor ratio is reduced without increasing the conversion cycles. For an N-bit (N-even) bipolar D/A converter, the capacitor ratio is $2^{N/2}$ and the total capacitance $C_{\text{total}} = [2^{(N/2+2)} - 1]C$, which corresponds to an improvement of $2^{N/2-1}$ over the direct method. The circuit is compatible with most process technologies and uses a single reference voltage for bipolar outputs.

6.4. HYBRID D/A CONVERTER STAGES

In the cascaded D/A converter shown in Fig. 6.19, the voltage corresponding to the LSBs propagates through two op-amps before reaching the output. The settling time of the two cascaded op-amp stages sets an upper limit on the maximum conversion speed. A more effective way to reduce component spread and achieve high precision is to combine the charge-mode and resistor-divider-mode DACs described earlier in the chapter. The most straightforward combination is to replace the LSB stage (first stage) of the cascaded D/A converter of Fig. 6.19 with one of the resistor-

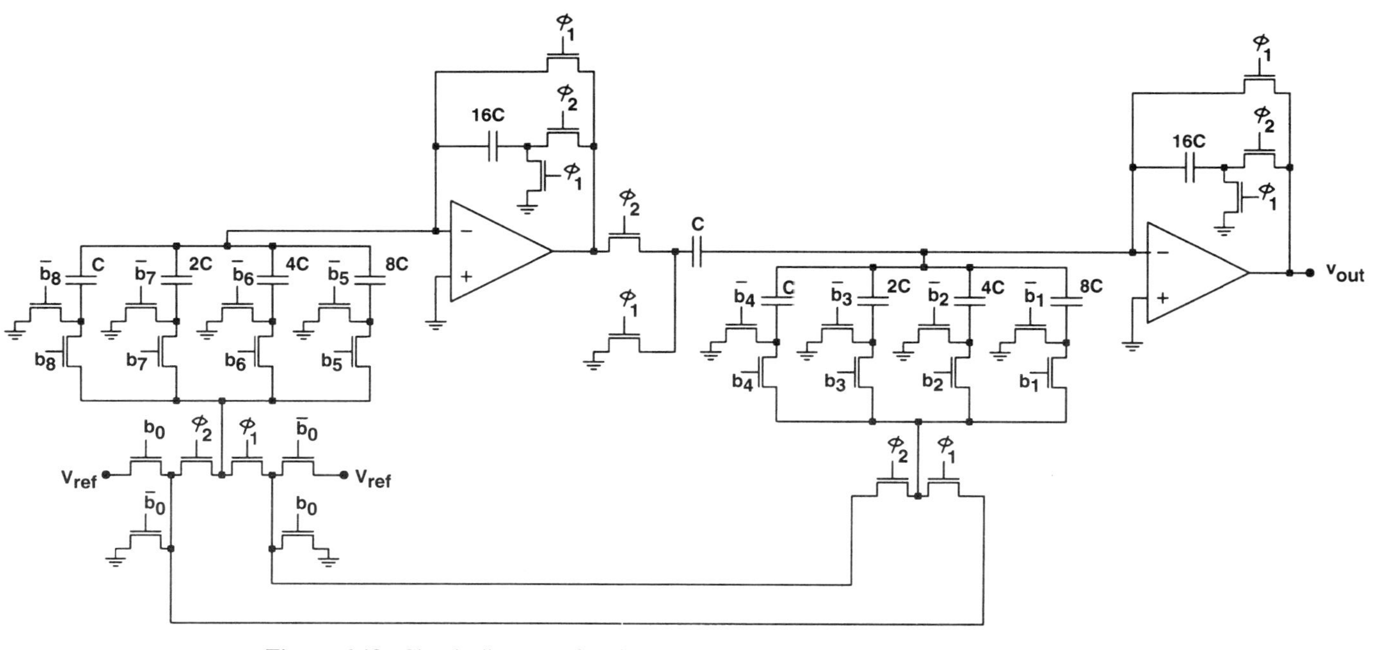

Figure 6.19. Circuit diagram of a bipolar output 8-bit cascaded D/A converter.

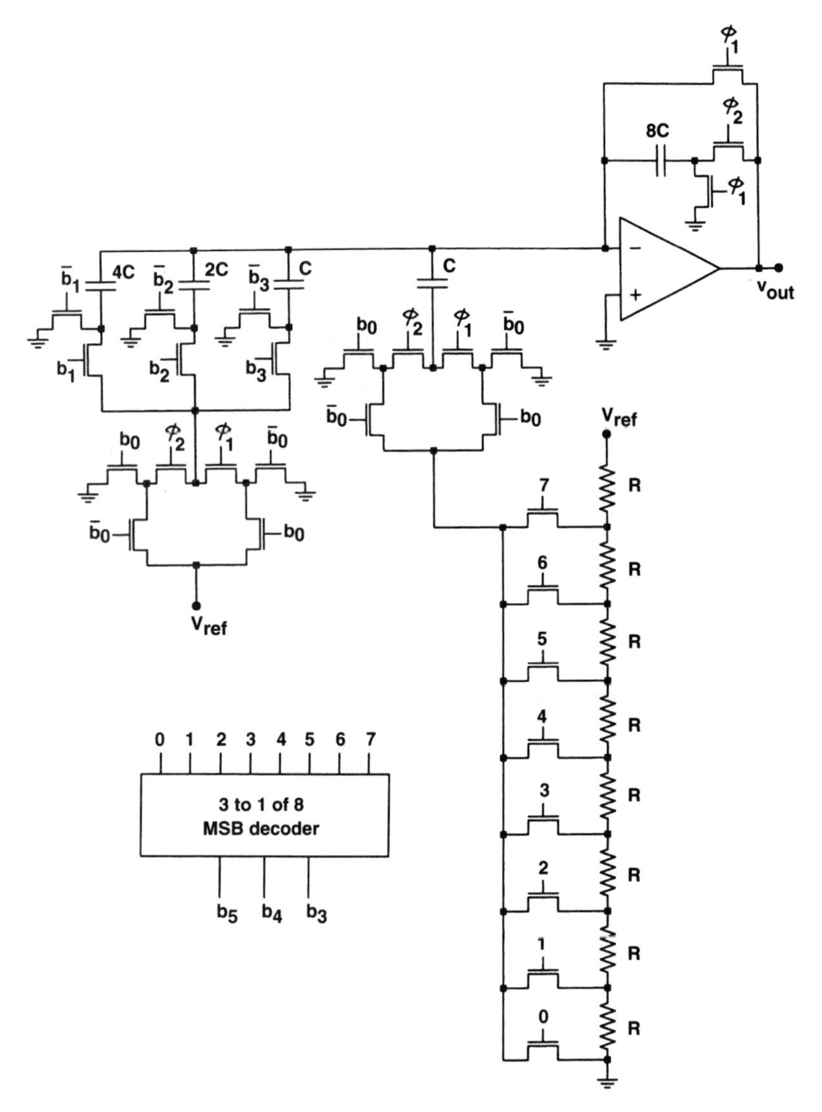

Figure 6.20. Seven-bit hybrid DAC stage with charge MSB and voltage-divider LSB DACs.

divider DACs of Fig. 6.5 or 6.6. One example of this approach is shown in Fig. 6.20, where a 7-bit bipolar-output DAC is realized as the cascade of a 3-bit charge-mode DAC and a 3-bit resistor divider-mode DAC. The MSB in this case is the sign bit and controls the polarity of the output. For a 16-bit unipolar DAC an 8-bit charge-mode DAC and an 8-bit resistor divider-mode DAC can be used. In this approach the LSBs are used to program the output of the resistor DAC while the MSBs control the capacitor array. The overall accuracy is determined largely by the

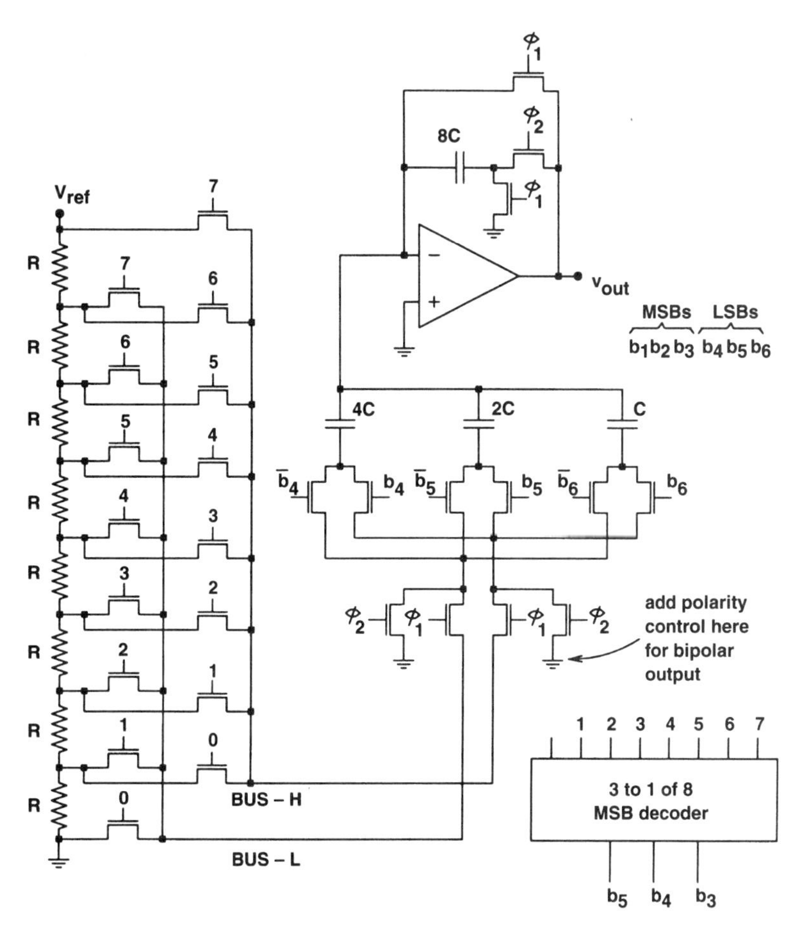

Figure 6.21. Six-bit hybrid unipolar DAC stage with resistor-divider MSB and charge LSB DACs.

MSB DAC. Monotonicity cannot be guaranteed because the charge-mode DAC is not accurate within one LSB of the overall DAC.

For inherently monotonic design, a two-stage approach can be used where the MSBs are used to program a resistor-divider DAC and the LSBs control a binary-weighted programmable capacitor array (PCA) [12]. This approach is similar to the one used in an inherently monotonic successive-approximation A/D converter [13]. A 6-bit two-stage unipolar D/A converter is shown in Fig. 6.21. For a 7-bit bipolar DAC, a sign bit can be added to control the clock phases of the switched-capacitor gain stage. The overall DAC consists of a 3-bit resistor-divider DAC and a 3-bit charge-mode DAC. The three MSB connect adjacent nodes of the resistor string to the two high (bus-H) and low (bus-L) buses. The three LSBs connect the binary-weighted capacitor array to the high and low buses through the two switches controlled by ϕ_1 and ϕ_2. If an LSB is a "1," the corresponding capacitor is connected to the high bus (bus-H); if the bit is "0" it is connected to the low bus (bus-L). For a positive output, when ϕ_1 is high, the bottom plates of the binary-weighted capacitor are connected to bus-H or bus-L. When ϕ_1 goes low and ϕ_2 goes high, the bottom plates of the capacitors switch to ground and a voltage corresponding to the digital code appears at the output of the gain stage. If a sign bit is used to control the sequence of the clock phases, a positive or negative output will be generated at the output of the DAC. The absolute accuracy and linearity of the entire DAC are limited by the accuracy of the voltage division of the resistor string. The monotonicity of the entire DAC is guaranteed as long as the capacitor array is monotonic. For a 16-bit bipolar output DAC, an 8-bit resistor DAC and a 7-bit capacitor DAC can be used. The sign bit will control the polarity of the output.

6.5. CURRENT-MODE D/A CONVERTER STAGES

All current-mode DACs are made of three basic blocks: a current reference generator, a controlled current switching matrix, and a current-to-voltage converter. The current reference generator is simply a voltage-to-current converter. One such circuit is shown in Fig. 6.22, where a reference voltage and resistor is used to generate the reference current. For the circuit of Fig. 6.22, the op-amp forces the voltage across the reference resistor to V_{ref}. So the reference current is given by

$$I_{\text{ref}} = \frac{V_{\text{ref}}}{R_{\text{ref}}}. \qquad (6.15)$$

The diode-connected p-channel device Q_1 generates a gate-to-source voltage that can be used as bias voltage to mirror the reference current into the current-switching

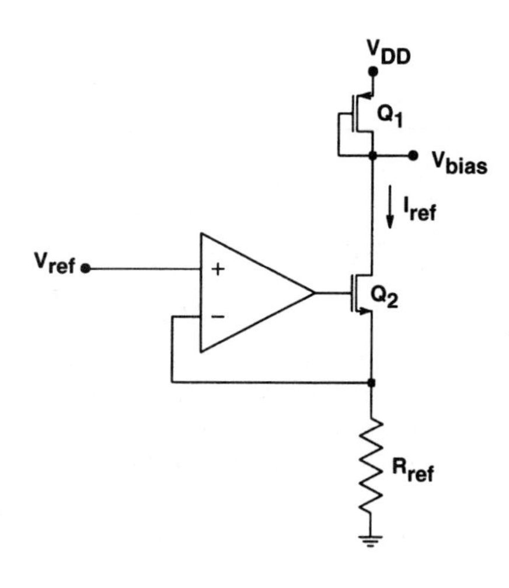

Figure 6.22. Current reference generator for the current DAC.

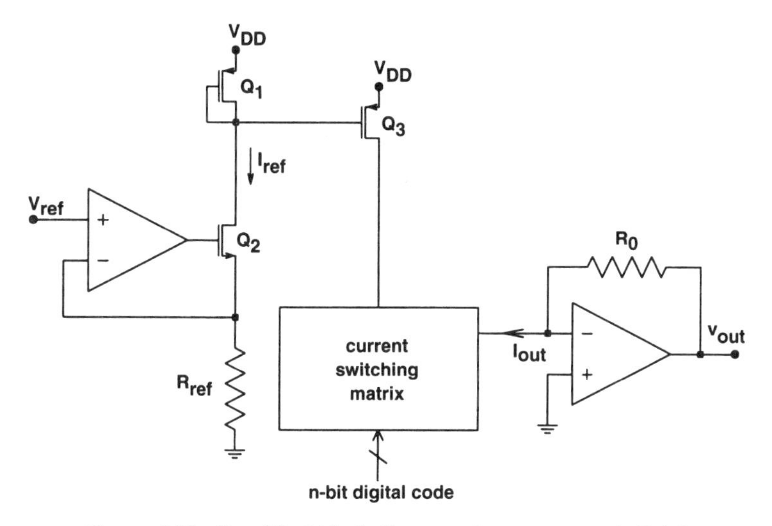

Figure 6.23. Simplified block diagram of a current-mode DAC.

matrix. The current-switching matrix under the control of the N-bit digital input code produces an output given by

$$I_{\text{out}} = I_{\text{ref}}\,(b_1 \cdot 2^{-1} + b_2 \cdot 2^{-2} + \cdots + b_N \cdot 2^{-N}), \tag{6.16}$$

where b_1 is the MSB and b_N the LSB.

The output current, I_{out}, flows into the current-to-voltage converter, which in its simplest form is a resistor. To provide a low-impedance output, an op-amp should be used. In this case the current-to-voltage converter is an op-amp with a feedback resistor. The simplified block diagram of a current-mode DAC is shown in Fig. 6.23, where the output v_{out} is given by

$$v_{\text{out}} = I_{\text{out}} R_o = V_{\text{ref}} \frac{R_o}{R_{\text{ref}}} (b_1 \cdot 2^{-1} + b_2 \cdot 2^{-2} + \cdots + b_N \cdot 2^{-N}). \tag{6.17}$$

One of the main applications of high-speed current-mode D/A converters is in raster-scan graphics monitors, which are used in most computer systems. High-speed DACs are also used in digital and high-definition television. These systems normally include three 8-bit high-speed DACs for red, green, and blue colors. In today's high-resolution color monitors, each DAC operates at speeds in excess of 200 MHz and is designed to have a current output that drives a 75-Ω doubly terminated line [14,15].

The basic architecture of a 3-bit current DAC is shown in Fig. 6.24. The DAC consists of $2^3 - 1 = 7$ identical current sources, where each current source, under the control of the input code, can switch between the output load and ground. For an inherently monotonic DAC with good differential nonlinearity (DNL), a thermometer-type decoder must be used. A 3-bit thermometer decoder converts the input 3

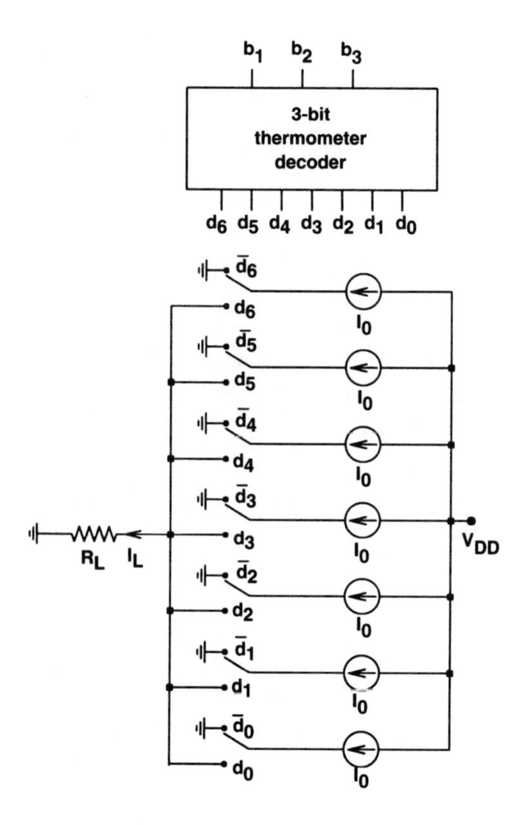

Figure 6.24. Three-bit current DAC using a thermometer decoder.

bits to $2^3 - 1 = 7$ output bits, where the number of 1's in the output code is equal to the decimal value of the binary code. Table 6.2 presents the truth table of a 3-bit thermometer-code decoder. If d_0 to d_6 are used to control the current sources of Fig. 6.24, then moving from one code to the next, one additional current source is turned on, which increases the total output current, hence guaranteeing monotonicity.

TABLE 6.2. Truth Table for 3-Bit Thermometer Decoder

b_1	b_2	b_3	d_6	d_5	d_4	d_3	d_2	d_1	d_0
0	0	0	0	0	0	0	0	0	0
0	0	1	0	0	0	0	0	0	1
0	1	0	0	0	0	0	0	1	1
0	1	1	0	0	0	0	1	1	1
1	0	0	0	0	0	1	1	1	1
1	0	1	0	0	1	1	1	1	1
1	1	0	0	1	1	1	1	1	1
1	1	1	1	1	1	1	1	1	1

The thermometer-decoding scheme also improves the glitch performance. A glitch occurs when, say, going from $3I_0$ to $4I_0$, for the output current, one set of three current sources turn off and another set of four current sources turn on. Any delay between turning the two groups on and off will result in a positive or negative glitch. This phenomenon is common in a binary-weighted current source DAC. In a thermometer decoder DAC, going from $3I_0$ to $4I_0$, the three current sources that supply $3I_0$ remain on and a fourth turns on to supply $4I_0$, eliminating any glitches.

As the number of bits increase, straight thermometer decoding becomes impractical. For more efficient implementation, a two-dimensional row–column decoding scheme can be used. For example, for a 6-bit DAC, a 3-bit row and 3-bit column decoder can be used. The DAC will consist of $2^6 - 1 = 63$ identical current sources arranged as a matrix. Figure 6.25 shows the basic architecture of the 6-bit DAC. For example, if the digital value for the 6-bit input code corresponds to a decimal number of 30, thirty current sources in the matrix are turned on and these outputs are summed to form the output current. In Fig. 6.25 the matrix consists of three types of rows: (1) rows in which all current cells are turned on, (2) rows in which all the current cells are turned off, and (3) a row in which the cells are partially turned on. Based on the three types of rows and the outputs of the row and column decoders, a logic is designed to control the individual current cells [14].

The architecture of Fig. 6.25 is such that in the actual physical layout of the

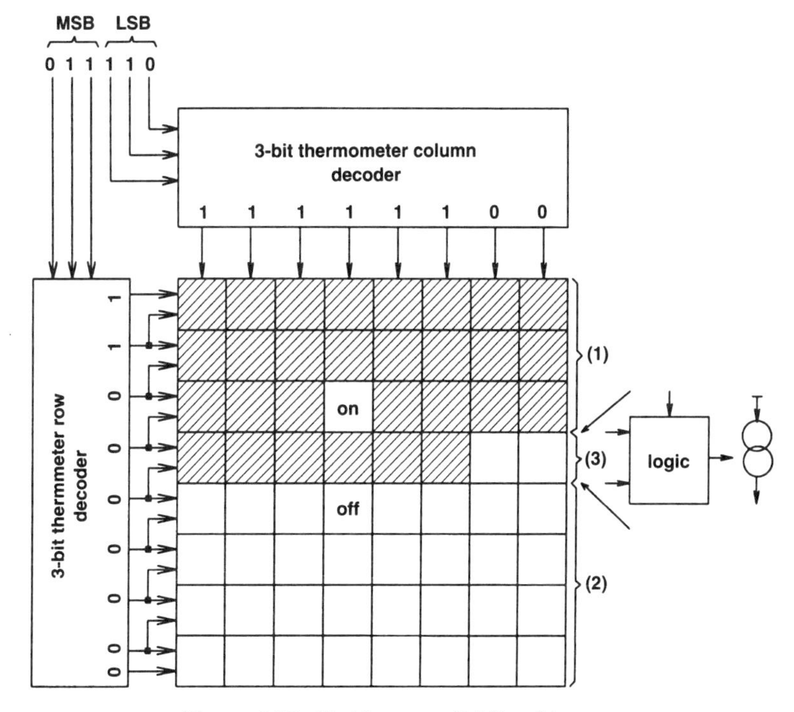

Figure 6.25. Six-bit current DAC architecture.

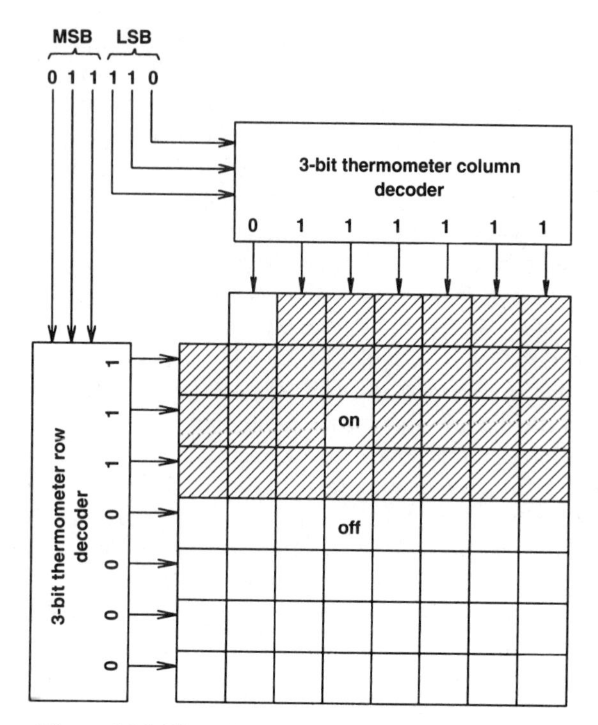

Figure 6.26. Simplified 6-bit current DAC architecture.

DAC, the control logic for each cell should be placed next to the current source. This requirement puts a limitation on the matching accuracy of the current cells and hence the linearity of the DAC. The architecture of the DAC can be greatly simplified if a slightly different decoding scheme is used. The block diagram of the simplified 6-bit DAC is shown in Fig. 6.26. It consists of one row of seven current cells and seven rows of eight current cells. The output of the column decoder controls the individual cells in the first row, and the seven outputs of the row decoder each control an entire row. In this arrangement, for a digital code corresponding to a decimal value of 30, three entire rows and six cells in the first row will turn on. As we increment the input digital code, the current cells of the first row turn on sequentially. When all seven current cells are turned on, one more increment will turn off all seven cells in the first row and turn on one entire row of eight current cells. Turning off one set of seven devices and turning on another set of eight devices may cause a slight glitch. Also, the architecture in not inherently monotonic. However, for a 6-bit DAC, monotonicity can be guaranteed as long as the DAC is 3-bit accurate. The physical layout of the DAC is very straightforward, however. Since the outputs of the thermometer decoders control the current cells directly, no additional logic is necessary and all current cells can be placed next to each other, improving the matching and hence the accuracy of the DAC.

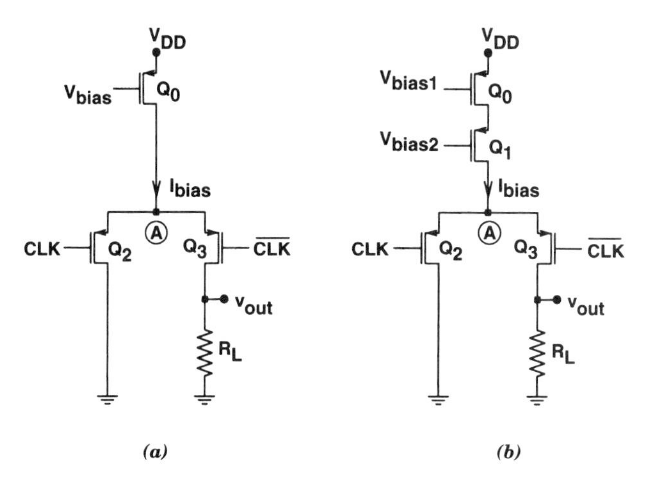

Figure 6.27. Configuration of current source: (*a*) single transistor configuration; (*b*) cascode configuration.

The individual current switch is shown in Fig. 6.27*a*. The gate of Q_0 is tied to a bias voltage such as the one shown in Fig. 6.22 and establishes the bias current. When the *Clk* signal is high Q_2 is off, Q_3 is on, and I_{bias} flows into the output load. When *Clk* goes low, Q_3 turns off, Q_2 turns on, and I_{bias} flows into ground. As shown in Fig. 6.27*a,* when the current flows into the load, the output voltage, v_{out}, modifies the voltage of node A, hence changing the drain-to-source voltage of Q_0 and consequently, the bias current. To increase the output impedance of the current source, the cascode-connected current mirror shown in Fig. 6.27*b* can be used. The output voltage swing of the current source can be improved by using the modified biasing schemes discussed in Chapter 3.

An alternative current switch with improved current regulation is shown in Fig. 6.28 [16]. In this circuit the gate of transistor Q_3 is not switched but is tied to a

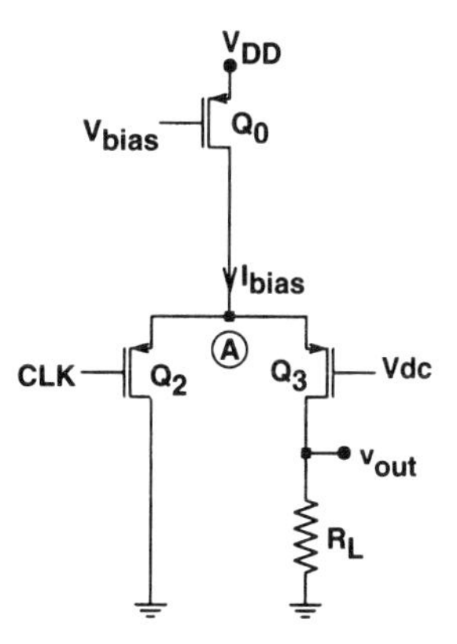

Figure 6.28. Configuration of improved current switch.

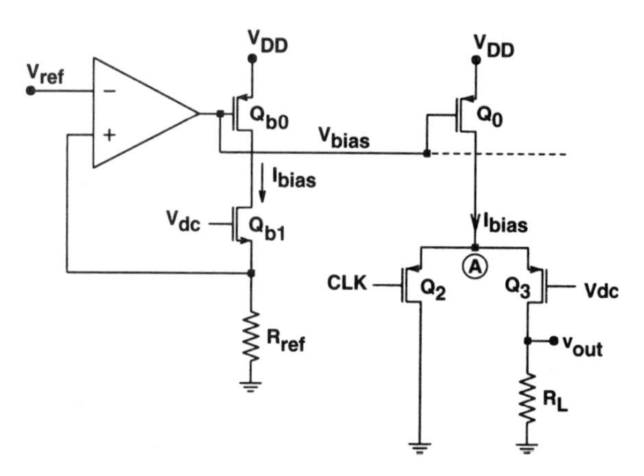

Figure 6.29. Current switch with a voltage-to-current converter.

constant voltage V_{dc}. This stage is essentially a fully switched differential pair. The current I_{bias} is steered either to ground or to the output load, depending on the polarity of the digital input signal that is applied to the gate of Q_2. In this circuit the potential of node A is independent of the potential of v_{out} and is determined by the bias voltage V_{dc}, the bias current, and the V_{GS} drop of Q_3. Since the drain-to-source voltage of Q_0 remains constant and independent of the output voltage, its drain current also remains unchanged. The simplified schematic of the voltage-to-current converter and the current switch is shown in Fig. 6.29 [17]. The transistor Q_{b1} is inserted in the feedback of the voltage-to-current converter to balance the drain-to-source voltage of all the current mirror transistors, hence equalizing their currents. Current output D/A converters using these types of switches exhibit very rapid settling times, typically on the order of 5 to 10 ns.

6.6. SEGMENTED CURRENT-MODE D/A CONVERTER STAGES

For high-resolution current-mode DACs, the methods described in Section 6.5 are not practical because the number of current elements rises exponentially and the silicon area necessary to implement the DAC becomes excessively large. A more efficient method of implementing high-resolution current DACs, similar to the voltage mode, is the segmented approach [18]. Figure 6.30 shows the basic block diagram of an N-bit segmented D/A converter. An array of M equal coarse current sources (I_u) is shown. One of the coarse current sources can be divided into more fine current levels by a passive current divider. Controlled by the value of the input digital code, a number of the coarse and fine currents are switched to the output terminal, and the remaining currents are dumped to signal ground. In the circuit of Fig. 6.30, a thermometer decoder that decodes the N_1 MSBs of the input code controls the coarse current sources. The fine current sources will be controlled by the remaining N −

N_1 LSBs. In this approach, use of the thermometer decoder guarantees monotonicity for the coarse current sources. The entire DAC is not inherently monotonic, however, because the last current source, which is divided into the fine currents, should match the other coarse current sources within one LSB of accuracy. To guarantee monotonicity, a three-way switch can be added to each coarse current source. For each code, the first $m - 1$ coarse current units are switched to the output, and coarse current unit number m is switched to the fine current divider. This method guarantees monotonic operation because the segment current selected, which is applied to the fine current divider, depends on the data input. Figure 6.31 shows the basic block diagram of an N-bit D/A converter. The N-bit input digital code is divided into C-*bit* coarse (MSBs) and F-*bit* fine (LSBs) codes. The C-*bit* MSBs are decoded by a binary-to-thermometer decoder that controls the coarse current sources. The remaining F bits directly control a binary-weighted current divider. The C coarse bits represent values from zero to $2^C - 1$. Therefore, 2^C unit coarse current sources are required, including one for the segmentation current. Figure 6.32 shows the output current of the converter as a function of the input code. Assume that point A on the graph represents the analog value corresponding to the input digital word. The unit coarse current sources 1 through $m - 1$ (of 2^C available unit current sources) are switched to the output line ($I_{0,\,\mathrm{coarse}}$) controlled by the MSBs, while the unit current source m denoted by I_{seg}, is divided into the fine current levels by a binary-weighted current divider and is switched to the output current line controlled by the LSBs. The output currents of the coarse network and fine current divider are added to form the total output current, expressed as

$$i_{\mathrm{out}} = \left[(m - 1) + D_f \frac{I_m}{2^F}\right] I_u, \tag{6.18}$$

where D_f represents the decimal value corresponding to the F LSBs and I_{u} is the unit coarse current source. An example of a 7-bit segmented DAC is shown in Fig. 6.33 where $C = 3$ and $F = 4$. There are a total of $2^3 = 8$ unit coarse currents and a 4-bit binary-weighted divider. The reference voltage V_{ref1} is used to bias the unit

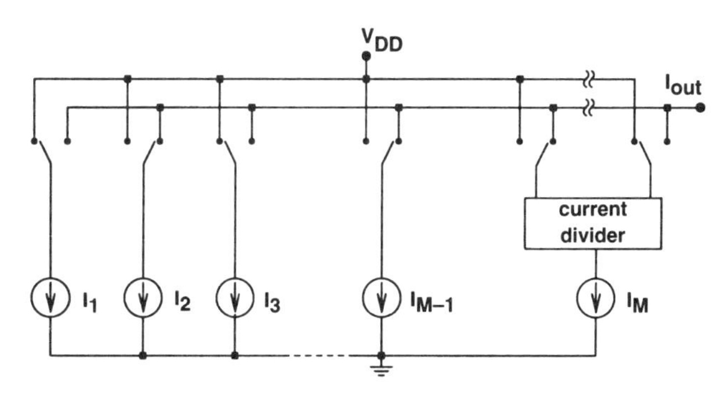

Figure 6.30. Basic block diagram of a segmented D/A converter.

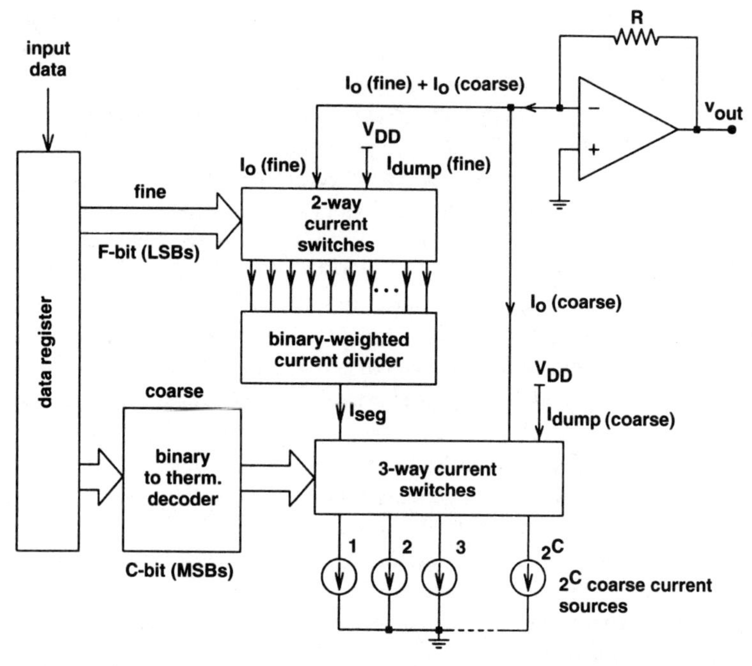

Figure 6.31. Basic block diagram of a segmented N-bit inherently monotonic current DAC.

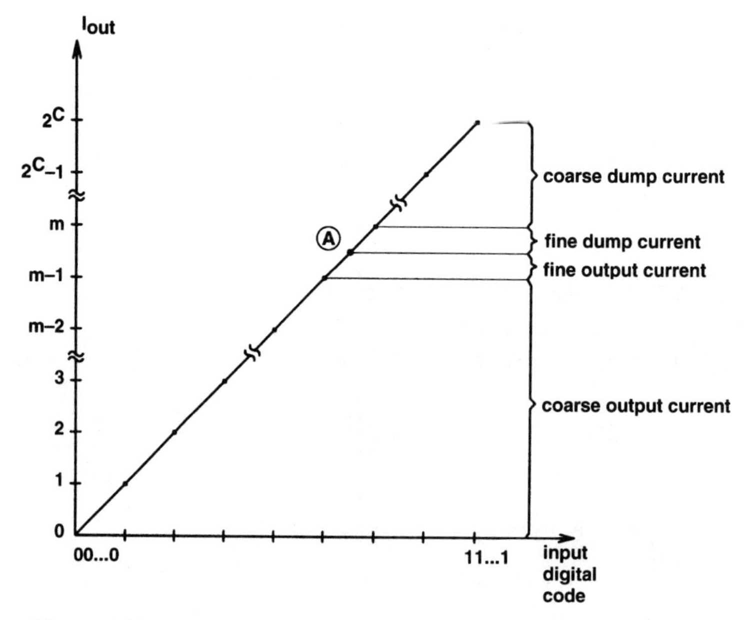

Figure 6.32. Segmented D/A output current as a function of input code.

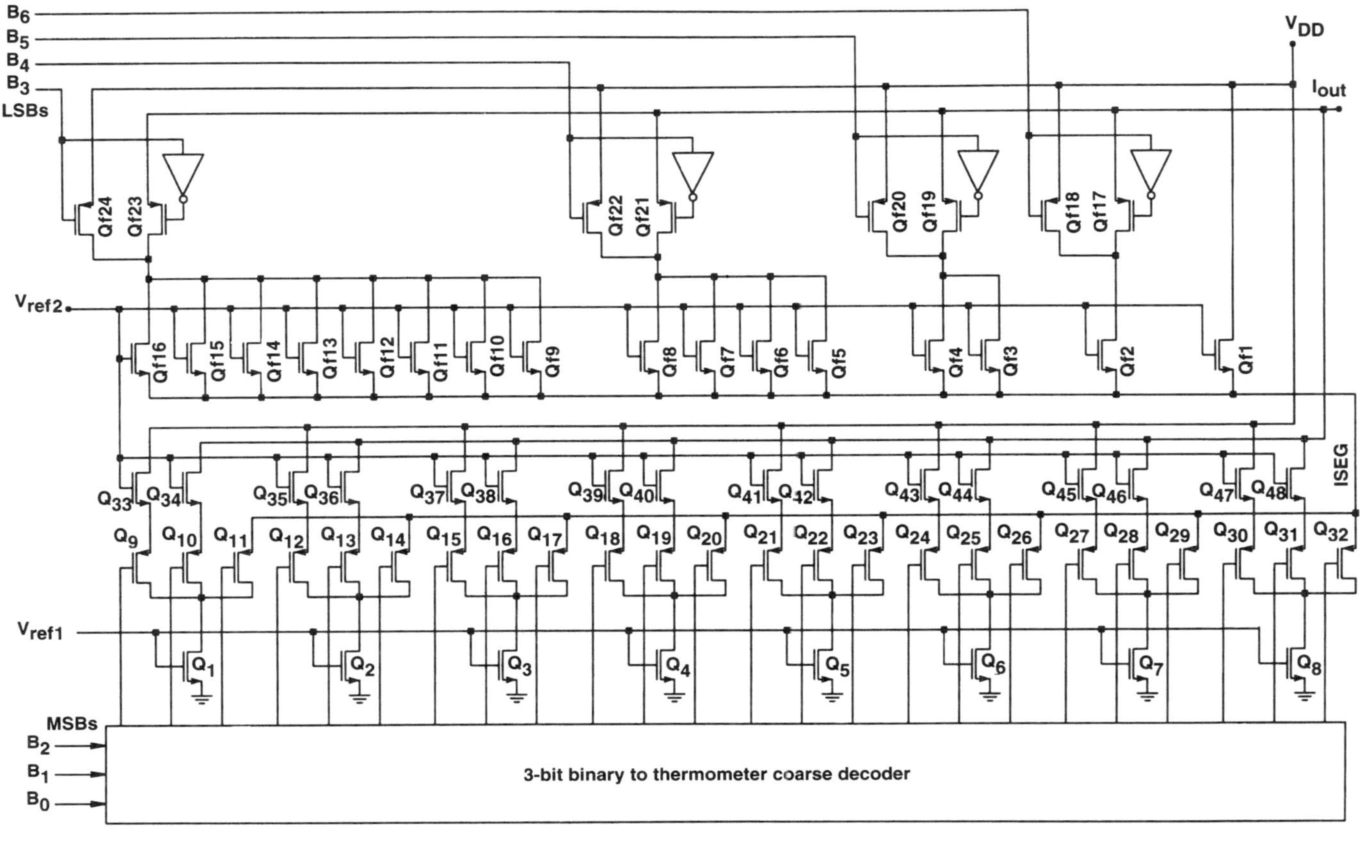

Figure 6.33. Seven-bit segmented current DAC.

coarse current transistors Q_1 to Q_8. The binary-to-thermometer decoder outputs control the gates of the three-way p-type MOS switches Q_9 to Q_{32} that operate in the linear region. Cascode devices Q_{33} to Q_{48} are added to improve the accuracy of the coarse currents. This is achieved by equalizing the potentials of the switches and the segment current line, I_{seg}, which is connected to the 4-bit binary-weighted current divider. The basic current divider consists of 16 equal-sized common-gate and common-source transistors Q_{f1} to Q_{f16}. The individual drains are combined in binary-weighted numbers, 1, 2, 4, and 8. The output current is controlled by a two-way switch, which consists of p-type MOS transistors Q_{f17} to Q_{f24}. The four LSBs directly control the gates of these transistors.

The accuracy of the segmented DAC is determined largely by the matching of the coarse unit current elements. Symmetrical layout techniques for the MOS transistors of the coarse current sources can improve the accuracy. However, the achievable precision based upon matching of components in a standard process is not sufficient. Therefore, additional calibration techniques are used to achieve high-resolution converters. Use of dynamically matched current sources is one of the self-calibration techniques that can be applied to the segmented current DAC of Fig. 6.30 to achieve well-matched current sources and hence a high-precision D/A converter [19]. To accomplish this, each unit coarse current source in Fig. 6.30 is continuously calibrated by a reference current I_{ref} in such a way that all coarse elements are matched precisely. Before describing the complete process, the calibration principle for one single current source will be explained.

The basic calibrated current cell is shown in Fig. 6.34. During the calibration cycle, the signal CAL goes to a high state, transistors Q_3 and Q_2 turn on, and Q_4 turns off. Consequently, the reference current I_{ref} is forced to flow through the diode-connected NMOS device Q_1 and establishes a voltage V_{gs} across its gate-to-source capacitance C_{gs}. The dimensions and parameters of the transistor determine the magnitude of this voltage. When CAL turns low, the calibration process is complete. Transistors Q_2 and Q_3 turn off and Q_4 turns on and the gate-to-source voltage V_{gs} of Q_1 remains stored on C_{gs}. Provided that the drain voltage of Q_1 also remains

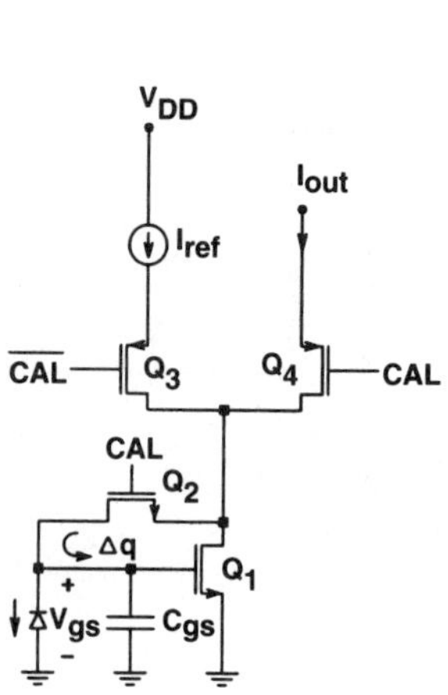

Figure 6.34. Calibration circuit for a single current cell.

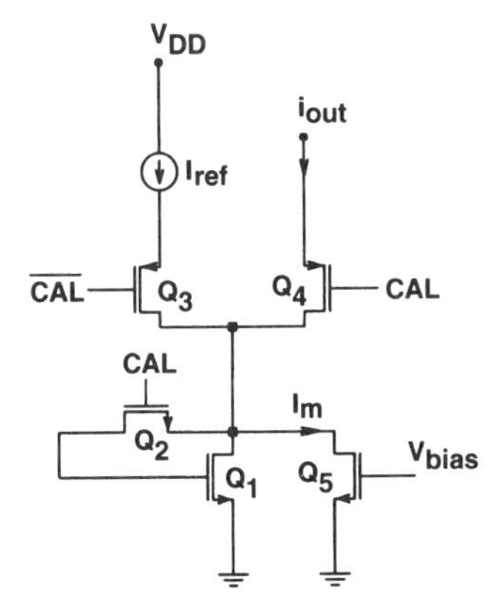

Figure 6.35. Improved calibration circuit for a single current cell.

unchanged, its drain current will still be equal to I_{ref}. This current is now available at the i_{out} terminal and I_{ref} is no longer needed for this current source.

Two nonideal effects degrade the calibration accuracy of the current cell. These effects, shown in Fig. 6.34, are the channel charge of Q_2 which when turned off is partially dumped on the gate of Q_1, and the leakage current of the reverse-biased source-to-substrate diode of Q_2. Both effects alter the charge that is stored on the gate-to-source capacitor C_{gs} and hence modify the drain current. It can be shown that decreasing the transconductance (g_m) of Q_1 can reduce the impact of both nonideal effects on the output (drain) current [19].

The calibration circuit of Fig. 6.34 can be modified by adding a fixed current source in parallel to the current-source transistor Q_1. The modified circuit is shown in Fig. 6.35, and the additional current source is represented by transistor Q_5, with its gate tied to a fixed bias voltage, V_{bias}. The added current source has a value of about 90% of the reference current I_{ref}. This reduces the value of Q_1's current to about 10% of its original value and decreases its transconductance by a factor of $\sqrt{10}$. Furthermore, since the current of Q_1 is lower, its W/L ratio can now be reduced by increasing the length L. In this way it is possible to reduce the transconductance of Q_1 further by a factor of 8 to 10 while increasing its C_{gs}.

For the calibration technique to be suitable for the DAC of Fig. 6.30, the principle must be extended to an array of current sources. A system that uses the continuous calibration technique for an array of current cells is shown in Fig. 6.36. The principle is characterized by using $N + 1$ current cells to generate N equal current sources. The current cell number $N + 1$ is the spare cell. An $(N + 1)$-bit shift register controls the selection of the cell to be calibrated. One output of the $(N + 1)$-bit shift register is a logic 1, while the other outputs are all zero. The cell corresponding to the register with the logic level 1 is selected for calibration and is connected to the reference current. Because this cell is now not delivering any current to its output terminal, the output current of the spare cell is switched to this terminal. After completion of a calibration cycle, the contents of the shift register is shifted by one place, and the next cell in the array is selected for calibration. This way, every current cell is sequentially calibrated and inserted back into the array. The switching network is responsible for taking the current source selected out of the array for calibration and replacing it with the spare cell. Since the output of the shift register is connected to its input, after all cells are calibrated sequentially, the first cell is calibrated again. By using one spare cell, no time is lost during the calibration period and there are always N equal current sources available at the output terminals. At this point it is worth mentioning that the purpose of calibration is not to make the value of the current sources precisely equal to I_{ref} but to make the current sources match accurately.

The coarse current array in the DAC of Fig. 6.31 can be replaced with the calibrated current array of Fig. 6.36. The basic block diagram of a 16-bit DAC is shown in Fig. 6.37. The DAC is segmented into 6-bit coarse and 10-bit fine DACs. The coarse current array now consists of 65 calibrated current cells shown in Fig. 6.35. The current outputs of 63 normally functioning cells are connected to 63 two-way current switches, and one cell is connected directly to the 10-bit binary-weighted current divider. The nonfunctioning current cell is connected to the reference current for calibration. In this arrangement since all current cells are calibrated, a unique current cell is dedicated to the fine current divider, unlike the DAC of Fig. 6.33. A simplified version of the calibration circuitry and the current cell is shown in Fig. 6.38. Of the 65 current cells, 63 supply currents to the coarse DAC, one supplies current to the fine current divider, and one cell is being calibrated. For a normally functioning cell, transistors Q_2, Q_5, and Q_7 are off and the current source comprised of transistors Q_1 and Q_4 supplies the current to the output terminal through the "on" device Q_6. For the cell that is being calibrated, devices Q_2, Q_5, and Q_7 are on and Q_6 is off. Since the cell is not operational, the spare cell is switched to the corresponding output terminal through device Q_7. Notice that terminals A_n and B_n of all 65 current cells are connected to nodes A and B of the calibration circuitry. For the cell selected, the reference current I_{ref} flows into Q_1 through Q_5, and the loop between the drain and gate of Q_1 is closed by the three transistors Q_5, Q_8, and Q_2. This process charges the gate of Q_1 to an appropriate voltage required for maintaining a

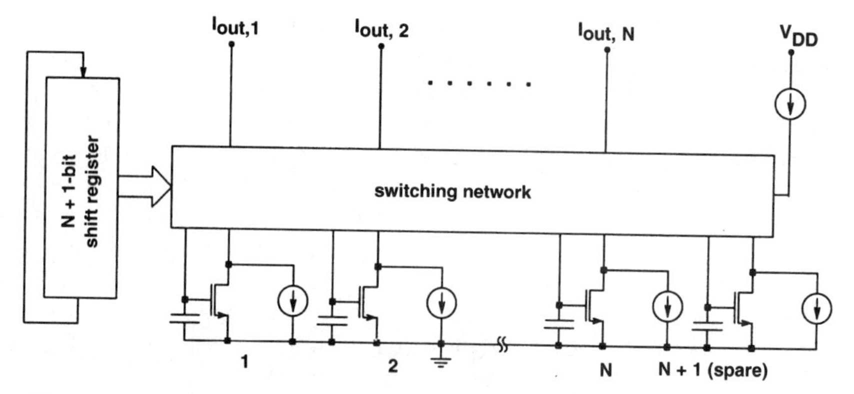

Figure 6.36. Block diagram of a continuously calibrated array of N current sources.

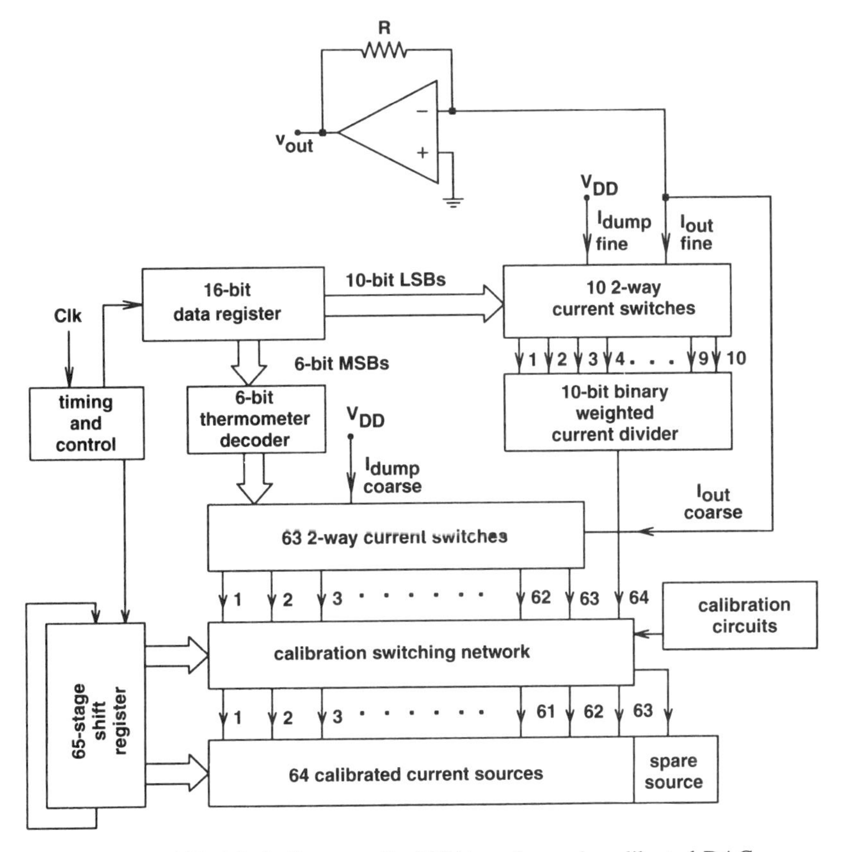

Figure 6.37. Block diagram of a 16-bit continuously calibrated DAC.

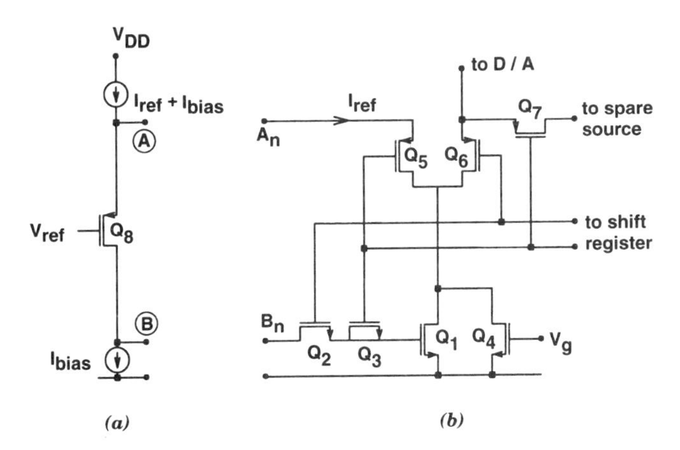

Figure 6.38. (*a*) Calibration circuitry; (*b*) current cell.

drain current of I_{ref}. At the end of the calibration period the cell returns to its normal operation and the next cell in the array becomes calibrated. In Fig. 6.38, transistor Q_3 has been added for channel charge cancellation. The gates of Q_2 and Q_3 are connected to the two opposite phases of the control clock. The transistors are identical except that Q_3 has half the channel width of Q_2. The charge transferred from the control signal to C from Q_2 during the falling edge of the clock is canceled by the charge transferred to C by Q_3 during the rising edge of the opposite clock.

Sixteen-bit DACs using the continuous calibration technique described in this section have achieved 0.0025% linearity at a power dissipation of 20 mW and a minimum power supply of 3 V [19].

PROBLEMS

6.1. What is the necessary relative accuracy of the resistor ratios in the 8-bit version of the resistive DAC of Fig. 6.4 to achieve 8-bit linearity?

6.2. Prove Eq. (6.4) for the folded resistor-divider DAC of Fig. 6.6.

6.3. Analyze the circuit of Fig. 6.13. Describe $v_o(nT)$ in terms of $v_{in}(nT)$. Assume first infinite, then finite op-amp gain.

6.4. Figure 6.39 shows an offset-compensated voltage amplifier that does not require a high-slew-rate op-amp [9]. Analyze the circuit for both choices (shown in parentheses and without parentheses) of the input-branch clock phases. How much does v_{out} vary between the two intervals ϕ_1 = "1" and ϕ_2 = "1"? Plot the output voltage v_{out} for both choices.

6.5. Calculate the total capacitance loading node X in an n-bit PCA as shown in Fig. 6.16.

6.6. What is the necessary relative accuracy of the capacitor ratios in the charge-mode D/A converter of Fig. 6.17 to achieve 11-bit linearity?

6.7. Design the two-stage cascaded D/A converter of Fig. 6.19 for 12-bit resolution. Determine the number of the bits in the first and second stages so that the total capacitance is minimized.

6.8. Design a 10-bit hybrid D/A converter using a charge-mode MSB and resistor-mode LSB structure (Fig. 6.20). Determine the relative accuracy of the capacitor and resistor ratios to achieve 10-bit linearity.

6.9. Derive an expression for the number of unit capacitors, resistors, and switches for the N-bit DAC of Fig. 6.21. Assume that $N = N_1 + N_2$, where N_1 is the number of LSBs assigned to the resistive DAC and N_2 is the number of MSBs assigned to the capacitive DAC.

6.10. Design an 8-bit current DAC according to the architecture of Fig. 6.25. Use

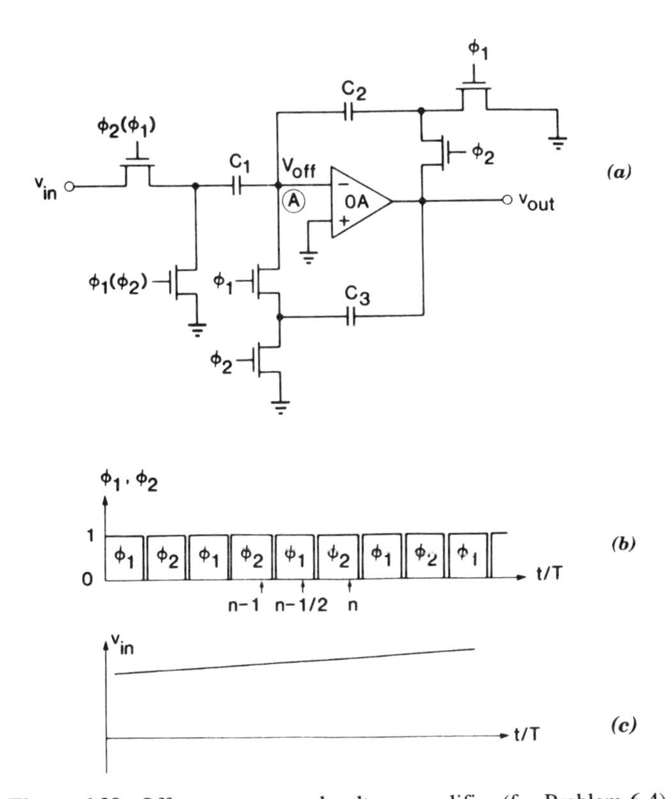

Figure 6.39. Offset-compensated voltage amplifier (for Problem 6.4).

an equal number of bits for the columns and rows. Design the column and row decoders and the current cell logic.

6.11. Repeat Problem 6.10 for the architecture of Fig. 6.26. Design the current reference and the unit currents in such a way that the full-scale output current generates 1 V of peak voltage across a 75-Ω load resistor.

6.12. For the circuit of Fig. 6.29, plot the waveform at node A when *Clk* goes from low to high and high to low. Assume that the low level is 0 V, the high level is V_{DD} (the positive supply voltage), and the *p*-channel threshold voltage is $V_{DD}/5$.

6.13. Design the 10-bit version of the segmented DAC of Fig. 6.31. Use 4 bits for the coarse current and 6 bits for the fine current DACs. If the feedback resistor of the current-to-voltage converter is 1 kΩ, find the value of the full-scale current if the maximum output voltage is 2 V. What is the value of each coarse current source and the current corresponding to one LSB?

6.14. Figure 6.31 shows a unipolar segmented current DAC where the output voltage varies between 0 V and $V_{FS} = RI_{FS}$ and I_{FS} is the full-scale output current.

In a single-supply application, assume that the positive input of the op-amp is connected to $V_{DD}/2$ (V_{DD} is the positive supply voltage). Modify the circuit of Fig. 6.31 so that the voltage output varies between $\frac{1}{4}V_{DD}$ (zero input code) and $\frac{3}{4}V_{DD}$ (full-scale input code). Assume that the DAC is 8 bits, $V_{DD} = 5$ V, and $R = 1\ \text{k}\Omega$. (*Hint:* Connect a fixed current source to the inverting input of the op-amp.)

REFERENCES

1. A. R. Hamadé, *IEEE J. Solid-State Circuits, SC-13*(6), 785–791 (1978).
2. A. Dingwall and V. Zazzu, *IEEE J. Solid-State Circuits, SC-20*(6), 1138–1143 (1983).
3. M. J. M. Pelgrom, *IEEE J. Solid-State Circuits, SC-25*(6), 1347–1352 (1990).
4. A. Abrial et al., *IEEE J. Solid-State Circuits, SC-23,* 1358–1369 (1988).
5. P. Holloway, *ISSCC Dig. Tech. Pap.*, pp. 66–67, Feb. 1984.
6. Y. A. Hague, R. Gregorian, R. W. Blasco, R. A. Mao, and W. E. Nicholson, *IEEE J. Solid-State Circuits, SC-14*(6), 961–969 (1979).
7. R. Gregorian, *Microelectron. J., 12,* 10–13 (1981).
8. R. Gregorian, K. Martin, and G. C. Temes, *Proc. IEEE, 71,* 941–966 (1983).
9. K. Huang, G. C. Temes, and K. Martin, *Proc. Int. Symp. Circuits Syst.*, pp. 1054–1057, 1984.
10. J. L. McCreary and P. R. Gray, *IEEE J. Solid-State Circuits, SC-10*(6), 371–379 (1975).
11. R. Gregorian and G. Amir, *Proc. Int. Symp. Circuits Syst.*, pp. 733–736, 1981.
12. K. W. Martin, L. Ozcolak, Y. S. Lee, and G. C. Temes, *IEEE J. Solid-State Circuits, SC-22*(1), 104–106 (1987).
13. B. Fotouhi and D. A. Hodges, *IEEE J. Solid-State Circuits, SC-14*(6), 920–926 (1979).
14. T. Miki, Y. Nakamura, M. Nakaya, S. Asai, Y. Akasaka, and Y. Horiba, *IEEE J. Solid-State Circuits, SC-21*(6), 983–988 (1986).
15. L. Letham, B. K. Ahuja, K. N. Quader, R. J. Mayer, R. E. Larson, and G. R. Canepa, *IEEE J. Solid-State Circuits, SC-22*(6), 1041–1047 (1987).
16. A. B. Grebene, *Bipolar and MOS Analog Integrated Circuit Design,* Wiley, New York, 1984.
17. D. A. Johns and K. Martin, *Analog Integrated Circuit Design,* Wiley, New York, 1997.
18. H. J. Schouwenaers, D. W. J. Greeneveld, and H. A. H. Tremeer, *IEEE J. Solid-State Circuits,* 23(6), 1290–1297 (1988).
19. C. A. A. Bastiaansen, *IEEE J. Solid-State Circuits, 24*(6), 1517–1522 (1989).
20. S. R. Norsworthy, R. Schreier, and G. C. Temes, *Delta-Sigma Data Converters Theory, Design, and Simulation,* IEEE Press, Piscataway, NJ, 1997.

CHAPTER 7

ANALOG-TO-DIGITAL CONVERTERS

The analog-to-digital converter (usually abbreviated ADC or A/D converter) is an essential building block in many digital signal-processing systems. It provides a link between the digital signal processor and the analog signals of a transducer. The A/D converter is considered to be an encoding device, where it converts an analog sample into a digital quantity with a prescribed number of bits. Numerous types of A/D converters have been designed for a wide variety of applications. The type of the application largely determines the choice of the A/D conversion technique. From the viewpoint of the implementation, analog-to-digital converters typically contain one or more comparators, switches, passive precision components, a precise voltage reference and digital control logic.

In this chapter the basic principles and performance metrics of the A/D converters are presented first. Following that, several types of Nyquist-rate A/D conversion techniques are examined, and their implementation in the CMOS technology is discussed.

7.1. ANALOG-TO-DIGITAL CONVERSION: BASIC PRINCIPLES

As was the case with Nyquist-rate digital-to-analog converters, there are a variety of algorithms and realizations available for analog-to-digital converters offering different advantages and disadvantages. The trade-offs between the conversion accuracy, speed, and economy (the latter measured by circuit complexity, chip area, power dissipation, etc.) offered by these options vary widely. As will be seen, practical converters exist for signal bandwidths ranging from 1 Hz to 5 GHz, with resolutions anywhere between a few bits to 24 bits. Different applications obviously require different parameters; Fig. 7.1 illustrates the approximate range of requirements for

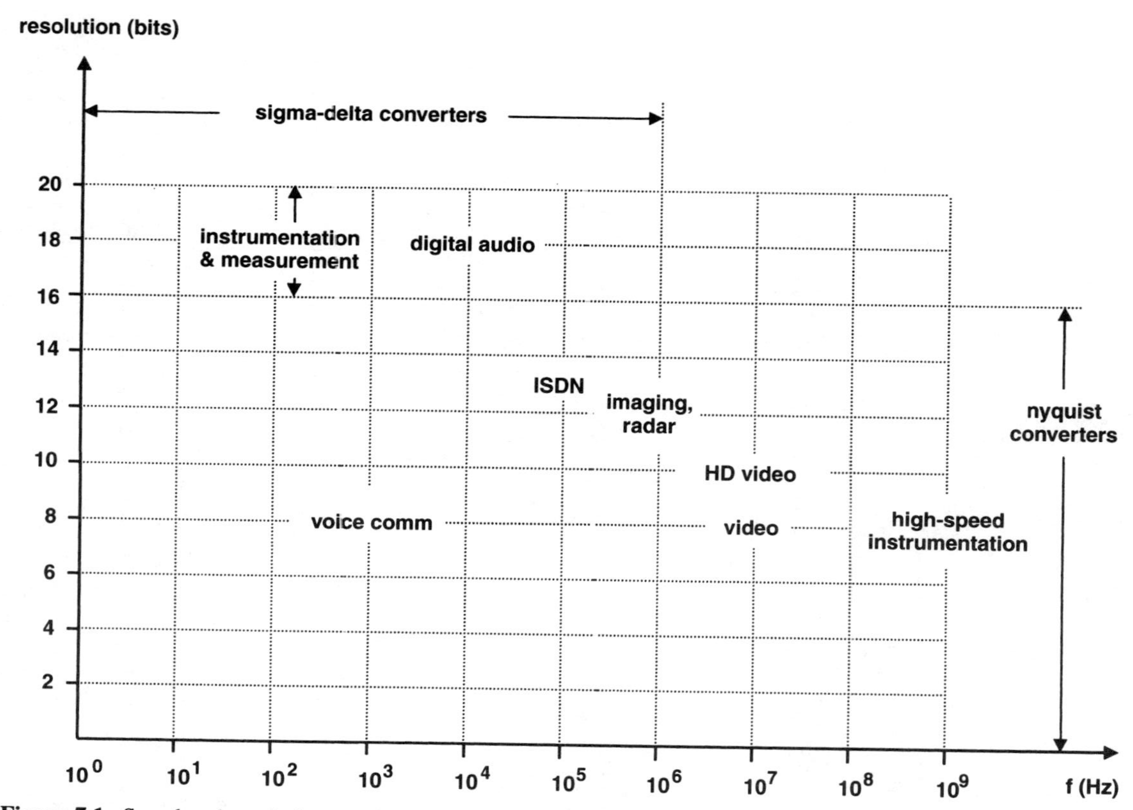

Figure 7.1. Speed and resolution requirements on ADCs in various applications.

some common systems containing ADCs. As we did for DACs, we classify Nyquist-rate ADCs according to their conversion speed into three categories:

1. *High-Speed ADCs.* In these devices the analog samples to be converted can be entered at a rate equal to the clock rate, or half the clock rate. Thus the throughput rate of data equals the clock rate, or half of it. There may, however, be a long constant delay (latency) between the time that the analog sample enters the converter and when its digital replica appears at the output. Typical examples of such high-speed ADCs are the flash, interpolating, two-step (or half-flash), pipeline, and time-interleaved converters. Such converters can achieve conversion rates in the range of 0.5 megasamples/second (MS/s) to 10 GS/s. Their accuracy ranges from 8 to 12 bits. Typical applications include video, imaging, and radar systems.
2. *Medium-Speed ADCs.* For an N-bit ADC, such converters require N clock periods for each analog sample. Thus their throughput is N times slower than the clock rate. Typical realizations include the various serial (successive-approximation) converters. These converters usually achieve a 10- to 14-bit resolution; their conversion speed may be in the range 0.1 to 5 MS/s. Their applications include telecommunication, control, and low- to medium-speed measurement systems.
3. *Low-Speed Converters.* For a resolution of N bits, these devices require approximately 2^N clock periods to convert an analog input sample. This can lead to very slow operation; for example, if $N = 16$, 65,536 clock periods are needed for each conversion. Clearly, these converters can only be used for constant, or very slowly varying signals. Converters in this categories include the various integrating and counting circuits, such as single- and dual-ramp converters. Typical accuracies range from 15 to 24 bits; typical applications include digital panel instruments, such as digital voltmeters and biomedical measurement instruments.

We discuss the most common A/D converter types and their key properties later in the chapter.

Next, the key parameters that characterize an A/D converter are discussed. The block diagram of an ADC is shown in Fig. 7.2. The analog input (typically, a voltage v_{in}), is normalized to a (voltage) reference V_{ref} and their ratio is converted into an N-bit digital output word B_{out} containing $b_1, b_2, \ldots, b_N$. Under ideal conditions, ignoring noise and any imperfections of the components, the relation between the three quantities is

$$B_{out} = b_1 \cdot 2^{-1} + b_2 \cdot 2^{-2} + \cdots + b_N \cdot 2^{-N} = \frac{v_{in} + v_q}{V_{ref}}. \tag{7.1}$$

Here v_q is the quantization error due to the finite number N of bits used in the conversion. This error is inherent in the process and can only be reduced by increas-

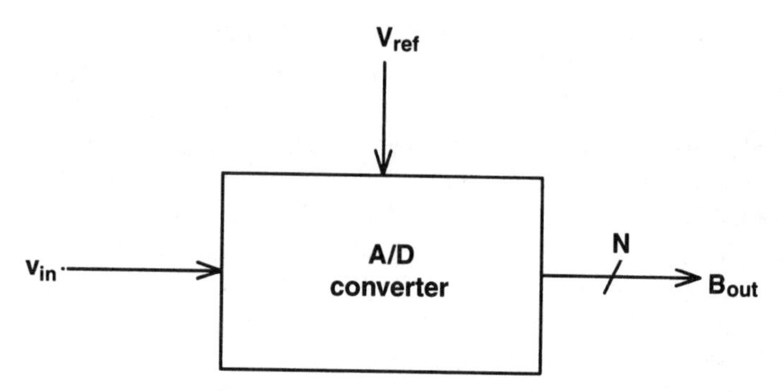

Figure 7.2. Block diagram of an analog-to-digital converter.

ing N or by reducing V_{ref}. The output–input characteristic of the conversion is illustrated in Fig. 7.3 for $N = 2$; the quantization error v_q is the difference between the solid staircase curve (the actual characteristic) and the dashed line (which represents the ideal curve for infinite N). We shall define the least-significant-bit (LSB) voltage $V_{lsb} = V_{ref}/2^N$. Here $V_{lsb} = V_{ref}/4$ for $N = 2$. Then, as Fig. 7.3 shows, the magnitude of v_q cannot exceed $V_{lsb}/2$ as long as v_{in} remains in the range 0 to $V_{ref} - V_{lsb}/2$. This is called the *linear conversion range* of the ADC; for v_{in} values outside this range, the converter overloads and the absolute value of v_q is not bounded

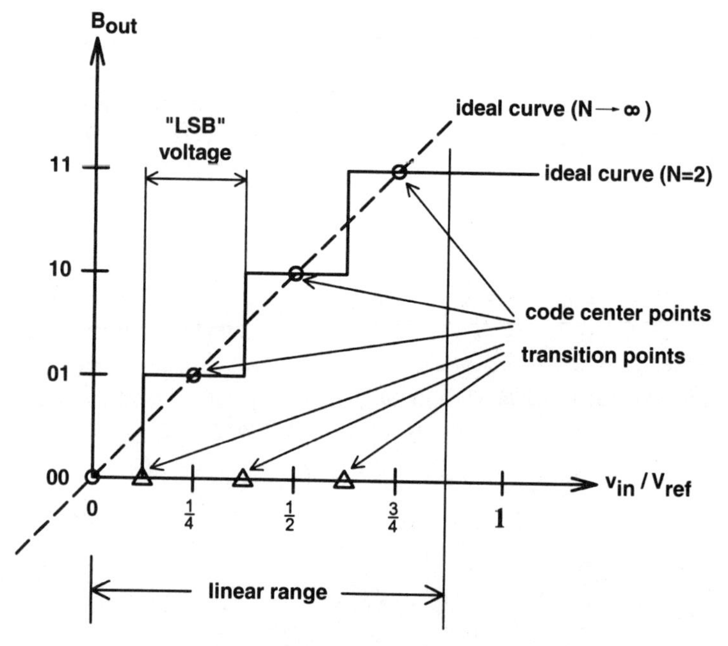

Figure 7.3. Input–output characteristics (transfer curve) for a 2-bit ADC.

by $V_{\text{lsb}}/2$. For reasonably large values of N, V_{ref} can approximate the upper boundary of the linear range. The figure also shows the transition (or threshold) voltages. These are the input voltage values where B_{out} changes its value; for our example, they are at $V_{\text{ref}}/8$, $3V_{\text{ref}}/8$, and $5V_{\text{ref}}/8$. Also, the code center points, located halfway between the transition voltages, can be observed. These are located on the ideal (infinite N) line characteristic.

For a randomly varying input signal $v_{\text{in}}(t)$ that stays within the linear range, we may often assume that the quantization error v_q is a random noise with a zero mean value and that the instantaneous value of v_q is a random variable with a uniform distribution probability between $-V_{\text{lsb}}/2$ and $+V_{\text{lsb}}/2$. Then its rms value can readily be found (Problem 7.1) to be

$$v_{q,\text{ms}} = \frac{V_{\text{lsb}}}{\sqrt{12}}. \tag{7.2}$$

The power of the quantization noise v_q is therefore

$$P_q = \frac{V_{\text{lsb}}^2}{12} = \frac{V_{\text{ref}}^2}{12 \times 2^{2N}}. \tag{7.3}$$

To illustrate the validity of this approach, Fig. 7.4 shows the spectrum of a *sine-wave* signal after an 8-bit quantization. The tall line at $\omega/\omega_c = 0.24$ represents

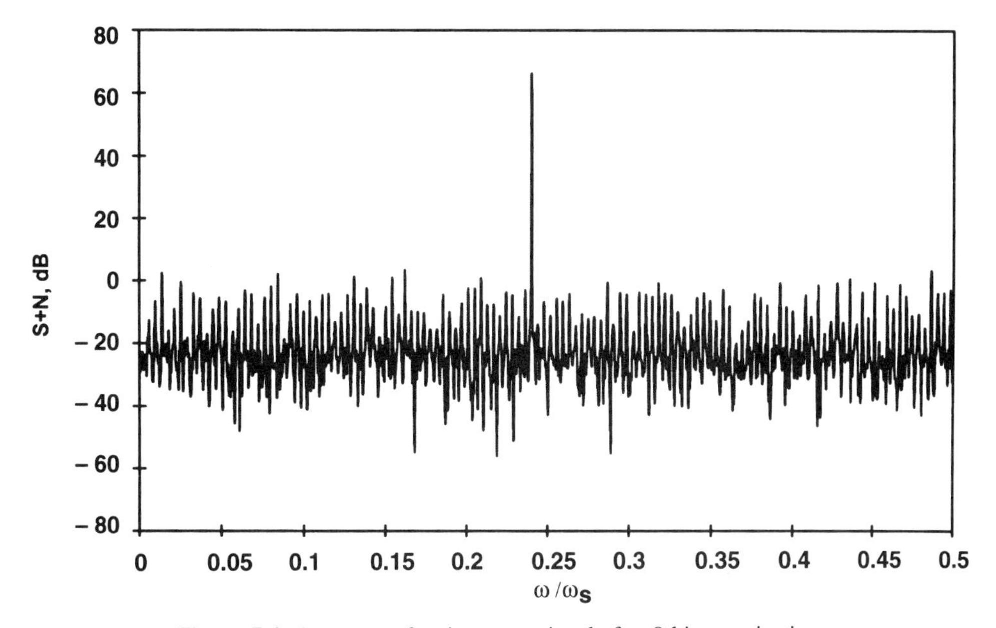

Figure 7.4. Spectrum of a sine-wave signal after 8-bit quantization.

the spectrum of the original sinusoid, and the random *white-noise-like spectrum* everywhere else belongs to v_q. Clearly, in this case the approximation of v_q by a white noise is reasonably accurate.

Next, let the input signal v_{in} be a *sine wave* with a peak-to-peak amplitude V_{ref}. This is the maximum input amplitude that the ADC can handle without overloading. The power of the signal v_{in} is, then, $P_{\text{in}} = V_{\text{ref}}^2/8$. Here we ignored the unimportant dc power due to the dc bias needed to center the input signal in the linear range of the ADC. Therefore, using the noise model of v_q, the signal-to-quantization-noise ratio of the N-bit ADC is

$$\text{SNR} = 10 \log \frac{P_{\text{in}}}{P_q} = 10 \log \frac{12 \times 2^{2N}}{8} = 6.02N + 1.761 \text{ dB}. \tag{7.4}$$

A similar calculation performed for a *sawtooth* input signal gives an SNR of $6.02N$ dB (Problem 7.2). Thus the SNR of an N-bit A/D converter with a given reference voltage V_{ref} is limited by quantization noise to about $6N$ dB; the SNR is increased by 6 dB for every added bit, since each extra bit reduces the amplitude of v_q by a factor of 2.

In practice, the ideal conversion curve of Fig. 7.3 cannot be achieved; the threshold voltages that are nominally at odd multiples of $V_{\text{lsb}}/2$ will in fact occur at different v_{in} values, and hence various errors will appear. Some of the commonly used error definitions are illustrated in Figs. 7.5 to 7.8. Figure 7.5 shows a transfer characteristic with offset error: the ideal converter curve is horizontally shifted. The offset error

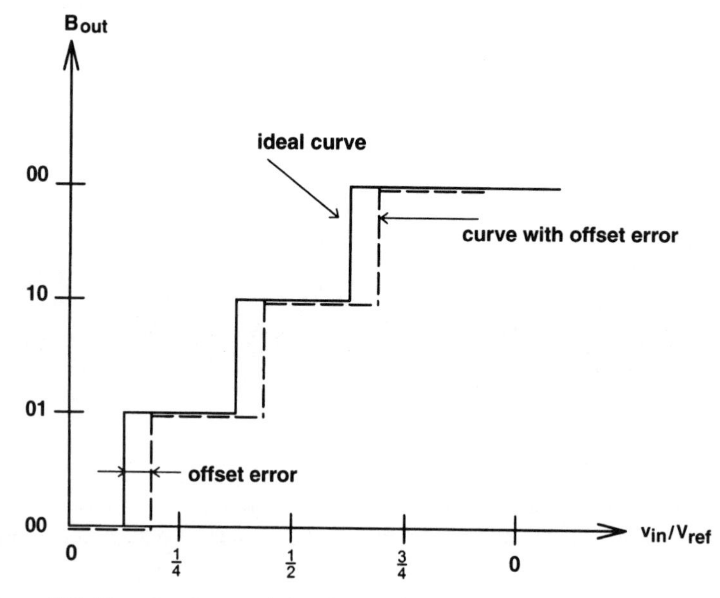

Figure 7.5. Transfer characteristics for a 2-bit ADC with and without offset error.

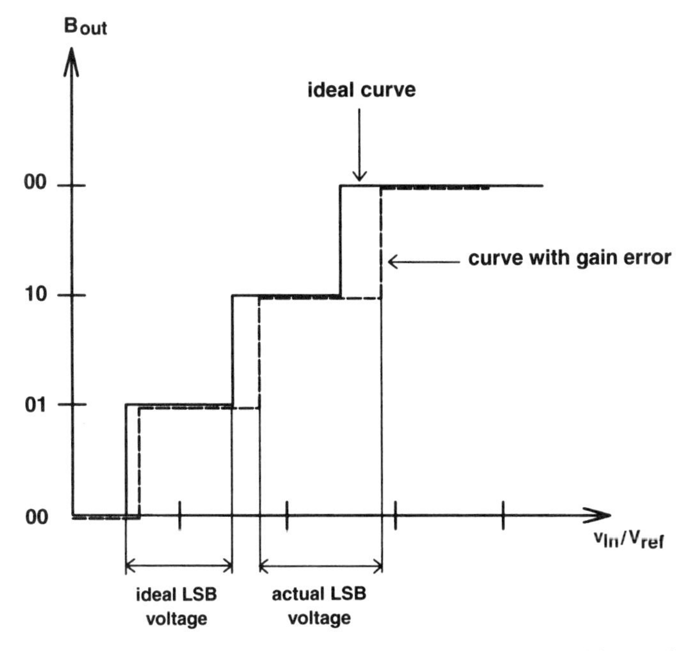

Figure 7.6. Transfer characteristics for a 2-bit ADC with and without gain error.

is simply the amount of shift. The gain error is illustrated in Fig. 7.6; the threshold voltages remain equally spaced, but the spacing is no longer the correct V_{lsb}. Both offset and gain errors are linear errors; they do not distort the input signal, only shift or scale it. Thus for a sine-wave input the quantization error remains a random-noise-like signal, and no harmonics of the input signal are generated. Hence such errors can usually be accepted and/or compensated elsewhere in the system. A more serious distortion results from the unavoidable unequal spacing of the threshold voltages, which causes nonlinearity errors. Two such nonlinear characteristics are shown in Figs. 7.7 and 7.8. In the former, the nonlinearity error is relatively small, and over the complete range of v_{in} the ADC generates every output code. In the latter, the distortion due to nonuniform variation of the threshold voltages is excessive, and one output code (010) cannot be generated at all. This effect, called *missing-code error,* is normally unacceptable in a practical ADC. As was the case for DACs, the nonlinearity errors of ADCs are usually quantified by the values of their *integral nonlinearity error* (INL) and *differential nonlinearity* (DNL). In ADCs, the INL is defined as the largest vertical difference (expressed in LSBs) between the code center points of the actual characteristic curve and the line connecting the endpoints on the curve (Fig. 7.9). The DNL is defined as the largest deviation between the actual differences of two adjacent threshold voltages and the ideal difference value (V_{lsb}), as shown in the ADC characteristics of Fig. 7.9. Here the largest error occurs when the transition-level difference equals d, and hence the INL expressed in LSBs is $(d - V_{lsb})/V_{lsb}$.

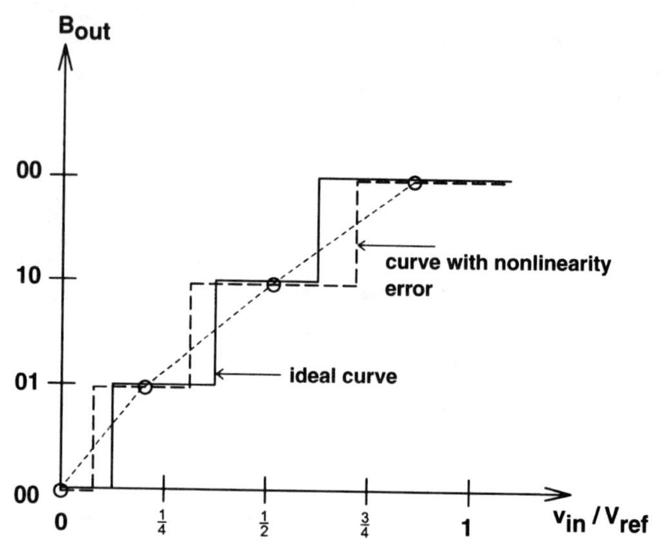

Figure 7.7. Transfer characteristics for a 2-bit ADC with and without nonlinearity.

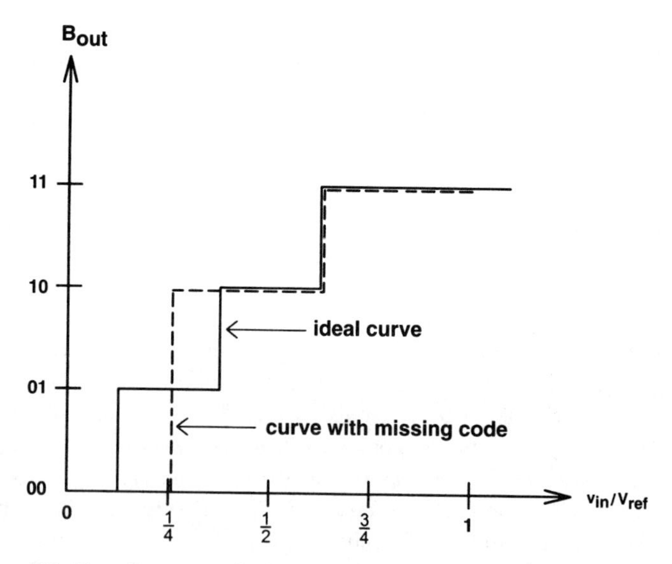

Figure 7.8. Transfer curve of a 2-bit ADC with and without missing-code error.

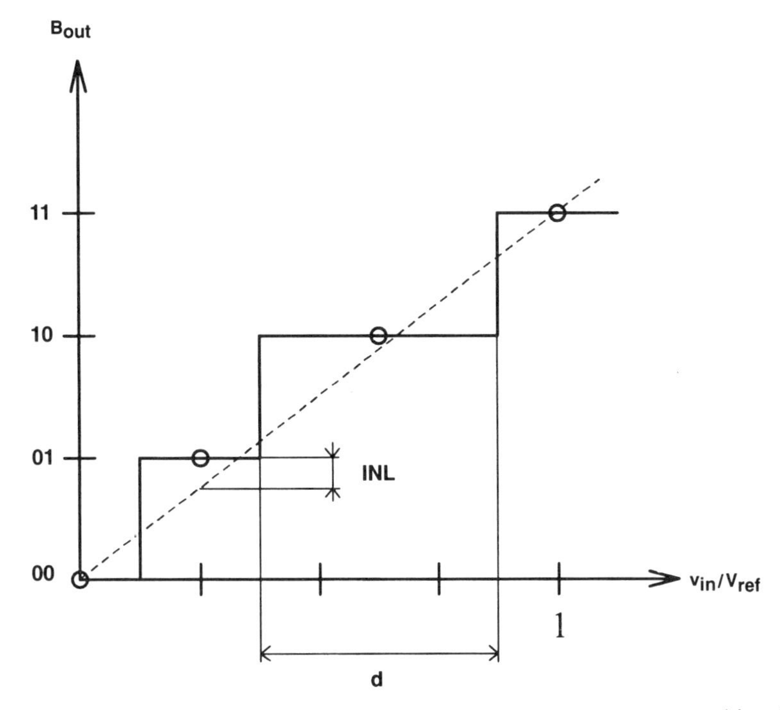

Figure 7.9. Nonlinear transfer curve showing the INL and the worst-case transition difference. The DNL is $|d - V_{\text{lsb}}|$.

7.2. FLASH A/D CONVERTERS

Flash analog-to-digital converters are the fastest and conceptually the simplest A/D converters. In an N-bit flash A/D converter, $2^N - 1$ separate analog comparators and reference voltages are used to convert the analog voltage to a digital word. Each one of the $2^N - 1$ reference voltages corresponds to one quantization level in the digital word. Figure 7.10 shows the conceptual diagram of an N-bit flash A/D converter. Here $2^N - 1$ comparators are used and the $2^N - 1$ quantization levels are generated by an N-segment resistive voltage divider. The outputs of the comparators are processed by the encoder/decoder logic to produce an N-bit digital word.

Figure 7.11 shows the structure of a 3-bit flash A/D converter in somewhat more detail. The positive input of the $2^3 - 1 = 7$ latching comparators is connected to a common analog bus, which is driven by the analog input voltage. The other input of each comparator is connected to a distinct analog decision level. For a given analog input level v_{in}, the comparators whose input reference levels are below v_{in} will have an output state of 1, and those with reference levels above v_{in} will have an output state of 0. Thus the reference voltages and comparator stages are used to convert the analog input to a digital thermometer output code. Table 7.1 shows the thermometer code and the corresponding digital binary code. With increasing analog

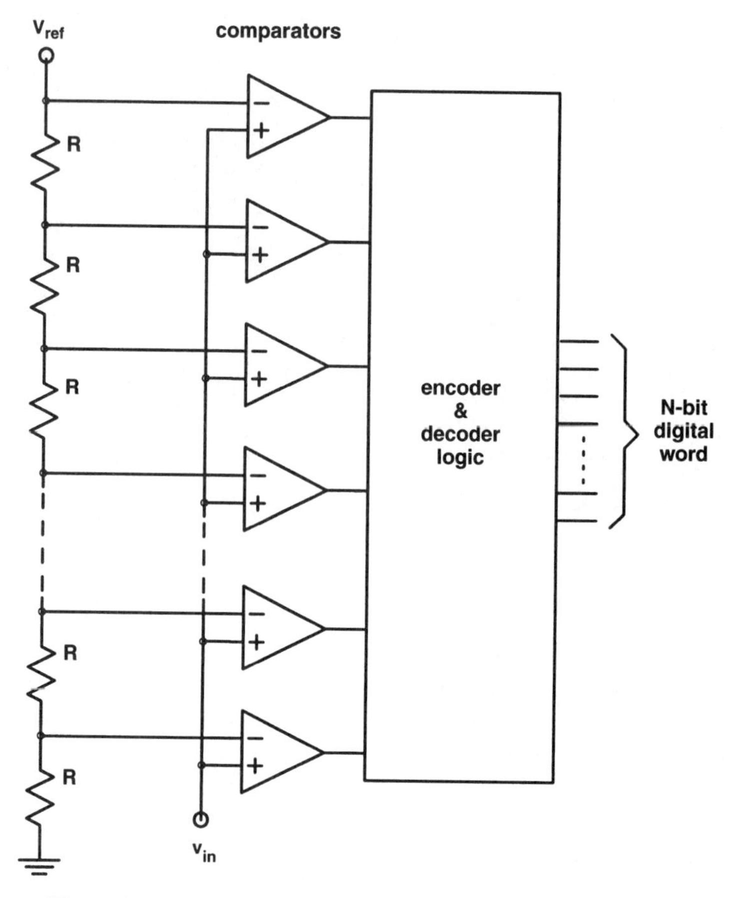

Figure 7.10. Conceptual diagram of an *N*-bit flash A/D converter.

voltage v_{in}, the number of the 1's in the thermometer output code will increase. Examining Table 7.1 reveals that the particular resistor segment in which the analog input lies can be determined by an encoder logic that compares the output logic levels of each comparator with the outputs immediately below it. As shown in Fig. 7.11 for the 3-bit flash A/D converter, the encoder logic consists of $2^3 - 1 = 7$ two-input AND gates. For example, for an analog input level corresponding to the binary digital word 100, the output of the AND gate with the input d$\bar{\text{e}}$ will be true. The output of the encoder is applied to a decoder logic to form a 3-bit word. As shown in Fig. 7.11, an additional comparator can be added which detects overflow conditions corresponding to input levels that exceed V_{ref}.

As mentioned earlier, the comparators in a flash converter generate a thermometer code. When everything is working ideally, the pattern of comparator outputs should resemble that of a thermometer, all 0's above the input level, all 1's below. The zero-to-one transition point rises and falls with the input level. However, under

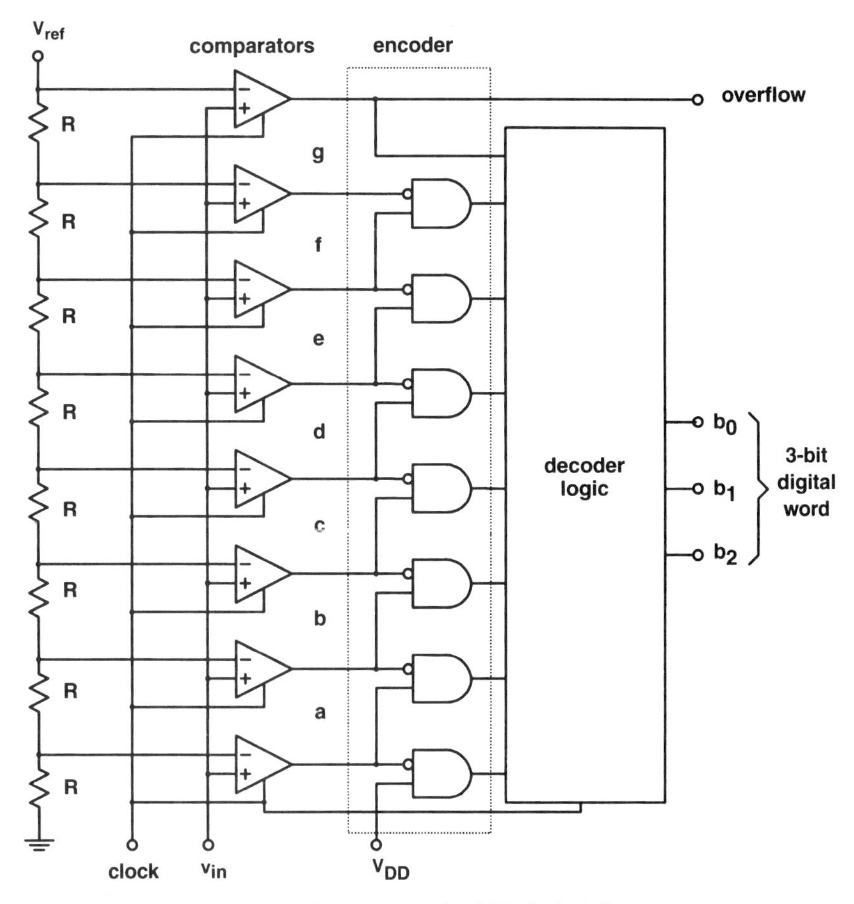

Figure 7.11. Basic structure of a 3-bit flash A/D converter.

extremely high input slew-rate conditions, timing differences between various signal paths or even slight differences between comparator response times can cause the effective strobe point of one comparator to be quite different from another. Consequently, a 1 may be found above a 0 in the thermometer code even though this cannot happen at dc. Errors of the type are sometimes referred to as *bubbles* because they resemble bubbles in the mercury of a thermometer. Table 7.2 shows a normal thermometer code and the corresponding encoder output, and a thermometer code that is contaminated by a bubble error, in this case a 0 surrounded by 1's. The encoder outputs of Fig. 7.11 would now generate two true outputs, which (depending on the design of the decoder) may produce grossly erroneous codes.

A common method of suppressing bubbles is to use a three-input AND gate that compares the logic output of each comparator with those immediately above and below. This would then require two 1's and a 0 to cause the encoder output to be true, and the bubble errors shown in Table 7.2 would be avoided. Figure 7.12 shows

TABLE 7.1. Binary–Thermometer Code Relationship

Decimal Number	Binary Code			Thermometer Code						
	b_2	b_1	b_0	a	b	c	d	e	f	g
0	0	0	0	0	0	0	0	0	0	0
1	0	0	1	1	0	0	0	0	0	0
2	0	1	0	1	1	0	0	0	0	0
3	0	1	1	1	1	1	0	0	0	0
4	1	0	0	1	1	1	1	0	0	0
5	1	0	1	1	1	1	1	1	0	0
6	1	1	0	1	1	1	1	1	1	0
7	1	1	1	1	1	1	1	1	1	1

the 3-bit flash A/D converter with a modified encoder to remove the bubble errors. However, this circuit will not remove bubble errors where two or more string of 0's are surrounded by 1's. More sophisticated encoders can be used to eliminate these and other possible bubble errors [1,2].

In bipolar technology, flash A/D converters operate in a continuous-time mode. Sample-and-hold circuits cannot be used because of the droop caused by the current of bipolar comparator input devices. Furthermore, analog switches are not easily implemented in bipolar technology. CMOS flash A/D converters, on the other hand, can operate in either continuous or discrete-time mode. In the discrete-time mode of operation, a sample-and-hold circuit can be combined with the comparator input stage. The comparator is normally implemented with one or more offset-canceling CMOS inverter stage. In an N-bit flash A/D converter, $2^N - 1$ offset-canceled

TABLE 7.2. Encoder Outputs for Thermometer Codes With and Without Bubble Error

Normal Thermometer Code	Encoder Output		Thermometer Code with Bubble Error	Encoder Output
0	0	—	0	0
0	0	—	0	0
0	0	—	1	1
0	0	Bubble →	0	0
1	1	—	1	1
1	0	—	1	0
1	0	—	1	0
1	0	—	1	0

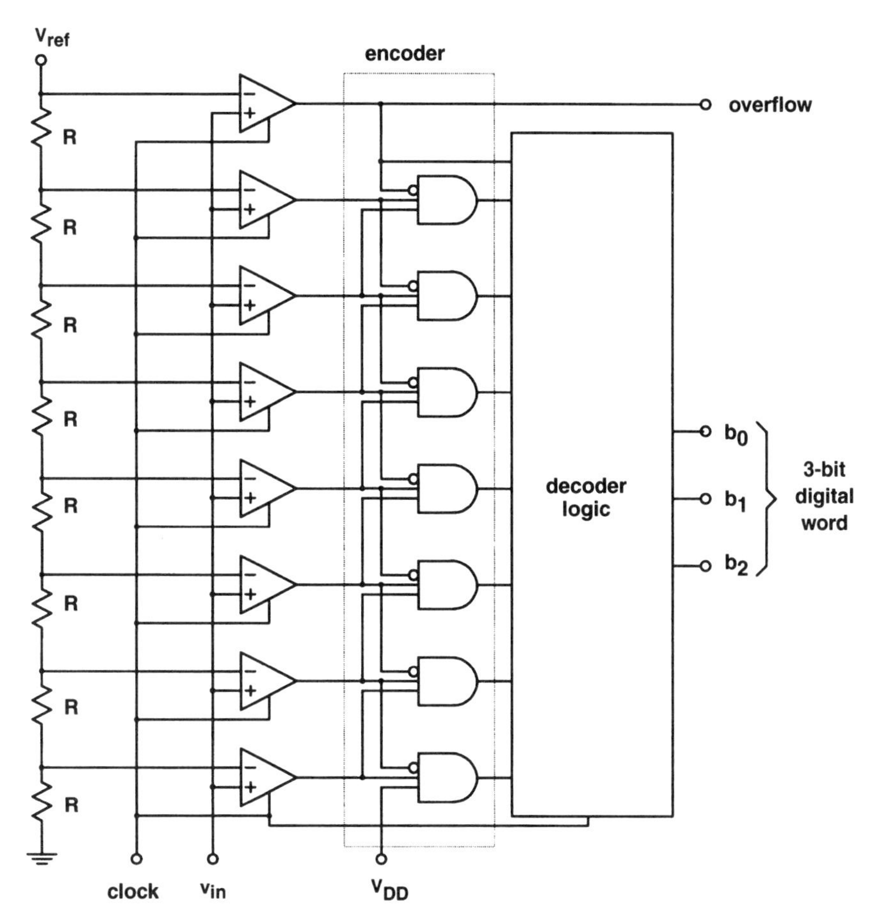

Figure 7.12. Structure of a 3-bit flash A/D converter with an encoder that eliminates simple bubble errors.

CMOS amplifier–comparator stages with sample-and-hold input capacitors will be used to perform the conversion.

Figure 7.13 shows the details of an auto-zeroed sequentially sampled differential input comparator stage [3]. It uses a CMOS inverter stage with three CMOS switches—S_1, S_2, and S_3—and capacitor C, which samples the input signal, as well as the input offset voltage of the inverter. The CMOS switches are controlled by the two complementary clock phases ϕ and $\bar{\phi}$. During the first half clock cycle, when ϕ is high, switches S_1 and S_2 are both on and switch S_3 is off. Switch S_1 shorts the input and output of the inverter stage and charges one side of the capacitor C to a bias voltage determined by the W/L ratios of the devices in the inverter stage. The PMOS and NMOS devices, with their gate and drain tied to each other, form a voltage divider between the positive supply and ground and are normally sized so that the bias voltage at the center node is approximately one-half of the positive

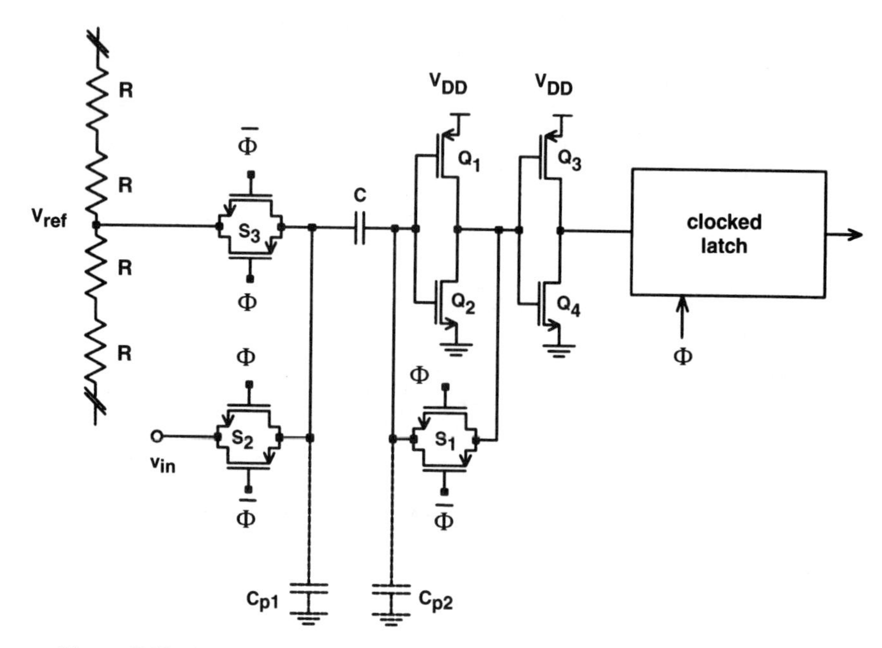

Figure 7.13. Auto-zeroed CMOS comparator with sample-and-hold input branch.

supply voltage. The other side of capacitor C is connected to the input voltage v_{in} via switch S_2. During this phase the comparator inverter is auto-zeroed to its toggle point by S_1, and the input voltage is connected to capacitor C through switch S_2. As soon as ϕ goes low, both S_1 and S_2 turn off, and the input voltage remains stored on capacitor C. With S_1 open, the inverter becomes a gain stage and its gate floats around its toggle point. During the second half-clock cycle, $\overline{\phi}$ goes high and the reference resistor tap is now connected to the capacitor C through switch S_3. Any difference between the sampled input and resistor tap voltages will be amplified by the gain of the inverter stage. The amplified signal will be applied to the following matched inverter stage, which has the same toggle point. This results in a forced digital "high" or "low" output level. This level is captured in the latch on the trailing edge of $\overline{\phi}$ clock signal.

As mentioned briefly in Chapter 5, the comparator of Fig. 7.13 has two potential problems. One problem is that the voltage gain of the single inverter stage is rather low, typically around 50. This puts a lower limit on the minimum voltage that the comparator can resolve. The other problem is that the channel charge injection and clock feedthrough of the switches modify the voltage stored on capacitor C, resulting in an offset voltage that is a function of the clock signal. To investigate the clock-feedthrough effect, consider the parasitic capacitors C_{p1} and C_{p2} shown in Fig. 7.13. When switches S_1 and S_2 turn off, a portion of their channel charges and the capacitively coupled clock-feedthrough charge are transferred to C_{p1} and C_{p2}. The parasitic charge transfer to C_{p1} has no adverse effects because when S_3 turns on, capacitor

C_{p1} will be connected to V_{ref}. So C_{p1} is always switched from one voltage source to another, and the intermediate voltage of C_{p1} when both S_2 and S_3 are off is unimportant. The charge coupling to C_{p2}, however, cannot be ignored, because the resulting change of voltage across C_{p2} modifies the bias voltage of the comparator that is stored on C. Using minimum-sized complementary switches with CMOS devices, so that the p- and n-channel control voltages are equal and opposite will result in some error cancellation. A more effective solution is to use the two-stage comparator shown in Fig. 7.14*a*. This circuit consists of two cascaded auto-zeroed comparator stages containing two capacitors, C_1 and C_2 and four switches, S_1, S_2, S_3, and S_4 [4]. During the first auto-zeroing half-cycle, when ϕ_1 and ϕ_2 are high, switches S_1, S_2, and S_4 are on and S_3 is off. Next, switches S_1 and S_2 turn off while S_4 remains on and C_2 stores the voltage change due to the charge injection of S_1. Subsequently, S_4 turns off and S_3 turns on, and the comparator enters the high-gain region. It amplifies the voltage difference between the sampled input signal and

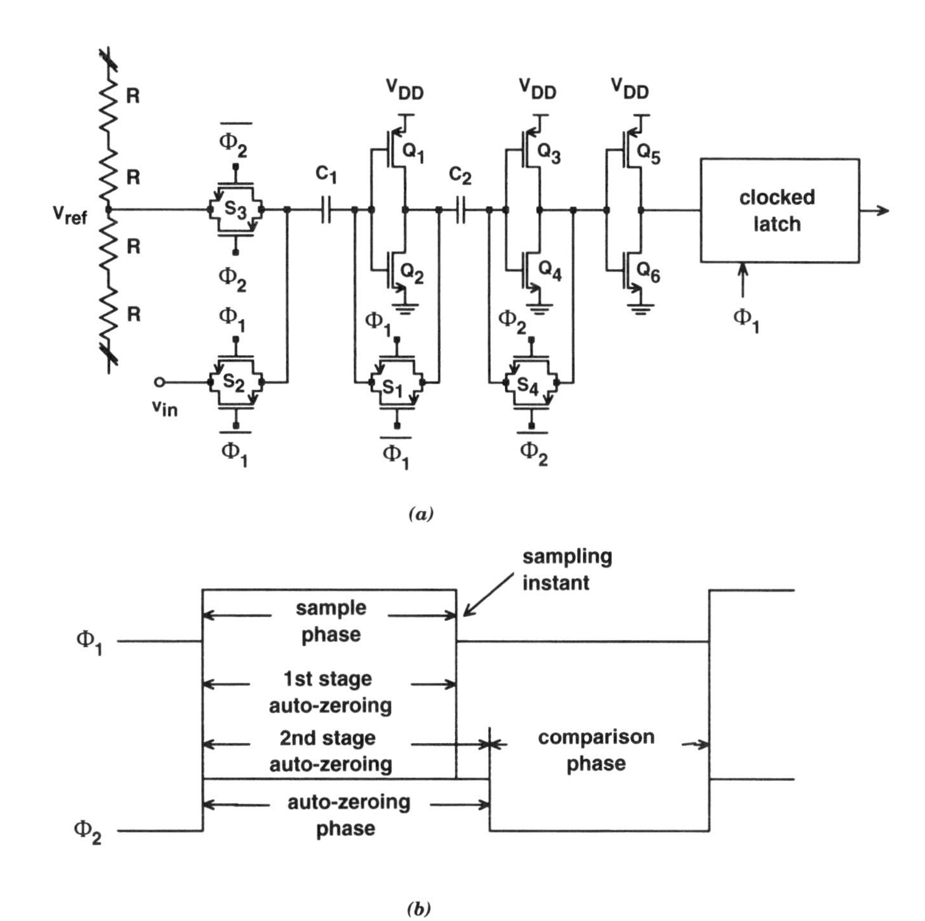

Figure 7.14. (*a*) Two-stage auto-zeroed comparator; (*b*) comparator timing.

the reference voltage. Figure 7.14*b* shows the timing of the clock waveforms. The comparator has a much higher gain and can respond accurately to much smaller input voltages. As before, a matching inverter and a clocked latch that captures the final digital output follow the comparator. The maximum sampling frequency of the flash A/D converter that uses the auto-zeroing comparator is determined by the speed of the comparator and the quantization voltage (absolute voltage of one LSB), which is a function of the magnitude of the reference voltage and the number of bits in the digital output.

7.3 INTERPOLATING FLASH A/D CONVERTERS

The concept of the flash A/D converter as presented in Section 7.2 is straightforward, and it can very easily be extended to higher-resolution systems. The complexity, however, increases very rapidly with the resolution (N bits). For an N-bit system, the architecture requires a minimum of $2^N - 1$ comparators and 2^N resistors. For example, an 8-bit A/D converter will require 255 comparators and 256 resistors. It is therefore difficult to realize high-resolution high-speed flash A/D converters and maintain low power dissipation and small die area. The physical layout of the flash converter poses another challenge. The comparators and long resistor strings should be located in a symmetrical fashion to avoid unequal propagation delays for the clock and input signals and to prevent uneven sampling instances of the input signal across the comparator array. As the number of bits increases, the input capacitance of the system increases linearly with the number of the auto-zeroed comparators. This large input capacitance causes high-speed current spikes in both the analog input and the reference taps. To overcome the effects of these dynamic transient signals, the user is required to provide a high-power signal buffer for driving the analog input terminal. One method for achieving lower input capacitance is to use interpolation technique [4]. Figure 7.15 shows a simple interpolation circuit that can reduce the number of the input comparators and consequently, the input capacitance by a factor of 2. For an N-bit flash A/D converter, the reference resistor divider is designed with 2^{N-1} taps, which is also a factor-of-2 reduction. Interpolating resistors are connected between the outputs of adjacent comparators, and the voltage at the common terminal of the two resistors is the average of the two comparator output signals:

$$V_{o,i+1/2} = \frac{V_{o,i} + V_{o,i+1}}{2} \tag{7.5}$$

To understand the operation of the circuit, assume that all inverter stages in the auto-zeroing input comparators and those that follow them are made of matching p- and n-channel transistors, so that they all have identical toggle voltage equal to V_{bias}. Assume also that the analog input voltage (once captured on the sample-and-hold capacitors) causes the output of inverter $i - 1$ and all those below it to be in

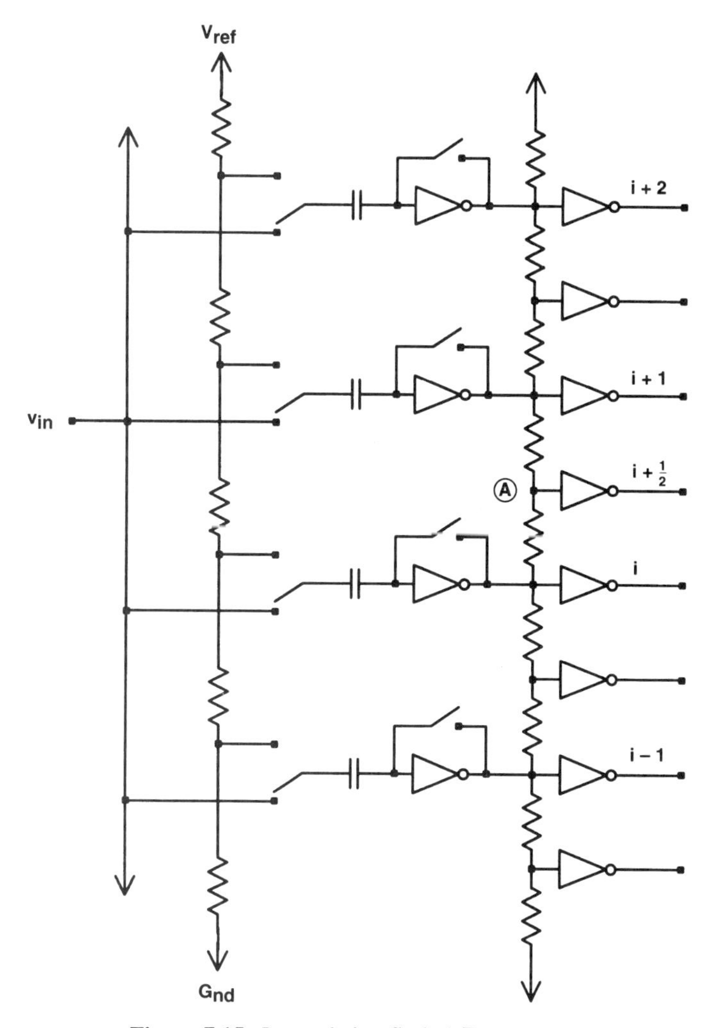

Figure 7.15. Interpolating flash A/D converter.

saturation, indicating a logic 1 and causes the outputs of inverter $i + 1$ and all those above it to be in saturation, indicating a 0. Comparators $i + 1$ and i will, however, be operating in their high-gain (linear) regions. They will hence be amplifying the difference between the analog input voltage and their respective tap voltages on the reference divider. In this case, the voltage that will appear at tap A on the second divider will be halfway between the voltages at the outputs of comparator i and $i + 1$ and will have the value given by Eq. (7.5). If this voltage is greater than V_{bias} (the toggle voltage of the inverter), the output of inverter $i + \frac{1}{2}$ connected to node A will be 0; otherwise, it will be a 1. The $2^N - 1$ outputs from the comparators and the averaging inverters are captured by an array of latches. Just as in all flash converters, the $2^N - 1$ thermometer code output data from the latches feed a decoder that converts them to an N-bit binary word.

Alternatively, active analog components can be used in an averaging circuit [5]. Figure 7.16 illustrates an analog averaging circuit employing CMOS inverters. The two inverters have their inputs tied to V_1 and V_2 (the voltages to be averaged), with the outputs connected to provide an amplified average of the two input signals. Assuming that the two inverter stages are identical and have equal gains given by A, the output will be the amplified average of the two input voltages V_1 and V_2:

$$V_o = A \frac{V_1 + V_2}{2}. \tag{7.6}$$

The implementation of the averaging function using active components has several advantages over the averaging circuits using passive components. First the circuit amplifies the averaged signal which improves resolution and the noise performance. Second, the circuit provides an active drive, which increases the speed of the circuit.

Figure 7.17 is a schematic of a flash A/D converter that uses the active analog averaging circuit shown in Fig. 7.16. In this circuit, as before, only half the number of required comparators is used in the input stage. The others are implemented by using the active averaging circuit placed between adjacent comparator outputs.

Interpolating flash A/D converters have a number of advantages over conventional

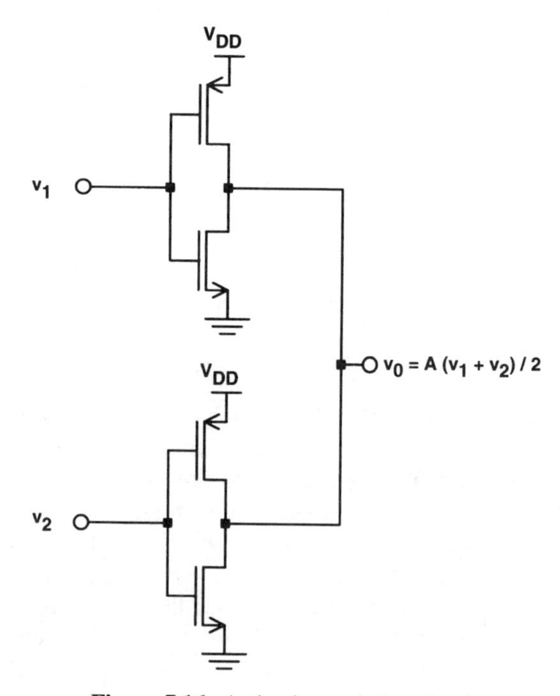

Figure 7.16. Active interpolating circuit.

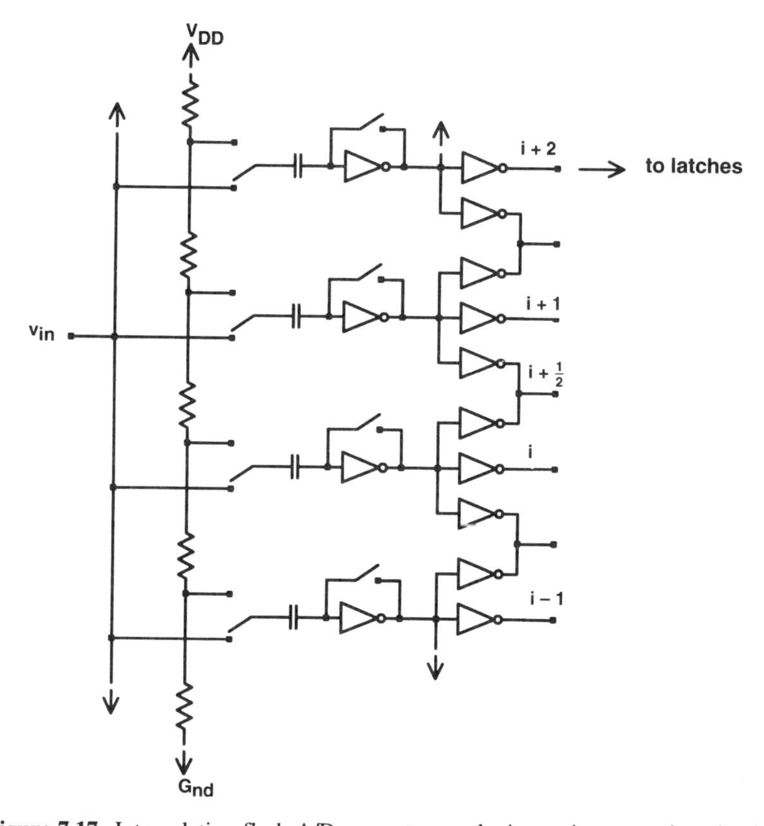

Figure 7.17. Interpolating flash A/D converter employing active averaging circuits.

full flash converters. Fewer auto-zeroing comparators are required, which significantly reduces the silicon area, power dissipation, and input capacitance. Hence the transient signals at the analog input and the reference voltage taps are also reduced. This also reduces the required number of the precision voltage taps, which leads to a significant reduction in silicon area and in improved accuracy. Active interpolation techniques, (Fig. 7.17) use simple inverter stages to amplify and average the effective residual signals. This increases the speed and accuracy of the converter significantly.

7.4. TWO-STEP A/D CONVERTERS [6–9]

The flash A/D converters described in Section 7.3 require $2^N - 1$ comparators to achieve an N-bit resolution. Thus the overall circuit complexity increases very rapidly with increasing N. The interpolating architectures reduce the total input capacitance by reducing the number of auto-zeroing comparators connected to the input line. This will increase the conversion speed but does not intrinsically reduce the total number of comparators and clocked latches. An alternative to a flash architecture is

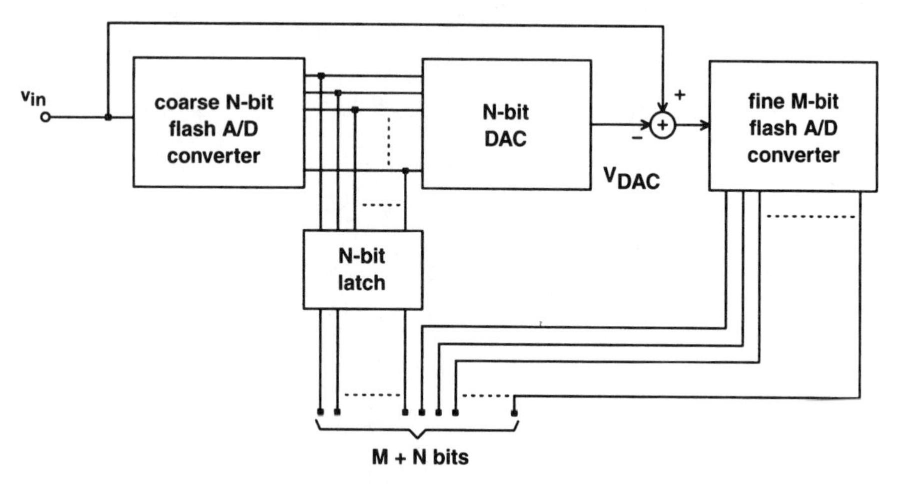

Figure 7.18. Block diagram of a two-step flash A/D converter.

the multistep A/D conversion technique, with the two-step flash being the most popular, due to its high speed and ease of implementation. The two-step architecture uses a coarse and a fine quantization to increase the resolution of the converter. Several two-step A/D conversion techniques are available. Figure 7.18 shows the block diagram of an $(N + M)$-bit two-step A/D architecture. The circuit operates on the input signal in two serial steps. First an N-bit ''coarse'' flash A/D converter determines the N most significant bits. After the coarse quantization has been performed, the N-bit digital data are reconverted into an analog value using an N-bit D/A converter. This analog value is then subtracted from the original analog input. The difference is subsequently applied to an M-bit ''fine'' flash A/D converter that generates the M least significant bits. In most cases, M and N are chosen to be equal to keep the circuit symmetrical and the complexity low. For example, in an 8-bit two-step flash, one would use two 4-bit flash A/D converters and a 4-bit D/A converter.

The two-step flash ADC (also known as *subranging* ADC) provides a powerful low-cost alternative to the flash A/D converter when maximum speed is not necessary. For an 8-bit A/D converter, the two-step flash approach reduces the comparator count from $2^8 = 256$ to 31. This requires a relatively small die size, with the input loading and power dissipation also greatly reduced. The major disadvantage of the two-step flash approach is the reduction in the throughput rate (one-half of that of flash): Two clock cycles are required for each conversion cycle, since the fine conversion cannot start until the coarse conversion is completed.

The block diagram of a 6-bit two-step flash A/D converter, which will be used to explain how the concept can be implemented, is shown in Fig. 7.19. The system contains two 3-bit flash A/D converters. Each flash subsection is made up of seven comparators, which compare the unknown input with the tap voltages of a string of reference resistors to get a 3-bit digital output. The reference string consists of eight

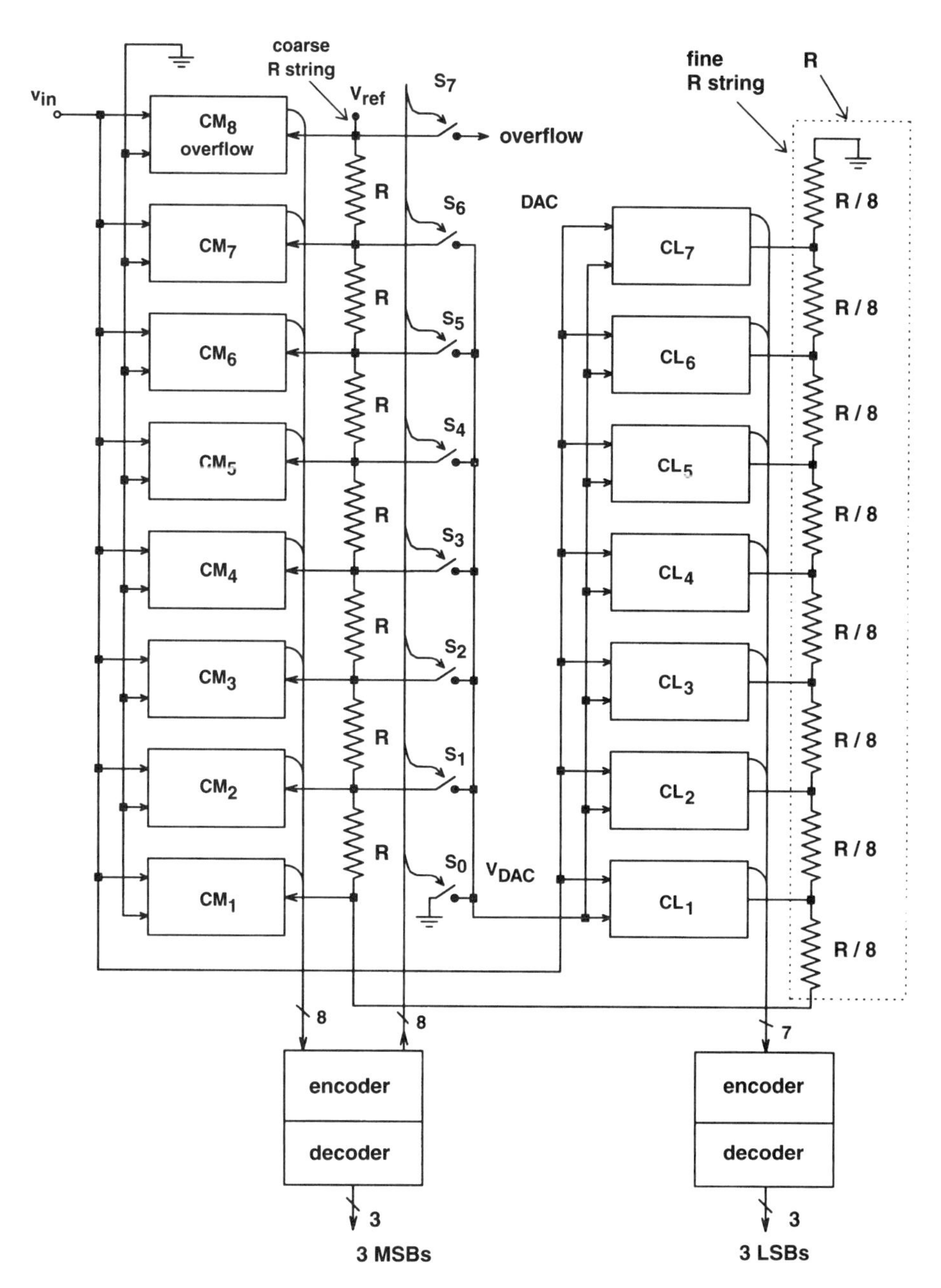

Figure 7.19. Six-bit two-step flash A/D converter.

"coarse" resistors, each of which has a value R. Each coarse resistor is made up of eight "fine" resistors, each with a value $R/8$. The total resistance of the entire string is thus $64 \times R/8 = 8R$. The coarse resistors provide seven taps for the "coarse" comparators CM_1 to CM_7, with voltages ranging from $V_{\text{ref}}/8$ to $7V_{\text{ref}}/8$ in increments of $V_{\text{ref}}/8$. The smaller resistors form a divider, with output voltages ranging from $V_{\text{ref}}/64$ to $V_{\text{ref}}/8$ in increments of $V_{\text{ref}}/64$. Only the taps of the lower resistor string are connected to the "fine" comparators. The A/D converter of Fig. 7.19 is a pipelined subranging architecture. It converts in two steps. The first 3-bit conversion subdivides the input range into eight segments. Then the subrange chosen is further quantized into eight segments (3 bits). The two 3-bit words are then merged to form a 6-bit output word.

The coarse and fine A/D converters each use seven auto-zeroing comparators, CM_1 to CM_7 and CL_1 to CL_7, respectively. An additional comparator CM_8 is used in the coarse A/D converter to detect overflow. The details of the coarse and fine auto-zeroing comparators are shown in Fig. 7.20*a* and *b*, respectively, and the timing diagram is illustrated in Fig. 7.20*c*.

Next, we discuss the operations of the various stages in the converter. The *coarse A/D converter* comparators (Fig. 7.20*a*) are auto-zeroed during the $\phi_1 = 1$ period: S_{M1}, and S_{M2} are closed, and hence C_1 in comparator CM_i charges to $V_{Mi} - V_{BM}$. Here $V_{Mi} = (i/8)V_{\text{ref}}$ is the ith tap voltage in the coarse R string, while V_{BM} is the self-bias (toggle) voltage of the coarse comparators. When next $\phi_1 \rightarrow 0$ and $\phi_2 \rightarrow 1$, the input voltage of comparator i changes from V_{BM} to

$$V_{BMi} = V_{BM} + \frac{C_1(v_{\text{in}} - V_{Mi})}{C_1 + C_2 + C_p}. \tag{7.7}$$

Here C_p is the parasitic input capacitance of CM_i and CL_i (Fig. 7.20*a*). Thus the output of CM_i is determined by the sign of $v_{\text{in}} - V_{Mi}$ at the time when $\phi_2 \rightarrow 1$. The coarse comparator outputs are then captured by an array of clocked latches and encoded/decoded into the three MSBs of the output word using the process illustrated in Fig. 7.12. The output of the encoder is also used to select as V_{DAC} the tap voltage V_{Mi} in the coarse resistor string, which is the largest $V_{Mi} < v_{\text{in}}$, by closing the appropriate switch S_i. This performs the 3-bit DAC operation (Fig. 7.18).

The *fine A/D converter* operates in a push-pull timing mode with the coarse converter. The ith comparator is auto-zeroed during $\phi_2 = 1$. In the following phase, S_{L1}, S_{L2}, and S_{L3} are opened and S_{L3} and S_{L4} closed. This causes the comparator input voltage to change from V_{BL} to

$$V_{BLi} = V_{BL} + \frac{C_2V_{Li} - C_1(v_{\text{in}} - V_{\text{DAC}})}{C_1 + C_2 + C_p}. \tag{7.8}$$

Since $C_1 = C_2 = C$ is used, the output of CL_i is determined by the sign of $V_{Li} - (v_{\text{in}} - V_{\text{DAC}})$. Therefore, $V_{BLi} < V_{BL}$ will occur if $v_{\text{in}} - V_{\text{DAC}} > V_{Li}$ and $V_{BLi} > V_{BL}$ if $v_{\text{in}} - V_{\text{DAC}} < V_{Li}$. Thus the CL_i comparator outputs give a thermometer-

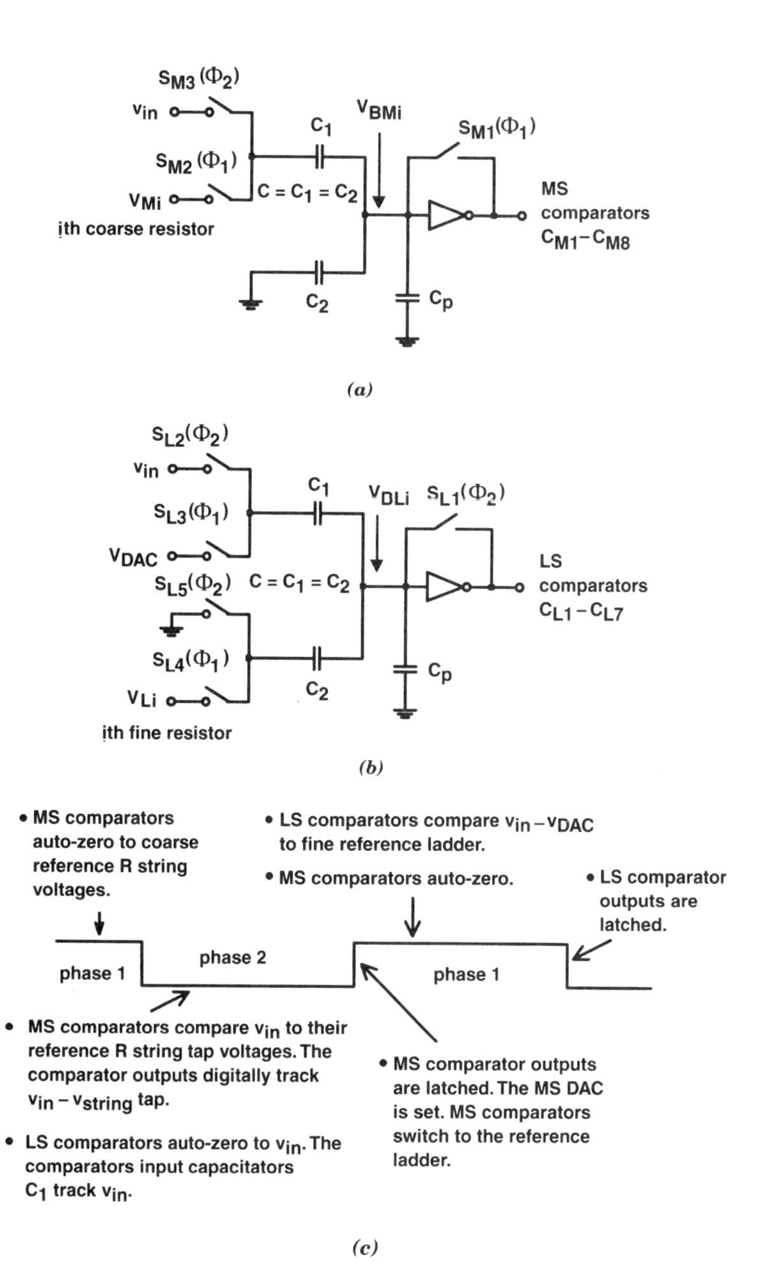

Figure 7.20. (*a*) Most significant auto-zeroing comparator; (*b*) least significant auto-zeroing comparator; (*c*) A/D timing diagram.

code representation of $v_{in} - V_{DAC}$, as required (Fig. 7.18). By recording these outputs, the three LSBs of the output word are obtained. These can then be combined with the three MSBs stored in the clocked latches to obtain the digital output of the ADC. It should be noted that v_{in} is acquired by the input capacitors C_1 of all comparators at the same time, when $\phi_2 = 1$. This generally makes use of an input sample-and-hold stage unnecessary.

Several alternative variations of the two-step flash A/D converters are available [10,11]. One architecture that avoids analog subtractions or DACs is the intermeshed ladder subranging architecture. The central feature of this approach is shown in Fig. 7.21. It is an intermeshed "coarse–fine" resistor network that provides the MSB and LSB reference levels against which the input is tested. As in the conventional architecture for an $(N + M)$-bit A/D converter, the coarse MSB string has 2^N low-resistance sections. A higher-resistance "fine" resistor string is intermeshed (paral-

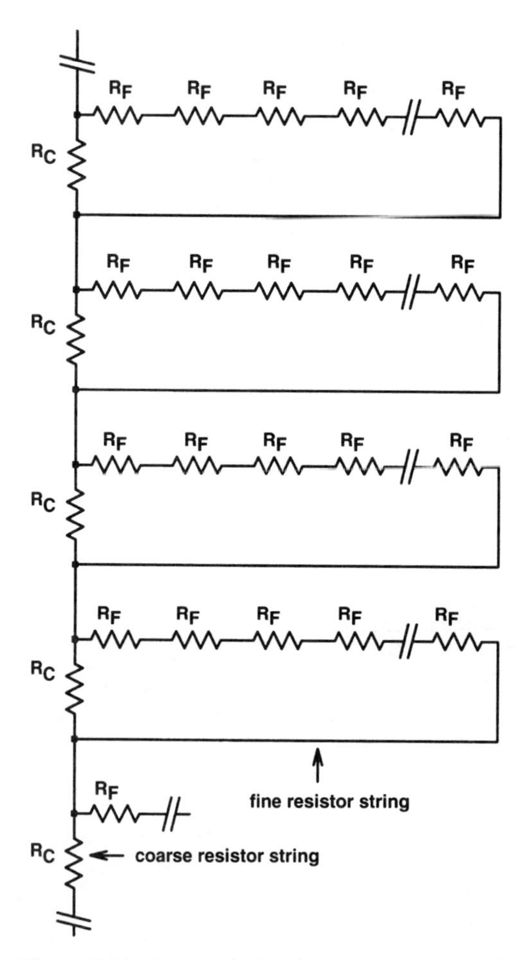

Figure 7.21. Intermeshed resistor segment network.

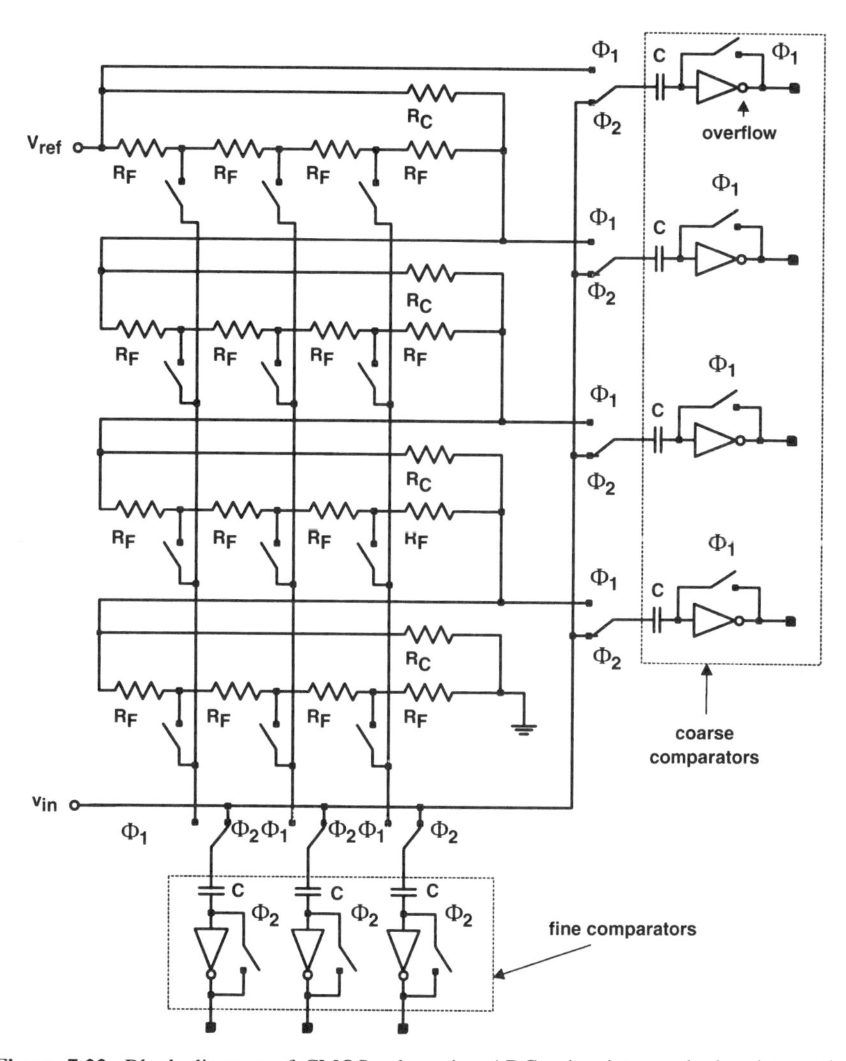

Figure 7.22. Block diagram of CMOS subranging ADC using intermeshed resistor string network.

leled across) each coarse resistor R_C; each of the 2^N fine sections is tapped at $2^M - 1$ nodes. As before, the system needs 2^N coarse comparators (one extra for overflow) and $2^M - 1$ fine ones.

Implementation of a 4-bit ADC example where the system is split into two 2-bit quantizers is shown in Fig. 7.22. Four comparators are connected to the taps of the "coarse" segment, while the three "fine" comparators are addressed by a block selection logic that obtains its information from the coarse quantizer. A two-phase clock with the timing shown in Fig. 7.20 can be used for the conversion. As before, during $\phi_1 = 1$ the coarse comparators are auto-zeroed and its input capacitors are charged to the coarse reference voltages. During $\phi_2 = 1$, the fine comparators are auto-zeroed and the input capacitors of both fine and coarse comparators are connected to the input voltage. At the end of phase 2, the outputs of the coarse comparators are latched, and the input voltage is sampled on the input capacitors of the fine comparators. The coarse quantization determines the two levels between which the fine quantization must take place. After decoding the thermometer code from the outputs of the coarse comparators, a block selection is performed and the fine reference levels between the previously determined coarse levels are applied to the input capacitors of the fine comparators. Then the fine conversion is performed. Notice that during phase 1, while the fine comparators are performing the fine conversion, the coarse comparators are auto-zeroed and their input capacitors are recharged to the coarse reference voltages.

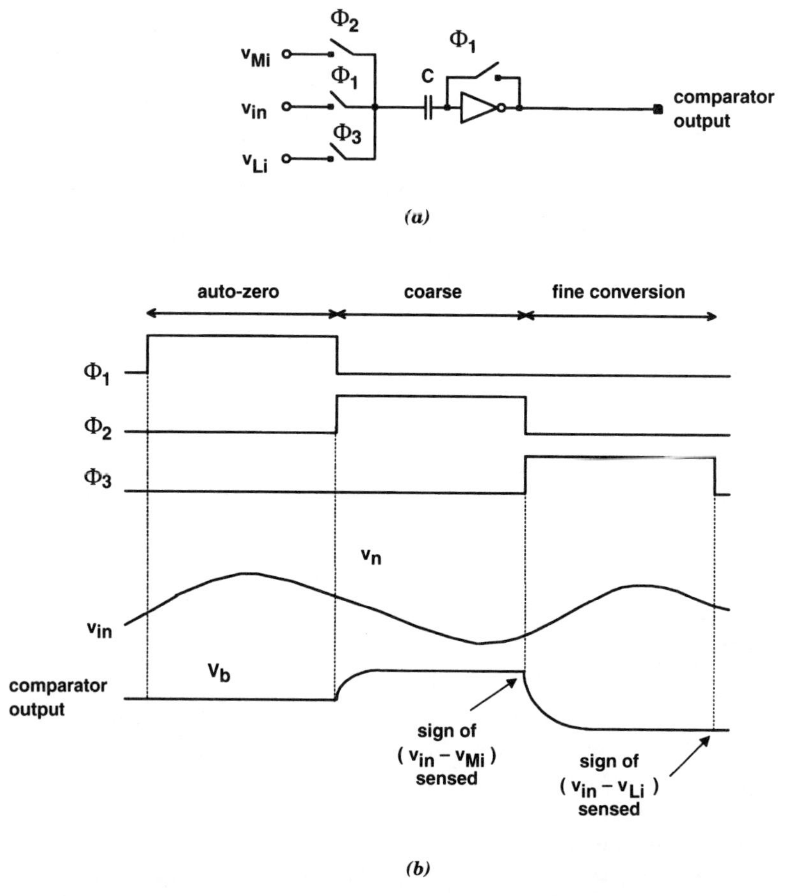

Figure 7.23. (*a*) Three-stage auto-zeroing comparator stage; (*b*) timing diagram of the conversion process.

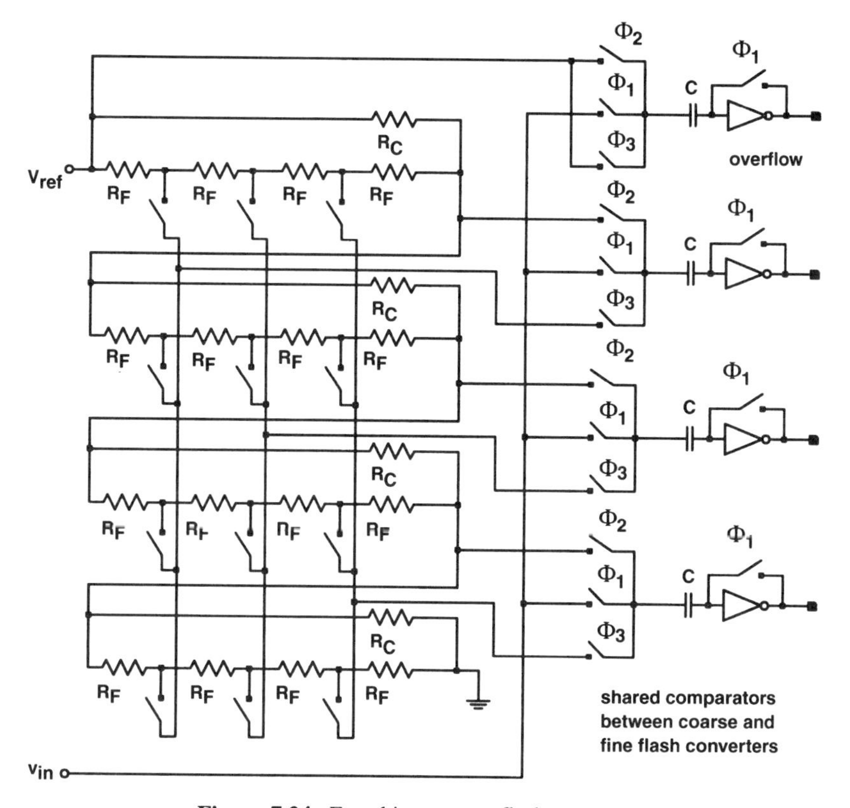

Figure 7.24. Four-bit two-step flash converter.

If three clock cycles are available during a conversion period, the three-input auto-zeroing comparator shown in Fig. 7.23*a* can be used. This allows sharing of the coarse and fine comparators. A two-step 4-bit flash A/D converter that uses four such comparators is shown in Fig. 7.24. The timing diagram of the conversion process is shown in Fig. 7.23*b*. During the first clock period ($\phi_1 = 1$) the comparators are auto-zeroed and the input signal is sampled and stored on the input capacitors. After completion of this sampling cycle, during the $\phi_2 = 1$ period a coarse comparison is performed with the four reference voltages equally spaced between V_{ref} and ground. The comparators produce a coarse thermometer code, which is decoded to produce the two output MSBs. The output of the encoder is also used to select the proper fine resistor network. Next, while $\phi_3 = 1$ the three fine-voltage taps between the previously determined coarse levels are applied to the same comparators and fine conversion is performed. Finally, the two newly obtained LSBs and the stored MSB data are combined to obtain the 4-bit output data. One of the major advantages of this architecture and the preceding one is the assured monotonicity, since the same circuit realizes the fine and coarse resistor strings.

7.5. SUCCESSIVE-APPROXIMATION A/D CONVERTERS

Successive approximation is one of the most popular A/D conversion techniques, because it offers the combination of high accuracy and moderate conversion speed. It is a feedback scheme that uses a trial-and-error technique to approximate each analog sample with a corresponding digital word. In Fig. 7.25, the basic block diagram of such a system is shown. It contains a sample-and-hold stage as well as a D/A converter, a successive-approximation register (SAR), and a voltage comparator. The successive-approximation register for an N-bit A/D converter contains an N-bit presentable register that is cleared prior to the start of the conversion process. The register bits are set one at a time to a high-state 1 and produce an input to the N-bit digital-to-analog converter. The output of the DAC is compared to the analog input voltage, and a decision is made to keep the last bit high or return it to zero. In the first step of the conversion, the MSB of the SAR is set to a high state ($b_1 = 1$). The MSB DAC voltage $V_{DAC} = V_{FS}/2$ (where V_{FS} is the full-scale voltage) is substracted from the analog input voltage. If the remainder ϵ is positive, the MSB remains high ($b_1 = 1$) for the rest of the conversion period. If the error signal is negative, the MSB is returned to zero ($b_1 = 0$) and remains zero for the rest of the conversion process. In the next clock cycle the next most significant bit of the register is set high ($b_2 = 1$) and the DAC output voltage $V_{DAC} = (b_1 V_{FS}/2 + b_2 V_{FS}/4)$ is subtracted from the input voltage. Again, if the remainder signal is positive, b_2 remains set ($b_2 = 1$); otherwise, it returns to zero ($b_2 = 0$). This process continues for each successive bit of the SAR for N clock cycles to complete the conversion process for one analog sample. In each cycle, if the error signal is positive, the bit

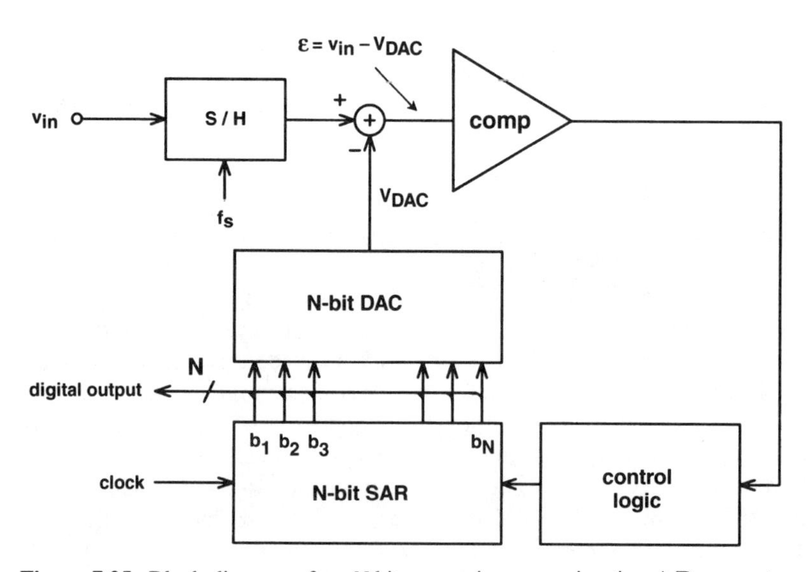

Figure 7.25. Block-diagram of an N-bit successive approximation A/D converter.

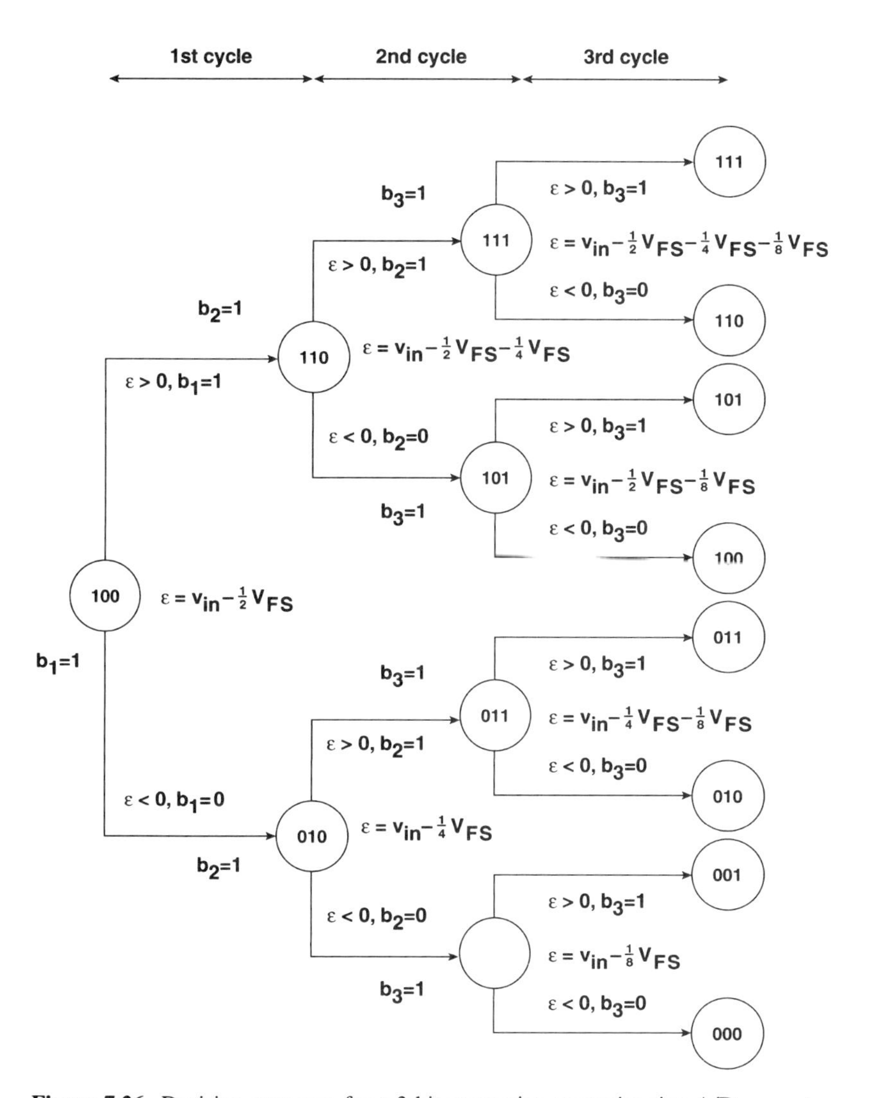

Figure 7.26. Decision sequence for a 3-bit successive-approximation A/D converter.

stays high; otherwise, it is returned to zero. The digital output, which corresponds to the data in the SAR, is not valid until the entire conversion cycle is completed and all bits (MSB through LSB) have been evaluated. The decision sequence for a 3-bit successive-approximation A/D converter is shown in Fig. 7.26. The corresponding input voltage, the sequential output of the DAC, and the remainder signal are shown in Fig. 7.27.

Successive-approximation A/D conversion techniques require analog comparators, digital logic, and precision analog components. A number of unique architec-

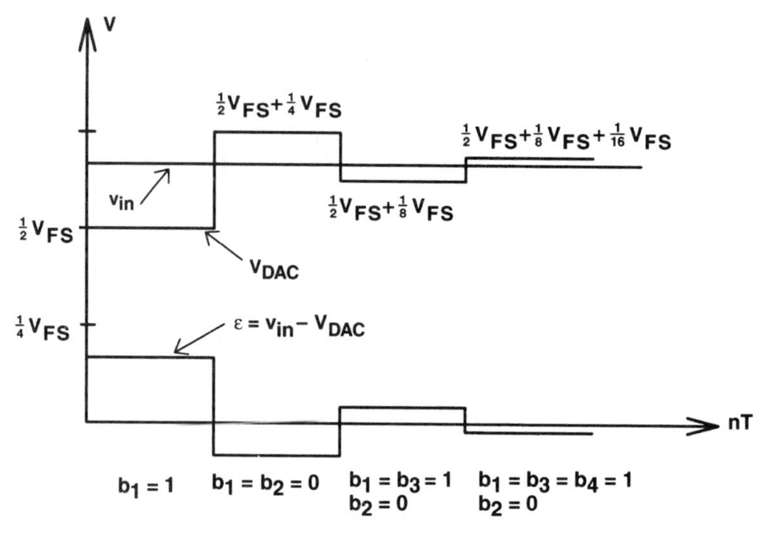

Figure 7.27. Remainder and DAC voltage timing diagram of a 4-bit successive approximation A/D converter.

tures have been developed for successive-approximation A/D converters in CMOS technologies. The most popular among these are the charge-redistribution capacitor circuit, the resister-string circuit, and resistor-capacitor (hybrid) circuit.

If capacitors are used as the precision components and MOS devices as switches, one can use charge rather than current or voltage to represent analog signals inside the converter. This technique, referred to as *charge redistribution,* has been used to implement monolithic A/D converters for many years [12]. A conceptual 5-bit version of a charge-redistribution A/D converter is shown in Fig. 7.28*a.* It consists of a comparator, a 5-bit binary-weighted capacitor array (plus one additional capacitor of a weight corresponding to the least significant bit), and MOS switches that connect the bottom plates of the capacitors to different voltages. The comparator is basically an inverting high-gain differential amplifier without feedback. Its output is thus normally latched to positive or negative supply, depending on whether its input voltage is negative or positive. The operation of the ADC is performed in three phases. In the first, S_0 is closed and the bottom plates of all capacitors are connected to v_{in}. This results in a charge proportional to v_{in} stored in all capacitors. In the second phase, S_0 is opened and all bottom plates are then grounded (Fig. 7.28*b*). This causes the top plate potential to become $-v_{in}$. In the final phase, the bits of the digital output are found one by one. To find the most significant bit (MSB) b_1, S_7 and S_1 switch the bottom plate of the largest capacitor (C) to V_{ref} (Fig. 7.28*c*). The top-plate potential v_x is now raised to $-v_{in} + V_{ref}/2$. If $v_{in} > V_{ref}/2$, this value will be less than 0. Therefore, the comparator output will be positive, corresponding to a logic 1, and this will be the value assigned to b_1; otherwise, $b_1 = 0$. Next, S_1

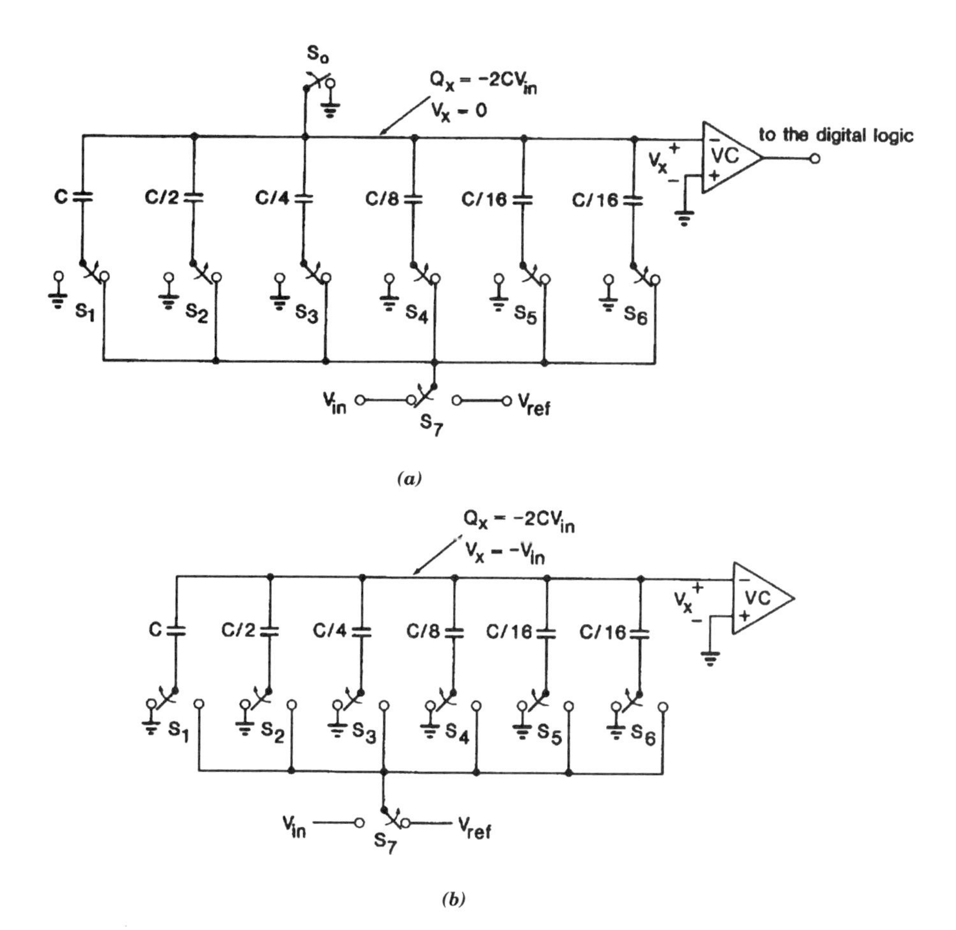

Figure 7.28. Successive-approximation analog-to-digital converter: (*a*) conceptual circuit diagram, shown in the first (sample) stage of operation; (*b*) circuit in the second (hold) stage; (*c*) approximation stage; (*d*) final configuration for the output 01001. (From Ref. 12, © 1975 IEEE.)

will return to ground if $b_1 = 0$ (or stay at V_{ref} if $b_1 = 1$), and S_2 will be switched to V_{ref}. The value of v_x then becomes

$$v_x = -v_{\text{in}} + \left(\frac{b_1}{2} + \frac{b_2}{4}\right)V_{\text{ref}}, \tag{7.9}$$

as can easily be shown (Problem 7.5). If $v_x > 0$, b_2 will be assigned the value of 0, and S_2 will return to ground: otherwise, $b_2 = 1$ and S_2 will stay at V_{ref}. Next, b_3, b_4, and b_5 are found in a similar manner. Figure 7.28*d* shows the final positions of S_1 to S_5 after obtaining the digital output 01001.

The circuit described represents the basic concept of the ADC. One of the factors

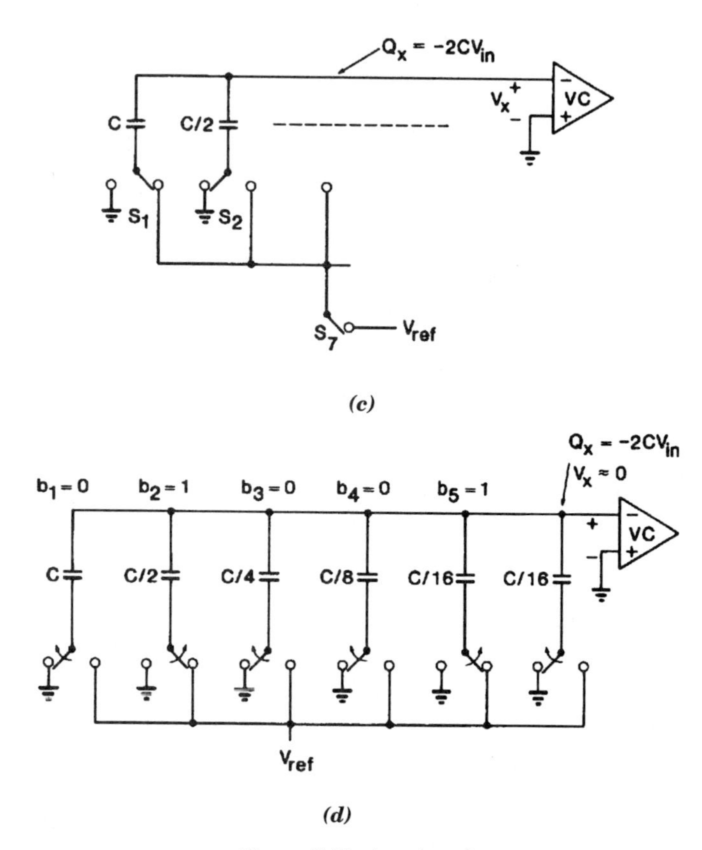

Figure 7.28. *(continued)*

ignored is the offset voltage of the comparator. This can be greatly reduced by using offset-compensation (auto-zeroing) circuits. Figure 7.29 illustrates the basic concept. When S_1 is closed and S_2 grounded, the capacitor C is charged up to the offset voltage V_{off}. The output v_{out} of the comparator will be high if $v_x < V_{\text{off}}$ and low if $v_x > V_{\text{off}}$ these conditions correspond to v_{out} being high if $v_{\text{in}} < 0$ and low if $v_{\text{in}} > 0$, as in an offset-free comparator. Another compensation method is described in Problem 7.7.

Comparators are often built by cascading several stages. If the gain of each stage is low, several stages may be required and the feedback path provided by S_1 may lead to instability. Then it is more expedient to connect S_1 to an intermediate stage (Fig. 7.30). The offset of A_1 is eliminated by the auto-zero circuit; the input offset of A_2 is not, but when it is referred back to the input of A_1, it gets divided by the gain of A_1 and hence greatly reduced (Problem 7.7). It is also possible to use separate auto-zeroing circuits for the two circuits.

The charge redistribution A/D converter of Fig. 7.28 is applicable for unipolar (e.g., only positive) input signals. A slightly modified version of this structure, shown

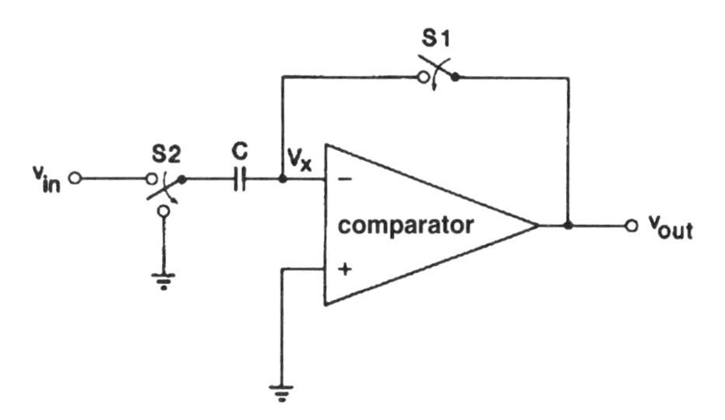

Figure 7.29. Comparator offset cancellation by auto-zeroing.

in Fig. 7.31, can be used for bipolar (positive or negative) inputs. The switching sequence has been altered to incorporate a sign-bit detection cycle. This architecture needs $+V_{ref}$ for positive inputs and $-V_{ref}$ for negative inputs.

The sign bit is detected during the second phase, when the top-plate grounding switch S_0 is opened and the bottom plates of the capacitors are connected to ground by S_1 to S_8. The output of the comparator corresponds to the polarity of $-v_{in}$, the top-plate voltage of the capacitor array. At the end of the second phase, the comparator output is stored in a clocked latch. It represents the sign of the analog sample. Next, this sign bit is used to switch the V_{ref}, with the appropriate polarity to the bottom plates of the capacitor array using S_7 and S_9. The reference voltage is used subsequently during the redistribution cycle to determine the bits of the digital word corresponding to the magnitude of the input.

The 5-bit charge-redistribution A/D converter of Fig. 7.31 also contains a capacitor $C/32$ that has a weight of $\frac{1}{2}$ LSB. Without this capacitor, the A/D transfer curve around zero would have a discontinuity similar to that shown in Fig. 7.32*a*. This problem is remedied by switching the capacitor $C/32$ from ground to the reference voltage after the sign bit is determined. This action will cause the top-plate voltage of the capacitor array to change from $-kv_{in}$ to $(-v_{in} + V_{ref}/64)k$, where $k = 1/(1 + 1/64 + C_p/2C)$ and C_p is the parasitic capacitance from the comparator input

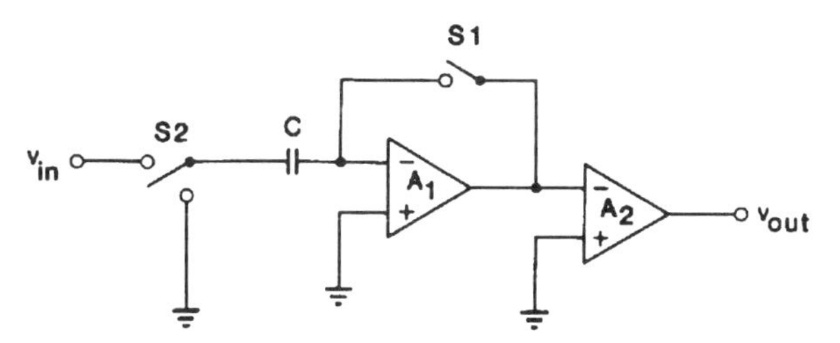

Figure 7.30. Reduction in input offset voltage by capacitive storage for a multistage comparator.

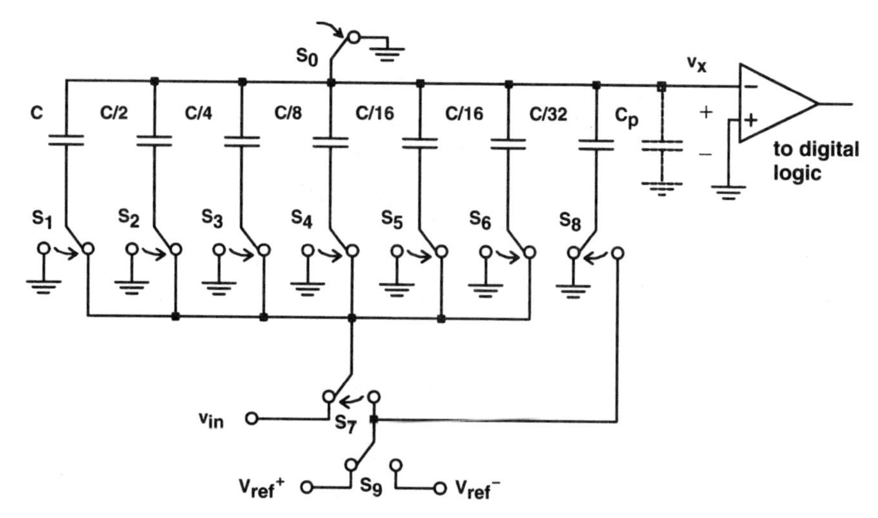

Figure 7.31. Five-bit signed charge-redistribution A/D converter.

to ground. This is equivalent to shifting the input voltage by $\frac{1}{2}$ LSB, and (as shown in Fig. 7.32*b*), it eliminates the discontinuity of the A/D transfer curve around zero.

High-performance data acquisition circuits make use of a fully differential architecture. The main motivation is to reject the noise from the substrate as well as from the power supply lines. Fully differential charge-redistribution A/D converters use a fully differential comparator, two binary-weighted capacitor arrays, and a positive and negative voltage reference. A fully differential 5-bit-plus-sign charge-redistribution A/D converter is shown in Fig. 7.33 [13]. There are two 5-bit binary-weighted capacitor arrays. The conversion process is similar to the single-ended case; it is performed in three phases: the sampling phase, the hold-and-sign-bit-determination phase, and the conversion phase. As before, two additional capacitors with weightings equal to $\frac{1}{2}$ LSB must be added to the two capacitor arrays to eliminate the conversion discontinuity around zero.

The resistor-string-based A/D converter structure consists of a voltage-output DAC, a comparator, and a successive-approximation register. For an N-bit system, the DAC contains a string of 2^N resistors connected in series and a switching matrix. A 3-bit version of a unipolar resistor-string A/D converter is shown in Fig. 7.34 [14]. The string of eight resistors acts as a voltage divider, and the voltage at any tap in the string is one LSB voltage higher than the voltage at the tap below it. The voltage at each tap defines a transition level of the A/D converter. The bottom resistor is chosen to be $R/2$ and the top resistor as $3R/2$, so that the decision levels are shifted to the midpoint of two adjacent transition levels. The comparator uses an offset-canceling capacitor C, which also acts as a sample-and-hold stage.

The conversion is again accomplished in three steps. In the first (sampling) step, ϕ_1 goes high, and the comparator enters the offset cancellation mode by closing the

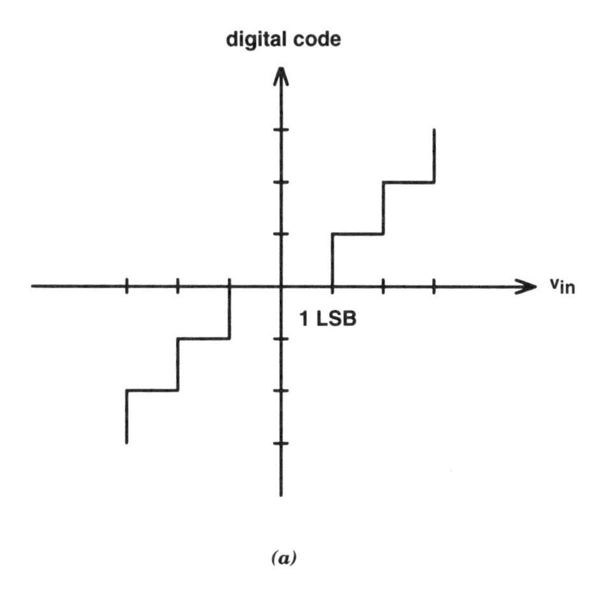

(a)

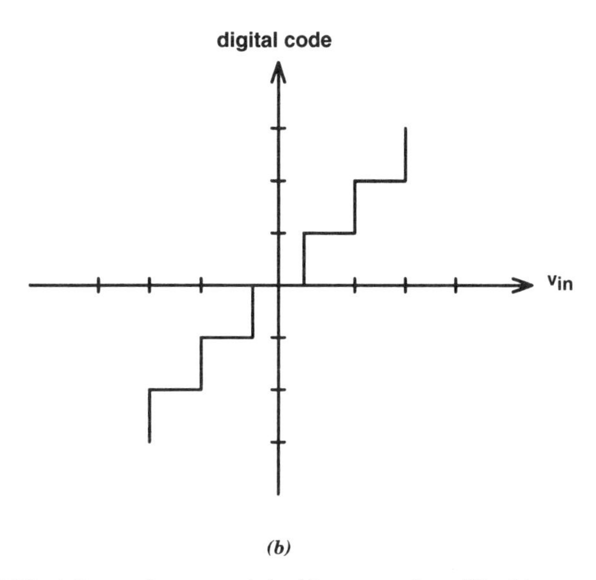

(b)

Figure 7.32. A/D transfer curve: (*a*) without capacitor; (*b*) with capacitor *C*/32.

MOS switch S_0 between the comparator output and its inverting input. During this phase the MOS switch S_1 is also turned on, and the sampling capacitor C is hence charged between the offset voltage V_{off} of the comparator and the input signal. When ϕ_1 goes low, both switches S_0 and S_1 turn off and the input signal is held across capacitor C. Also, the feedback path across the comparator is opened. During the

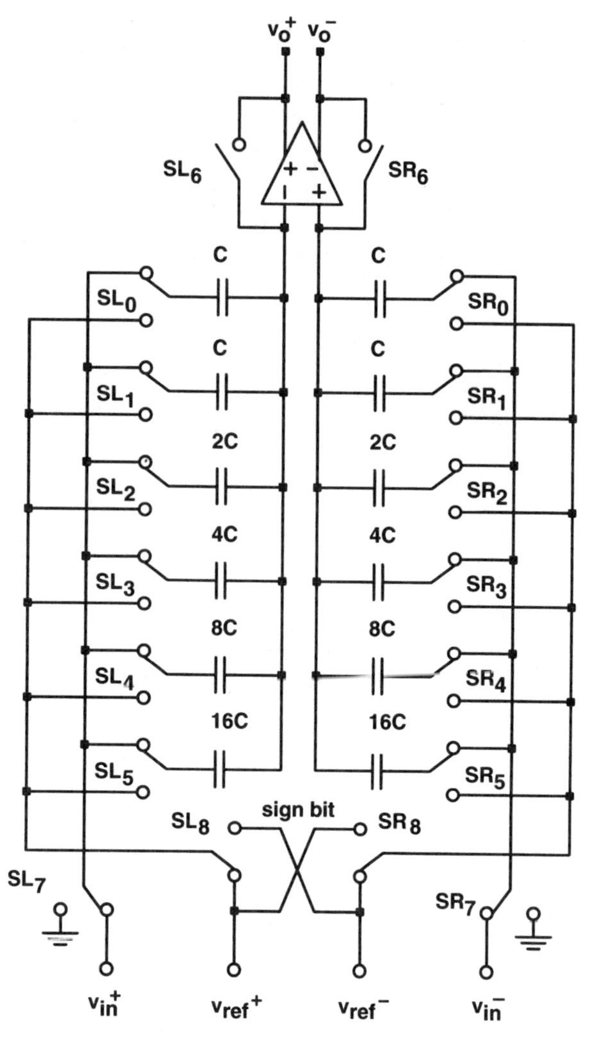

Figure 7.33. Five-bit charge-redistribution fully differential A/D converter.

subsequent hold cycle, ϕ_2 goes high and the capacitor is connected to the output of the DAC. The voltage at the input of the comparator thus becomes

$$v_x = -(v_{\text{in}} - V_{\text{off}} - V_b) + V_{\text{DAC}}. \tag{7.10}$$

Here V_b represents an appropriate bias voltage, normally midway between the positive supply and ground, connected to the positive input of the comparator. During the third phase, the usual successive-approximation search is performed by placing a trial value 1 in the MSB position and making a decision by examining the state

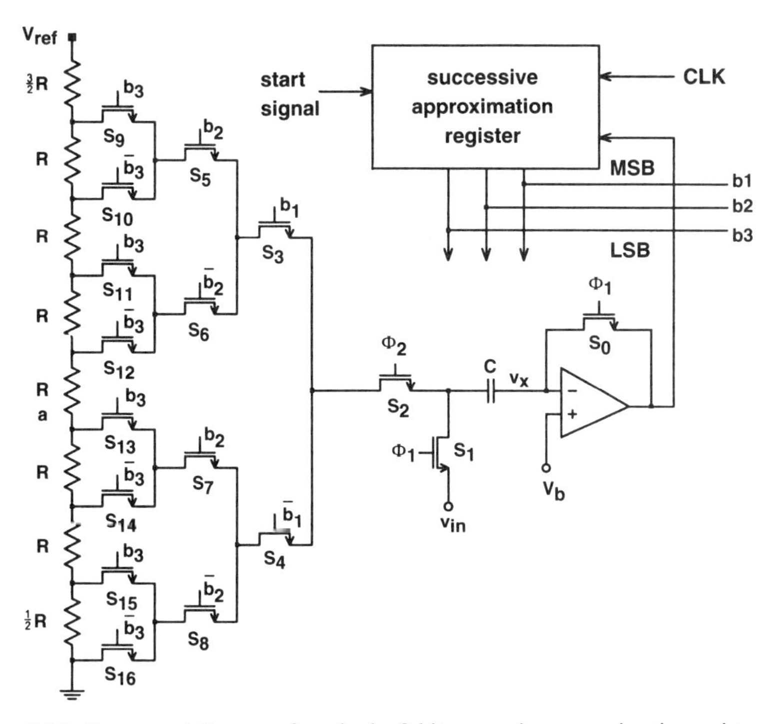

Figure 7.34. Conceptual diagram of a unipolar 3-bit successive-approximation resistor-string A/D converter.

of the comparator output. For example, in the first step when $b_1 = 1$ and $b_2 = b_3 = 0$, switches S_3, S_6, and S_{12} close and the voltage at node A of the resistor string is applied to the capacitor C. From Eq. (7.10), voltage v_x is now given by

$$v_x = -(v_{\text{in}} - V_{\text{off}} - V_b) + \frac{7}{16} V_{\text{ref}}. \tag{7.11}$$

v_x is next compared to the voltage on the positive input of the comparator given by $V_{\text{off}} + V_b$. If v_{in} is greater than $\frac{7}{16}V_{\text{ref}}$ the output of the comparator goes high and the bit b_1 remains 1. However, if v_{in} is less than $\frac{7}{16} V_{\text{ref}}$, the output of the comparator goes low and the bit b_1 is changed to 0. This process is then repeated in descending sequence for each of the lesser significant bits.

Since the voltage at any given node of the resister string is always greater than the voltage at the node below it, the D/A converter is inherently monotonic, and the A/D converter will have no missing codes. For higher-resolution A/D converters, the folded-resister ladder-string DAC described in Chapter 6 can be used to simplify and improve the performance of the circuit.

As the bit count N of the A/D converter is increased, the complexity of the simple resistor-string or charge-redistribution A/D converters increases rapidly. For

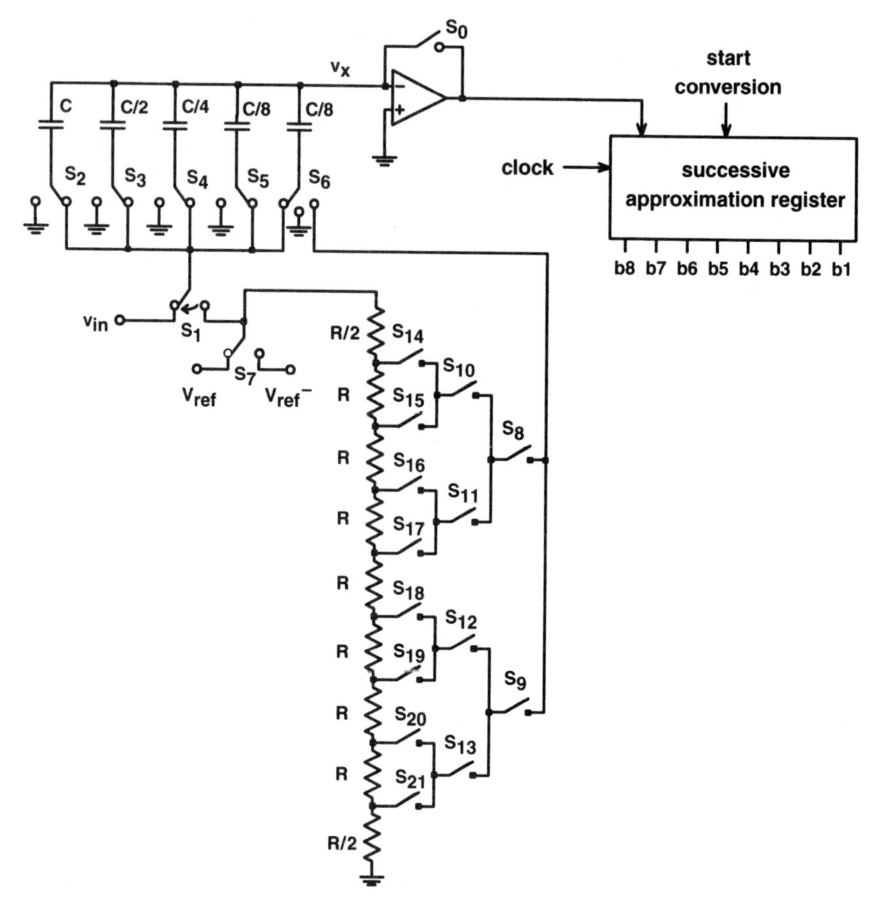

Figure 7.35. Successive-approximation hybrid 8-bit A/D converter with sign.

example, for an N-bit converter, 2^N resistors or capacitors will be needed. Major simplifications become possible if one uses hybrid architectures by combining the resister-string techniques with the charge-redistribution methods [15,16]. An example of an 8-bit (7-bit plus sign) hybrid successive-approximation bipolar A/D converter is shown in Fig. 7.35. It contains a 4-bit binary-weighted capacitor array for the MSBs and a 3-bit resistor string for the steps. The conversion is again performed in three phases. In the first (sample-and-auto-zero) phase, switch S_0 in the feedback of the comparator is closed and switches S_2 to S_6 connect the bottom plates of the capacitors to the input voltage through switch S_1. Thus the capacitors are charged to $v_{in} - V_{off}$ where V_{off} is the offset voltage of the comparator. In the second (hold-and-sign-determination) phase, S_0 is opened and the bottom plates of all capacitors are connected to ground. This will cause v_x, the voltage on the top plate of the capacitors, to charge to $-v_{in} + V_{off}$. The output state of the capacitor now represents the polarity of the input sample. At the end of the second step the state of the comparator output is latched and it is used to control switch S_7, which supplies the reference voltage with the correct polarity to the capacitor array and to the resister string. In the third step a successive-approximation search is performed to find the 7 bits. First, the bottom plate of capacitor $C/8$ is switched to the point on the resistor string that corresponds to $V_{ref}/16$. This in effect shifts the top-plate voltage by $\frac{1}{2}$ LSB to $-v_{in} + V_{off} + V_{ref}/512$, which is necessary to maintain continuity around zero. Next, the four MSBs are determined in the capacitive DAC, one at a time, by setting the bits to a binary 1 state and examining the state of the comparator output. Finally, the three LSBs are determined by switching the bottom plate of capacitor $C/8$ to the output of the resistive D/A converter and performing a successive-approximation search to find the correct tap voltage on the resistor string, which resets the top-plate voltage to zero (within $\pm\frac{1}{2}$ LSB). During this cycle, the top-plate voltage is given by

$$v_x = -(v_{in} + V_{off}) + \left(\frac{b_2}{2} + \frac{b_3}{4} + \frac{b_4}{8} + \frac{b_5}{16}\right)V_{ref} + \frac{V_{DAC}}{16}. \tag{7.12}$$

The total conversion takes nine cycles: one cycle to sample the input and auto-zero the comparator, one cycle to determine the sign bit, and seven cycles to determine the 7 bits.

The accuracy of the A/D converter of Fig. 7.35 is determined largely by the matching accuracy of the capacitors. Since monotonicity cannot be guaranteed for binary-weighted capacitor arrays, the A/D structure is not inherently monotonic. An alternative architecture for a hybrid A/D converter that uses a resistor DAC for the MSBs and a capacitor array for the LSBs is shown in Fig. 7.36 [16]. Operation of the circuit is again performed in three phases. In the first phase, S_F is closed and the bottom plates of all capacitors are connected to v_{in}. Thus all capacitors are charged to $v_{in} - V_{off}$, where V_{off} is the offset (threshold) voltage of the comparator. Next, S_F is opened and a search is performed among the resistor string taps to find the segment within which this stored voltage sample lies. Nodes A and B are then switched to the terminals of the resistor R_i that defines this segment. In the final stage, the bottom plates of C_{k+1}, C_k, . . . , C_1 are switched successively back and forth between A and B until the input voltage of the comparator converges to V_{off}. The sequence of comparator outputs during the successive approximations gives the binary code for v_{in}. Due to the first step in which V_{off} was subtracted from v_{in}, the offset voltage of the comparator does not affect its output. Since the resistor string DAC is inherently monotonic, as long as the capacitor array is monotonic, the A/D converter will also be monotonic and there will be no missing codes.

The successive-approximation A/D conversion techniques using single arrays of capacitors or resistors are suitable for up to 10 bits of resolution. The capacitive charge-redistribution technique exhibits excellent temperature stability due to the very low temperature coefficient of MOS capacitors. Typical conversion times for 8- to 10-bit resolution are in the range 10 to 40 μs. The hybrid technique using a combination of resistors and capacitors can easily be extended to resolutions greater

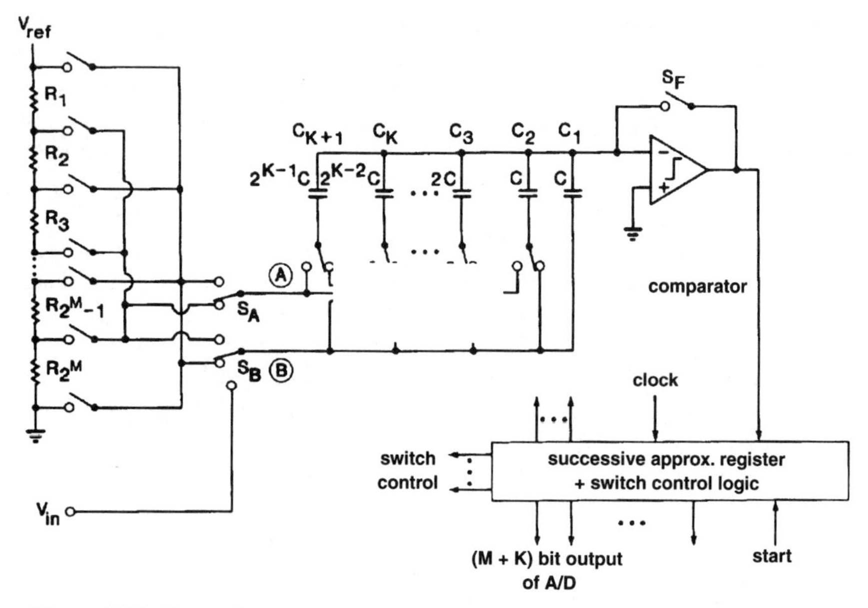

Figure 7.36. Successive-approximation hybrid ADC. (From Ref. 16, © 1979 IEEE.)

than 16 bits. Unless self-calibration is used for high accuracy [17], typical capacitor and resistor matching can guarantee up to 10-bit linearity by following careful layout techniques.

7.6. COUNTING AND TRACKING A/D CONVERTERS

Counting and tracking A/D converters are the two main classes of digital-ramp converters. They contain a binary digital counter and a D/A converter in a feedback loop around a voltage comparator. Figure 7.37 shows the block diagram of a counting A/D converter. In this system, a digital ramp input is applied to a D/A converter. The analog output of the DAC is then compared with the analog input. In each conversion step, first the counter is cleared and then it starts counting the input pulses. The output of the digital counter is applied to the D/A converter, which creates a staircase analog output signal. The counting continues until the D/A output exceeds the input value. At this point, the output of the digital counter is the required output word and is dumped into a storage register. The counter is subsequently cleared, and the circuit is ready to perform the next conversion. Although quite simple in concept, this converter, has the disadvantage of being very low speed for a given resolution. Also, the conversion time is a function of the analog input signal level. For an N-bit converter and a clock period of T, the conversion time for a full-scale input is equal to $T_c = 2^N \times T$. For example, for $N = 12$ and $1/T = 10$ MHz, the maximum input sample rate is $F_c = 2441$ Hz, very low.

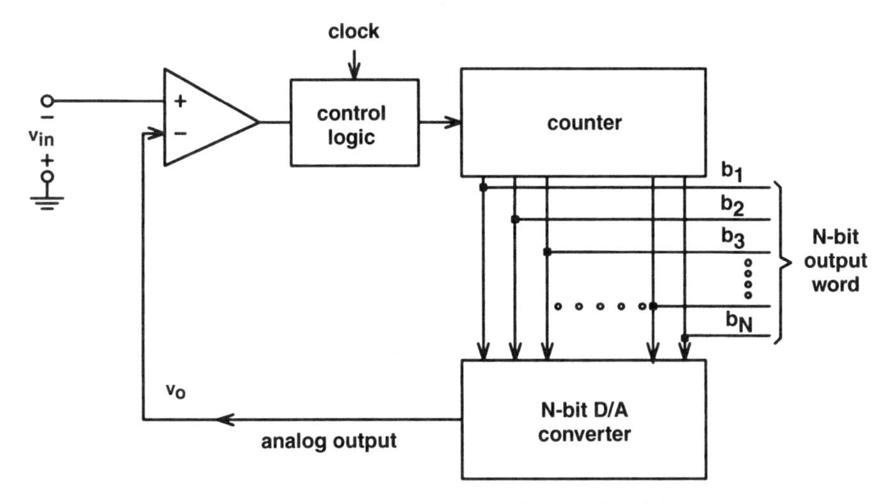

Figure 7.37. Functional block diagram of a counting A/D converter.

An alternative form of the counting A/D converter is the *tracking* or *servo converter,* which is realized by replacing the simple unipolar digital counter of the system shown in Fig. 7.37 with an up-down counter. In this converter, if the input voltage is higher than the output of the D/A converter, the counter counts up. Conversely, if the input voltage is lower than the D/A output, the counter counts down. The analog output of the D/A converter is compared continuously with the analog input. The end of the conversion occurs at the clock period immediately following a change in the state of the comparator, at which point the counter stops counting and its state is dumped into an output register. Unlike the counting A/D converter discussed previously, the counter does not get cleared at the beginning of a conversion cycle. Instead, it retains the previous count and it counts up or down from there based on the polarity of the error signal and the output of the comparator. The conversion is again complete when the output of the D/A reaches the input signal level and the comparator changes state.

If the difference between two consecutive input samples is large, close to full scale, the conversion time for the tracking A/D converter is quite slow, similar to the counting converter, which is on the order of 2^N clock cycles. However, for slowly-varying input signals (i.e., for highly over sampled signals), the conversion time can be quite fast and the digital output can track the analog input signal within a few clock cycles. It is this property of these A/D converters that gave rise to the adjectives *tracking* or *servo.*

7.7. INTEGRATING A/D CONVERTERS [18,20]

Integrating A/D converters are used when high accuracy is needed and low conversion speed is acceptable. In integrating-type high-resolution A/D converters, an indirect conversion is performed by first converting the analog input signal to a time interval whose duration is proportional to the input signal and then converting the time duration into a digital number using an accurate clock and a digital counter. The conversion time is slow because of the lengthy counting operation during the time-to-digital conversion cycle. Single- and dual-ramp ADCs are examples of integrating A/D converters. The dual-ramp ADC is specially suitable for digital voltmeter applications, where the relatively long conversion time provides the benefit of reduced noise due to signal averaging.

A single-slope integrating A/D converter is shown in Fig. 7.38*a*. The circuit consists of an integrator with a resetable feedback capacitor, a comparator, a digital counter, and control logic. The integrator generates the accurate reference ramp signal. Figure 7.38*b* shows the timing waveforms associated to a conversion cycle. Prior to starting the conversion cycle, the digital counter is reset to zero and the feedback capacitor of the integrator is discharged by closing the switch S_F. At the moment the conversion starts, the switch S_F is opened, the input signal v_{in} is applied to the noninverting input of the comparator, and the integrator starts generating the ramp function applied to the inverting input. In the meantime, the control logic enables the digital counter, which starts counting the clock pulses. The counter continues counting the clock pulses, and the count accumulates until the ramp voltage reaches the analog input level v_{in}. At this time the

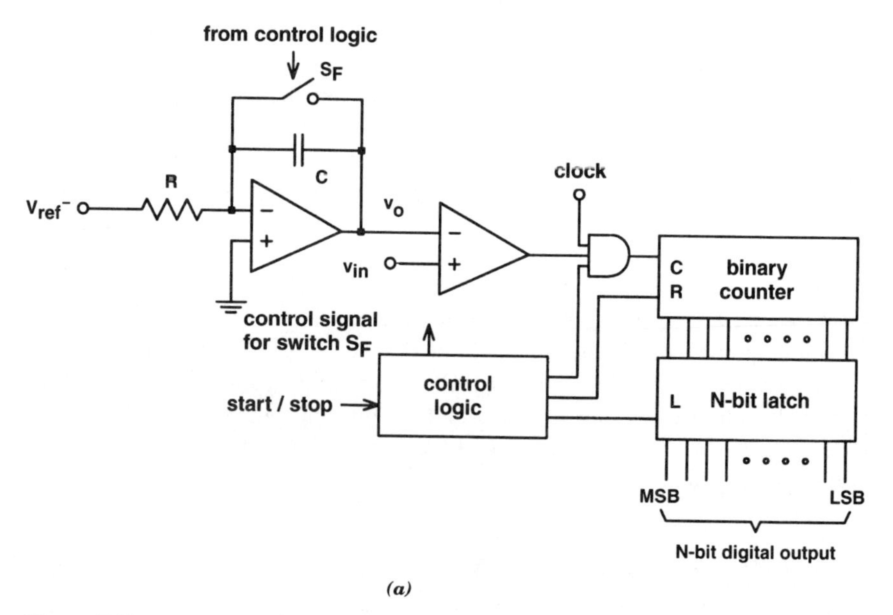

(*a*)

Figure 7.38. (*a*) Basic structure of a single-slope A/D converter; (*b*) timing waveforms.

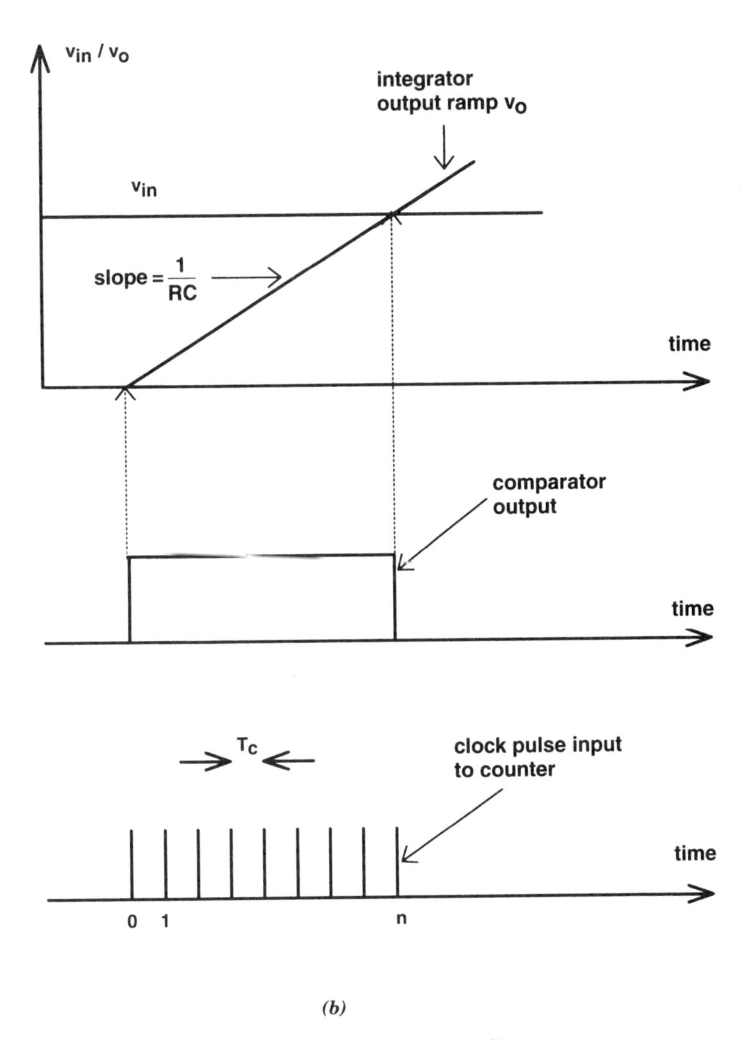

(b)

Figure 7.38. *(continued)*

comparator changes state, stops the counter, and terminates the conversion cycle. This process converts the analog input level to a time-interval duration that is obtained by counting the input clock pulses. If $-V_{\text{ref}}$ represent the negative reference voltage and RC the time constant of the integrator, the time t_1 needed for the ramp to reach the input signal level is

$$t_1 = RC \frac{v_{\text{in}}}{V_{\text{ref}}}. \tag{7.13}$$

The time t_1 as a function of the input clock period and the accumulated count n is given by

$$t_1 \simeq nT_c. \tag{7.14}$$

Of course, n must be an integer number, the nearest to t_1/T_c. For N-bit resolution, the full scale must be at least $n_F = 2^N$, and for a given input level v_{in}, the accumulated count n in the counter is

$$n \simeq 2^N \frac{v_{in}}{V_{ref}}. \tag{7.15}$$

Thus n is the quantized value and digital representation of the input signal. The accuracy of the converter is a function of the time constant RC, the reference voltage V_{ref}, and the clock period.

Several nonideal effects are associated with the basic single-slope A/D converter of Fig. 7.38. These include the offset voltage of the integrator op-amp and comparator, the inaccuracies associated with the initial ramp startup point, and the accuracy of the reference voltage and RC time constant. Improved circuit techniques can be used to avoid these problems. One such circuit technique uses the principle of dual-slope integration. This is one of the most popular techniques for high-accuracy A/D converters and has been used extensively in practice.

The basic block diagram of the dual-slope converter is shown in Fig. 7.39*a*, together with the integrator's output voltage waveform. This type of converter offers a significant advantage over the single-slope converters in that the conversion scale factor will be independent of the integrator time constant and of the clock frequency. The operation of the system is as follows. Prior to starting the conversion, the switch S_F is closed, discharging the integrating capacitor C. At the same time, the counter is reset and the switch S_1 connects the input of the integrator to the input voltage. At the beginning of the conversion, switch S_F is opened and at the same time the counter starts counting clock pulses. The integrator starts from a reset state and integrates the input voltage for a predetermined period of time, normally equal to 2^N clock cycles, which corresponds to the full count of the counter. At the end of the first phase, the voltage at the output of the integrator is given by

$$v_o = 2^N \times T_C \times \frac{v_{in}}{RC}, \tag{7.16}$$

where T_C is the clock period of the counter and N in the resolution of the A/D converter in bits.

In the second phase the counter is reset to zero and the input of the integrator is switched to a reference voltage $-V_{ref}$ which has the opposite polarity with respect to the input signal. The integrator output is now ramping down while the counter again counts the clock pulses starting from zero. When the integrator output reaches

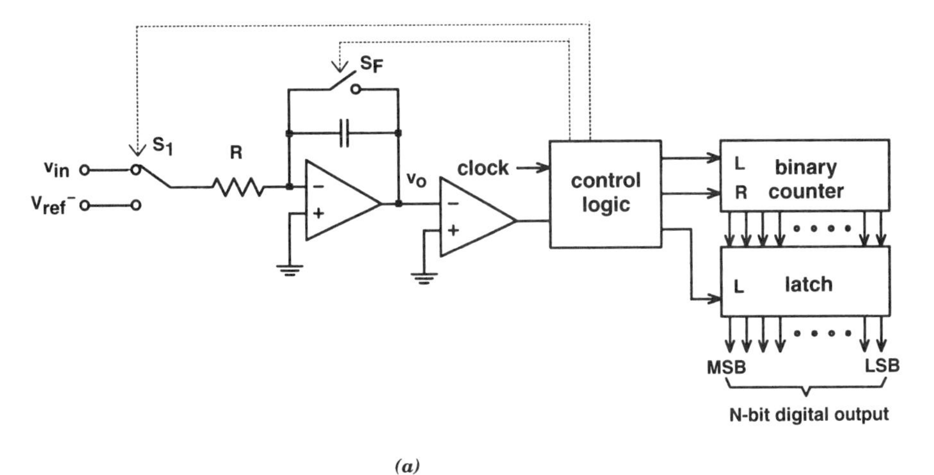

(a)

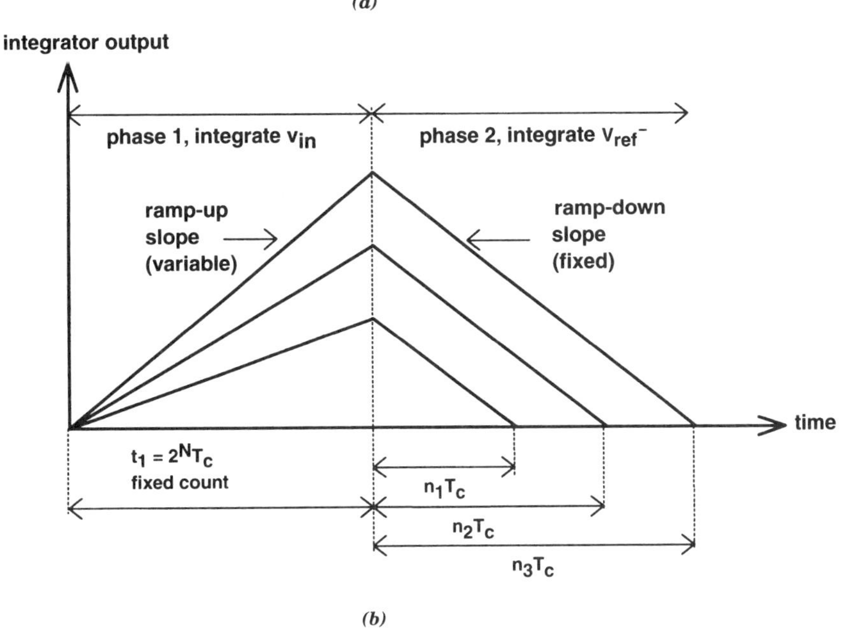

(b)

Figure 7.39. (*a*) Basic block diagram of dual-slope A/D converter; (*b*) integrator output waveform.

zero, the comparator stops the counter. The counter content n is then stored in a latch. Since in the second phase the integrator discharges from its initial voltage v_o to zero in a period $t_2 = nT_C$, t_2 is given by

$$t_2 = nT_C = -\frac{v_o RC}{-V_{\text{ref}}}. \tag{7.17}$$

Combining eqs. (7.16) and (7.17), we have

$$n = 2^N \times \frac{v_{\text{in}}}{V_{\text{ref}}}. \tag{7.18}$$

Therefore, the count stored in the digital counter at the end of phase 2 is the digital equivalent of the analog input.

Figure 7.39*b* shows the output waveform of the integrator during phases 1 and 2. As noted earlier, the duration of phase 1 is fixed at $2^N T_C$, while the integrator output slope varies with the magnitude of the input signal. The duration of phase 2, however, is variable, and the ramp slope is a constant determined by the fixed value of the reference voltage.

The dual-slope A/D converter has several advantages over the single-slope converter. The accuracy is independent of clock frequency and the integrator time constant RC because they affect the up- and down-ramp voltages in the same way. The differential linearity is excellent, because the digital counter generates all codes and all codes are inherently present. The integration provides a rejection of high-frequency noise and noise coupled through the power supplies. For example, if one wishes to reject a specific frequency, such as 60 Hz and its harmonics, the integration time should be 16.67 ms, which sets an upper limit of 30 Hz on the conversion rate. This makes the dual-slope A/D converter suitable for high-accuracy applications with low conversion rates.

The basic dual-slope converter shown in Fig. 7.39 has a major shortcoming: It is sensitive to the comparator and operational amplifier offset voltages, which show up as errors in the output digital word. Introducing auto-zeroing phases into the conversion cycle can reduce the effect of these errors [19]. Using these techniques, multiple-slope integrating A/D converters can be used in the 12- to 14-bit resolution and linearity range.

PROBLEMS

7.1. For a unipolar 10-bit A/D converter, if $v_{\text{in}} = 2.376$ V and $V_{\text{ref}} = 5$ V, determine the digital output code and the quantization error.

7.2. Design a 4-bit flash A/D converter similar to the one shown in Fig. 7.12. Design the decoder logic for the 4-bit digital output code.

7.3. For the two-stage auto-zeroed comparator of Fig. 7.14*a*, draw the timing waveforms at the output of the first, second, and third inverters, based on the timing shown in Fig. 7.14*b*. Assume that $v_{\text{in}} = 1$ V and $V_{\text{ref}} = 1.1$ V.

7.4. Prove Eq. (7.8) for the comparator of Fig. 7.20.

7.5. By analyzing the circuit of Fig. 7.28*a* five times, calculate the values v_x in the five conversion cycles of $v_{\text{in}} = 0.7V_{\text{ref}}$.

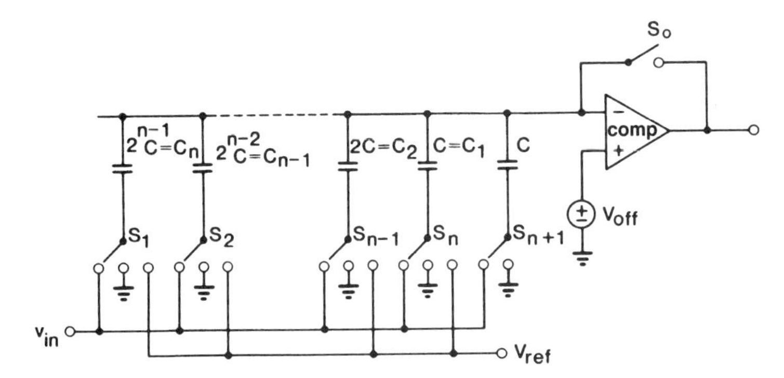

Figure 7.40. Comparator offset elimination method (for Problem 7.7).

7.6. Analyze the effects of the stray capacitances C_{st} and C_{sb} between the top and bottom plates of the capacitors and ground in Fig. 7.28*a* on the digital output.

7.7. The circuit of Fig. 7.40 can be used to eliminate the effect of the comparator offset voltage in the D/A converter of Fig. 6.28*a*. Analyze the operation of the circuit. Find the input referred offset voltage.

7.8. For the hybrid 8-bit A/D converter of Fig. 7.35, draw the timing waveforms of the successive approximation. Include the sampling phase and eight consecutive cycles. Assume that $|V_{ref}| = 5$ V and $v_{in} = 3.1$ V. Find the digital output code.

7.9. For the *N*-bit resistive-string A/D converter of Fig. 7.34, find an expression for the number of switches as a function of *N*.

7.10. For the 16-bit single-slope A/D converter of Fig. 7.38*a*, what is the worst-case conversion time if the clock frequency is 2 MHz?

7.11. Repeat Problem 7.10 for the dual-slope A/D converter of Fig. 3.39*a*.

7.12. For the 16-bit A/D converter of Fig. 7.38*a*, determine the integrating resistor *R* so that the output of the integrator never exceeds 5 V. Assume that $0 < v_{in} < 5$ V, $V_{ref} = 5$ V, $C = 100$ pF, and the clock rate is 2 MHz.

REFERENCES

1. J. G. Peterson, *IEEE J. Solid-State Circuits, SC-14* (6), 932–937 (1979).
2. C. W. Mangelsdorf, *IEEE J. Solid-State Circuits, SC-25* (1), 184–191 (1990).
3. A. G. F. Dingwell, *IEEE J. Solid-State Circuits, SC-14* (6), 926–932 (1979).
4. R. Van de Plassche, *Integrated Analog-to-Digital and Digital-to-Analog Converters*, Kluwer Academic Publishers, Dordrecht, The Netherlands, 1994.

5. J. Caruso, Active analog averaging circuit and ADC using same, U.S. patent 5,298,814, Mar. 29, 1994.
6. N. Fukushima, T. Yamada, N. Kumazawa, Y. Hasegawa, and M. Soneda, *ISSCC Dig. Tech. Pap.,* pp. 14–15, Feb. 1989.
7. M. Ishikawa and T. Tsukahara, *ISSCC Dig. Tech. Pap.,* pp. 12–13, Feb. 1989.
8. T. Shimizu, M. Hatto, and K. Maio, *ISSCC Dig. Tech. Pap.,* pp. 224–225, Feb. 1988.
9. R. J. Van De Plassche and R. E. Van De Grift, *IEEE J. Solid-State Circuits, SC-14* (6), 938–943 (1979).
10. A. G. F. Dingwall and V. Zazzu, *IEEE J. Solid-State Circuits, SC-20* (6), 1138–1143 (1985).
11. T. Sekino, *ISSCC Dig. Tech. Pap.,* pp. 46–47, Feb. 1981.
12. J. L. McCreary and P. R. Gray. *IEEE J. Solid-State Circuits, SC-10* (6), 371–379 (1975).
13. C. C. Shih, K. K. Lam, K. L. Lee, and R. W. Schalk, *IEEE J. Solid-State Circuits, SC-22* (6), 990–995 (1987).
14. A. R. Hamadé, *IEEE J. Solid-State Circuits, SC-13* (6), 785–791 (1978).
15. T. P. Redfern, J. J. Connolly, Jr., S. W. Chin, and T. M. Fredriksen, *IEEE J. Solid-State Circuits, SC-14* (6), 912–920 (1979).
16. B. Fotouhi and D. A. Hodges, *IEEE J. Solid-State Circuits, SC-14* (6), 920–926 (1979).
17. H. S. Lee, D. A. Hodges, and P. R. Gray, *IEEE J. Solid-State Circuits, SC-19* (6), 813–819 (1984).
18. D. H. Sheingold, *Analog–Digital Conversion Notes,* Analog Devices, Norwood, Mass., 1977.
19. F. H. Muse and R. C. Huntington, A CMOS Monolithic $3\frac{1}{2}$ digit A/D converter *ISSCC Dig. Tech. Pap.,* pp. 144–145, 1976.
20. A. B. Grebene, *Bipolar and MOS Analog Integrated Circuit Design,* Wiley, New York, 1984.

CHAPTER 8

PRACTICAL CONSIDERATIONS AND DESIGN EXAMPLES

The design of low-cost and high-performance mixed-signal VLSI systems requires compact and power-efficient analog and digital library cells. While the digital library cells benefit from downscaling of the CMOS process technology, in contrast, analog library cells, such as op-amps and comparators, cannot be designed using minimum-length components, for reasons of gain, dc offset voltage, and other factors. Furthermore, the reduction of the power supply voltage in the tighter geometry technologies does not necessarily result in lower power dissipation in the analog cells, mainly because lower-voltage analog cells are more complex to design and they often require a larger quiescent current.

In Chapters 4 and 5 design techniques for high-performance CMOS op-amps and comparators were discussed in detail. In this chapter the design principles presented earlier in the book are employed to work out several design examples to acquaint the reader with the problems and trade-offs involved in op-amp and comparator designs. Practical considerations in CMOS op-amp design such as dc biasing, systematic offset voltage, power supply, and substrate noise coupling are discussed in some detail.

8.1. PRACTICAL CONSIDERATIONS IN CMOS OP-AMP DESIGN

In Section 4.1, several nonideal effects, which can degrade the performance of practical op-amps, were listed. The minimization of these effects is an important aspect of op-amp design. The corresponding considerations are discussed briefly next for the most important nonideal effects.

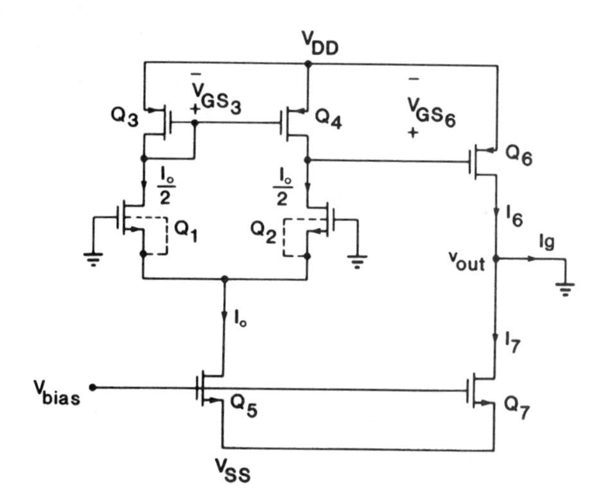

Figure 8.1. Two-stage CMOS op-amp.

1. *Finite Gain.* Earlier we discussed the available gain for various CMOS gain stages and the special circuits that may be utilized, such as cascode devices, to enhance the voltage gain without reducing the bandwidth.
2. *Finite Linear Range.* This question was also discussed briefly and circuits were introduced (see, e.g., Figs. 3.15 and 3.16) for maximizing the allowable signal swing.
3. *Offset Voltage.* As defined in Section 4.1, the input-referred offset voltage $v_{\text{in,off}}$ is the differential input voltage needed to restore the output voltage v_{out} to zero. It contains two components: a *systematic offset,* which is due to improper dimensions and/or bias conditions, and a *random offset,* which is due to the random errors in the fabrication process, resulting, for example, in the mismatch of ideally symmetrical devices.

   To illustrate the generation of systematic offset (and ways to avoid it), consider the simple two-stage CMOS op-amp shown in Fig. 8.1. The first state (Q_1 to Q_5) is the differential-input/single-ended-output input stage introduced in Fig. 3.42; the output stage is a single-ended gain stage with a driver Q_6 and a current-source load Q_7. Clearly, if the circuit has no systematic or random offset, grounding both input terminals (as shown) results in $v_{\text{out}} = 0$. Then if the output terminal is also grounded (as shown in Fig. 8.1), the current I_g in the grounding lead will also be zero. Thus the condition for zero offset is equivalent to requiring that $I_g = 0$ for grounded input and output terminals; this in turn requires that $I_6 = I_7$.

   Assuming symmetry in the input stage, $(W/L)_1 = (W/L)_2$ and $(W/L)_3 = (W/L)_4$. Then all currents and voltages will also be symmetrical, and hence $V_{DS3} = V_{DS4}$. Then also $V_{GS3} = V_{GS6}$. If this value of V_{GS6} results in I_6 being equal to the source current I_7 when $V_{DS6} = 0 - V_{DD} = -V_{DD}$, then $I_g =$

0, as required. If this is *not* the case, then $I_g \neq 0$ and systematic offset exists. Specifically, let $\hat{V}_{GS6}$ denote the value of V_{GS6} needed to make I_6 equal to I_7. Then the input offset voltage is clearly

$$v_{\text{in,off}} = \frac{V_{GS6} - \hat{V}_{GS6}}{A_d} = \frac{V_{GS3} - \hat{V}_{GS6}}{A_d}. \tag{8.1}$$

where A_d is the voltage gain of the *input* stage. For example, an error of 0.1 V in the bias voltage of Q_6 will result in a 1-mV input offset if $A_d = 100$.

Assuming that all devices are in saturation, and neglecting channel-length modulation effects, the voltages of Q_3 and Q_4 can be expressed as

$$V_{GS3} = V_{DS3} = V_{DS4} = V_{Tp} + \sqrt{\frac{I_0/2}{k_p'(W/L)_3}}. \tag{8.2}$$

Here V_{Tp} is the threshold voltage of the p-channel devices Q_3 and Q_4, and k_p' is the constant transconductance factor $\mu_p C_{\text{ox}}/2$ of the drain-current equation for PMOS devices. Similarly, for Q_6,

$$V_{GS6} = V_{Tp} + \sqrt{\frac{I_6}{k_p'(W/L)_6}}. \tag{8.3}$$

Substituting $V_{GS6} = V_{GS3}$ and the required condition $I_6 = I_7$ into Eq. (8.3) yields

$$V_{GS3} = V_{Tp} + \sqrt{\frac{I_7}{k_p'(W/L)_6}}. \tag{8.4}$$

From Eqs. (8.2) and (8.4), the condition

$$\frac{(W/L)_3}{(W/L)_6} = \frac{I_0/2}{I_7}$$

is obtained for zero offset.

Turning to Q_5 and Q_7, since they have equal gate-to-source voltages, neglecting channel-length modulation, we obtain

$$\frac{(W/L)_5}{(W/L)_7} = \frac{I_0}{I_7}. \tag{8.5}$$

Combining the equations above, the design relations are

$$\frac{(W/L)_3}{(W/L)_6} = \frac{(W/L)_4}{(W/L)_6} = \frac{1}{2}\frac{(W/L)_5}{(W/L)_7} = \frac{I_0}{2I_7}. \tag{8.6}$$

Physically, if Eq. (8.6) is satisfied, the current I_7 induced in Q_7 by its gate-to-source voltage V_{GS7} and the current I_6 induced in Q_6 by the gate-to-source voltage $V_{GS6} = V_{DS4}$ are the same, and hence $v_{out} = 0$ is possible when Q_6 and Q_7 are in saturation. If the gate-to-source voltages are *not* compatible and the output terminal is open-circuited, v_{out} assumes a nonzero value such that the drain voltages of Q_6 and Q_7 will compensate for the discrepancies of the gate voltages. This may also result in Q_6 or Q_7 operating out of saturation. It usually represents a major systematic offset voltage effect and may reduce the gain and the bandwidth of the op-amp.

To minimize the effects of random process-induced channel-length variations on the matching of the devices, and thus the random offset, the channel lengths of Q_3, Q_4, and Q_6 should be chosen equal. Then the current density I_d/W is the same for these devices when Eq. (8.6) is satisfied, and the required current ratios are determined by the ratios of the widths. If ratios as large as (or larger than) two are required, the wider transistor can be realized by the parallel connection of two (or more) "unit transistors" of the size of the narrower one. Note, however, that this process is in conflict with the guidelines for the minimization of noise established in Section 4.7: According to those rules, the transconductances of the load devices Q_3 and Q_4 should be low while that of Q_6 should be high for high gain and good high-frequency response. The channel lengths for all three devices Q_3, Q_4, and Q_6 should be long for high output impedance and high gain. While this helps to reduce the noise of Q_3 and Q_4 and increase the gain of the differential stage, it decreases the transconductance of Q_6 and hence reduces the gain of the second stage. Clearly, the optimum trade-off among these conflicting requirements will vary from application to application.

The random offset voltage will be affected by several factors, including mismatch between the (ideally symmetrical) input devices Q_1 and Q_2 and/or between the load devices Q_3 and Q_4. This can be caused either by geometrical mismatch or by a process gradient causing different threshold voltages. Assume first that the current mirror (Q_3–Q_4) is imperfect, so that

$$I_3 = \tfrac{1}{2}(1 - \epsilon_1)I_0 \neq I_4 = \tfrac{1}{2}(1 + \epsilon_1)I_0. \tag{8.7}$$

The differential voltage $v_{G1} - v_{G2}$ needed at the input terminal to restore symmetry is clearly

$$v_{\text{off1}} = \frac{\epsilon_1 I_0}{g_{mi}}. \tag{8.8}$$

Thus v_{off1} can be reduced by increasing the transconductance g_{mi} of the input devices or by reducing the bias current I_0.

Assume next that the dimensions and the threshold voltages of the input devices are mismatched while the load devices are symmetrical. Thus let

$$(W/L)_1 = (1 - \epsilon_2)(W/L)_2, \tag{8.9}$$

$$V_{T1} = V_{T2} - \Delta V_T. \tag{8.10}$$

Clearly, it requires an input offset voltage $v_{\text{off2}} = \Delta V_T$ to cancel the effect of the threshold voltage mismatch. The geometric mismatch causes a current imbalance $\Delta I_1 \simeq -\epsilon_2 I_1 \simeq -\epsilon_2 k_1(V_{GS1} - V_{T1})^2$. This can be balanced by a change v_{off3} in v_{G1} such that

$$\begin{aligned} g_{mi} v_{\text{off3}} &\simeq 2k_1(V_{GS1} - V_{T1}) v_{\text{off3}} = \Delta I_1 \\ &\simeq \epsilon_2 k_1 (V_{GS1} - V_{T1})^2. \end{aligned} \tag{8.11}$$

Here, we used Eq. (2.18) to express g_{mi}, with $v_{SB}^0 = 0$ and $\lambda \simeq 0$. From Eq. (8.11),

$$v_{\text{off3}} = \frac{\epsilon_2}{2}(V_{GS1} - V_{T1}) = \frac{\epsilon_2}{2}\sqrt{\frac{I_0/2}{k'(W/L)_1}}. \tag{8.12}$$

Therefore, v_{off3} can be reduced (as could v_{off1}) by increasing $(W/L)_1$—and thus g_{mi}—or by reducing I_0. Both will reduce $V_{GS1} - V_{T1}$.

The variation of the threshold voltage ΔV_T is independent of I_0 or W/L; it depends only on process uniformity. It can be reduced by building Q_1 and Q_2 from unit transistors arranged in a common-centroid structure [1, Chap. 6; 2].

4. *Common-Mode Rejection Ratio* (CMRR). As defined in Section 4.1, CMRR $= A_D/A_C$, where A_D is the differential gain and A_C is the common-mode gain. For the op-amp of Fig. 8.1, the common-mode rejection is provided by the input stage. The value of the CMRR for this stage was found earlier and was given by Eq. (3.88) as

$$\text{CMRR} \simeq 2\frac{g_{mi} g_{ml}}{g_0 g_{di}}. \tag{8.13}$$

As explained in Section 2.4, both g_{mi} and g_{ml} are proportional to $\sqrt{I_0}$, while g_0 and g_{di} are proportional to I_0. Thus the rejection ratio is inversely proportional to I_0. Values of 10^3 to 10^4 can readily be achieved, as Eq. (8.13) shows.

If a mismatch exists between Q_1 and Q_2 such that $g_{mi} = (1 + \epsilon)g_{m2}$, a common-mode voltage v_c will cause as much differential output voltage as a differential input voltage $(g_{di}/g_{mi})\epsilon v_c$ (Problem 8.1). Thus now we have CMRR $\simeq g_{mi}/\epsilon g_{di}$. This again illustrates the importance of making the input devices symmetrical, using common-centroid geometry if necessary: A 1% mismatch may lower the CMRR to 60 dB!

5. *Frequency Response, Slew Rate, Biasing, Power Dissipation.* The requirements on the speed (i.e., high-frequency gain and slew rate) of a CMOS amplifier depend very much on its application. In switched-capacitor circuits, most op-amps drive capacitive loads only. The speed requirement is then that the op-amp must be able to charge the load capacitance C_L and settle to within a specified accuracy (usually, 0.1% of the final voltage) in a specified time interval.

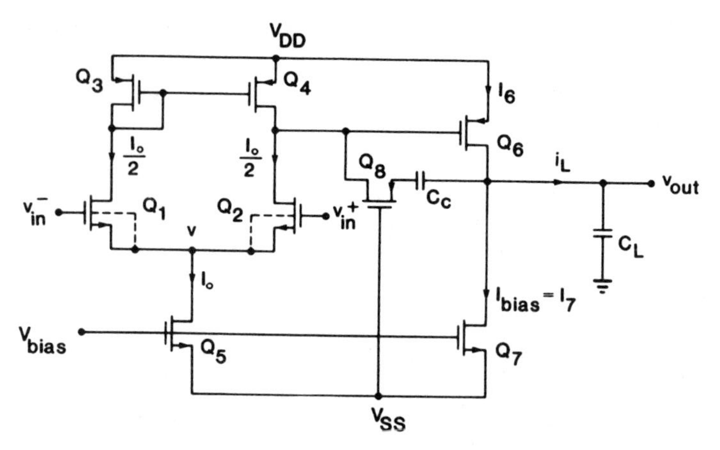

Figure 8.2. CMOS op-amp with capacitive load.

Figure 8.2 shows the simple op-amp of Fig. 8.1 supplemented by a feedback branch (Q_8, C_c) for compensation and driving a capacitive load C_L. As explained in Section 4.6, the unity-gain bandwidth of the stage is given by $\omega_0 = g_{mi}/C_c$. A detailed analysis of the performance of a compensated op-amp used to charge or discharge a capacitor reveals [3,4] that (in linear operation) the condition on the unity-gain bandwidth,

$$\omega_0 = \frac{g_{mi}}{C_c} \geq \frac{15}{T_{ch}} \tag{8.14}$$

is usually sufficient to guarantee adequate speed. Here T_{ch} is the time available for recharging C_L; for a two-phase circuit, usually $T_{ch} \simeq \frac{1}{2f_c}$, where f_c is the clock frequency.

The minimum value of g_{mi} is usually determined by the required dc gain and by noise considerations. In addition, there is also an upper bound on g_{mi}/C_c, based on the requirement that the second pole frequency $|s_{p2}|$ must be considerably higher than ω_0. Often, $|s_{p2}| = 3\omega_0$ is chosen. This, as derived in (4.57) to (4.62), requires for our circuit that

$$C_c \sim 3C_L \frac{g_{mi}}{g_{m6}}. \tag{8.15}$$

A rule of thumb that usually results in a good compromise among all requirements is to choose $C_c = C_L$. Then Eq. (8.14) gives a lower bound for g_{mi} and Eq. (8.15) a lower bound for g_{m6}.

For large-signal operation, the slew rate S_r of the input stage must also be considered. From Eq. (4.94), $S_r = I_0/C_c$. For $C_c = C_L$, the minimum value of I_0 (the bias current of the input stage) is thus determined. In addition, for the circuit of Fig. 8.2, the slew rate of the output also needs attention. For *positive-going* v_{out}, the output current i_L is supplied by Q_6. The magnitude of i_L is limited only by the size of Q_6, and that of v_{out} only by the v_{DS6} needed to keep Q_6 in saturation. For *negative* v_{out}, by contrast, the output stage must *sink* the load current i_L. This is performed by reducing I_6 below I_{bias} so that $I_6 = I_{bias} - |i_L|$.

The maximum value $|i_L| = I_{bias}$ is obtained when $I_6 = O$, that is, when Q_6 is cut off. The negative-going slew rate due to the output stage is then

$$S_{r0} = \left|\frac{dv_{out}}{dt}\right| = \frac{|i_L|}{C_L} = \frac{I_{bias}}{C_L}. \tag{8.16}$$

This establishes the minimum value of I_{bias}. Note that by Eqs. (4.56) and (4.61), the transconductances of the output devices (here Q_6 and Q_7) must be large for stability. A large I_{bias} will help to satisfy this with moderate-sized Q_6 and Q_7.

The dc power dissipated by the circuit of Fig. 8.1 in the quiescent state is thus

$$\begin{aligned}(V_{DD} - V_{SS})(I_0 + I_{bias}) &\geq (V_{DD} - V_{SS})(S_r C_c + S_{r0} C_L) \\ &= (V_{DD} - V_{SS})(S_r + S_{r0}) C_L,\end{aligned} \tag{8.17}$$

where $C_c = C_L$ has been set. Therefore, the higher the slew rates and the larger C_L, the more dc standby power is needed by the stage.

Using a class AB stage can often reduce the standby current of the output stage. The resulting circuit was discussed earlier (cf. Fig. 4.15) and is reproduced in Fig. 8.3. The level-shifter stage (Q_9 and Q_{10}) should be dimensioned

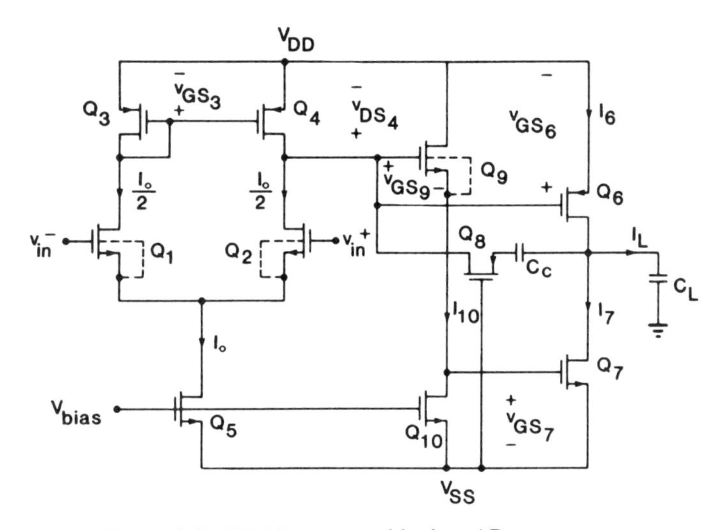

Figure 8.3. CMOS op-amp with class AB output stage.

such that the voltages v_{GS6} and v_{GS7}, and hence the quiescent drain currents of Q_6 and Q_7, are not too large. Since signal voltages drive the gates of both Q_6 and Q_7, the load current can now be much larger than the bias current for either positive- or negative-going v_{out}. The minimum value of the output-dc bias current is thus determined only by the requirement on the transconductances g_{m6} and g_{m7} needed for a good phase margin.

The value of the standby current I_{10} will determine the locations of the zero and the pole of the source follower Q_9–Q_{10}, as given by Eq. (3.103). Obviously, the pole (which is at a lower frequency) will result in a positive phase shift, while the higher-frequency left-half-plane zero will cause a negative phase, resulting in a dip in the phase characteristic. The location of the dip is determined by the frequency of the pole, and its depth and width by the separation between the pole and the zero. The minimum value of I_{10} can be thus determined such that the dip moves to a sufficiently high frequency where it affects only slightly the phase shift of the stage at ω_0. This requirement gives the minimum value of g_{m9}, and thus of the bias current I_{10}. Forcing the zero as close to the pole as possible can reduce the depth of the dip. It is worth mentioning that the source follower's time response, due to the presence of the pole–zero pair, will contain an exponential term. In this term the time constant is determined by the pole frequency, while the amplitude (residue) is determined by the difference between the pole and zero frequencies. Increasing the pole frequency and reducing the distance between the pole and zero will therefore improve the time response as well.

The voltage V_{bias} for the op-amps of Figs. 8.2 and 8.3 can be obtained using the circuits of Section 3.1. In particular, for the op-amp of Fig. 8.1 a supply-independent bias is desirable. This will keep I_0 and I_{bias} independent of the supply voltages, and hence the parameters, which affect the stability, unchanged. (The dc power dissipation will, of course, vary with the supply voltages.) Thus V_{bias} may be obtained from the circuit of Fig. 3.5.

For the circuit of Fig. 8.3, by contrast, supply-independent biasing can cause problems. In particular, if I_0 remains constant with supply voltage variation, V_{GS3} does also; if I_0 remains constant, V_{GS9} does too. Hence in the expression for the gate-to-source voltage of Q_7,

$$V_{GS7} = V_{DD} + V_{DS4} - V_{GS9} - V_{SS} \simeq (V_{DD} - V_{SS}) + (V_{GS3} - V_{GS9}), \quad (8.18)$$

$(V_{GS3} - V_{GS9})$ is invariant of V_{DD} and V_{SS}. Therefore, all power supply changes appear directly in V_{GS7}, and Q_7 can cut off if $V_{DD} - V_{SS}$ drops significantly.

A suitable bias circuit for the op-amp of Fig. 8.3 is shown in Fig. 8.4. When this circuit is used, the dimensions of the NMOS devices Q_5, Q_{10}, and Q_{13} are obviously related by

$$\frac{(W/L)_5}{(W/L)_{13}} = \frac{I_0}{I_{ref}}, \qquad \frac{(W/L)_{10}}{(W/L)_{13}} = \frac{I_{10}}{I_{ref}}. \quad (8.19)$$

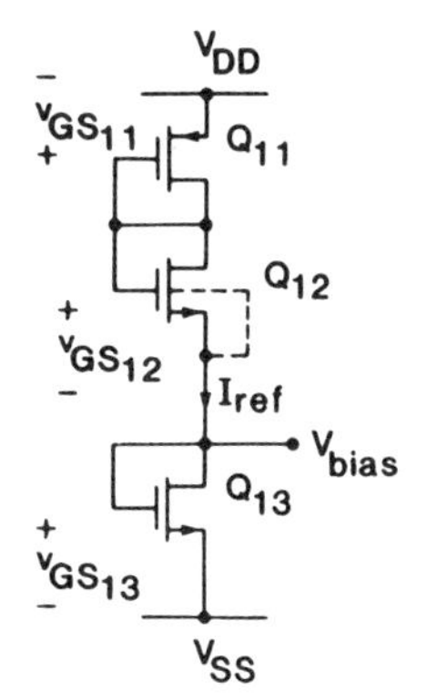

Figure 8.4. Bias circuit for the op-amp of Fig. 8.3.

In addition, the PMOS transistors Q_3, Q_4, and Q_{11} can be dimensioned such that

$$\frac{(W/L)_3}{(W/L)_{11}} = \frac{(W/L)_4}{(W/L)_{11}} = \frac{I_0/2}{I_{\text{ref}}} \tag{8.20}$$

holds. Then we have

$$V_{GS3} = V_{DS3} = V_{GS4} = V_{DS4} = V_{GS11}. \tag{8.21}$$

In addition, we can dimension the NMOS devices Q_9 and Q_{12} so as to satisfy

$$\frac{(W/L)_9}{(W/L)_{12}} = \frac{I_{10}}{I_{\text{ref}}} = \frac{(W/L)_{10}}{(W/L)_{13}}. \tag{8.22}$$

This will cause $V_{GS9} \simeq V_{GS12}$, and using Eqs. (8.18) and (8.21), we find that

$$\begin{aligned} V_{GS7} &= V_{DD} + V_{DS4} - V_{GS9} - V_{SS} = V_{DD} + V_{GS11} - V_{GS12} - V_{SS} \\ &= V_{\text{bias}} - V_{SS} = V_{GS13} = V_{GS5} = V_{GS10}. \end{aligned} \tag{8.23}$$

This matching of voltages is independent of V_{DD} and V_{SS}.

Thus if I_7 is the desired value of the output bias current, we must choose the dimensions of the NMOS devices Q_7 and Q_{13} to satisfy

$$\frac{(W/L)_7}{(W/L)_{13}} = \frac{I_7}{I_{\text{ref}}}. \tag{8.24}$$

Finally, since clearly

$$V_{GS6} = V_{DS4} \simeq V_{DS3} = V_{GS3}, \tag{8.25}$$

we should also choose

$$\frac{(W/L)_6}{(W/L)_{11}} = \frac{I_7}{I_{ref}}. \tag{8.26}$$

Note that this choice of dimensions will establish the desired bias currents without introducing any systematic offset voltage. Assume now that the supply voltages vary in the circuit. Then I_{ref} will change, and so will v_{GS11}, v_{GS12}, and v_{GS13}. However, Q_{11}, Q_{12}, and Q_{13} will certainly continue to conduct; in fact, they will all also remain in saturation since their gates and drains are connected. But since the currents and gate-to-source voltages of Q_5, Q_{10}, and Q_7 mirror those of Q_{13}, they too will conduct and remain in saturation. Also, since the currents and gate-to-source voltages of Q_3, Q_4, and Q_6 mirror those of Q_{11}, these transistors will conduct and remain saturated. Finally, the conduction and saturation conditions of Q_9 follow from those of Q_{12}. Thus all transistors have stabilized dc bias conditions, and the operation of the circuit will be insensitive to process variations.

6. *Nonzero Output Resistance.* This is usually important only for the output amplifier of the overall circuit, which may have to drive a large capacitive and/or resistive load. The low-frequency output resistance R_{out} of the op-amp *without* negative feedback (i.e., open loop) is on the order of $r_d/2$ for an unbuffered circuit, such as that shown in Fig. 8.2 or 8.3. Here r_d is the drain resistance of the output devices, on the order of 0.1 to 1 MΩ. For a buffered circuit (such as that shown in Fig. 4.12), the output impedance is around $1/g_m$, where g_m is the transconductance of the output device; hence $R_{out} \sim 1$ kΩ. In closed-loop operation, the effective output impedance is $R_{out}(1 - A_{CL})/A$, where A is the open-loop gain and A_{CL} is the closed-loop gain (Problem 8.3). Since usually $A > 1000$, the effective closed-loop impedance is around 1 kΩ for unbuffered op-amps, and very low (on the order of a few ohms) for buffered ones. This value is sufficiently low for most applications.
7. *Noise and Dynamic Range.* These subjects were discussed in some detail in Section 4.7 for CMOS op-amps. Hence they are not analyzed here.
8. *Power Supply Rejection.* This is one of the most important nonideal effects in MOS analog integrated circuits, for several reasons. First, several circuits (some analog, some digital) may operate off the same power supply. Hence a number of analog and digital signal currents can enter the supply lines. Since these lines have nonzero impedances, digital and analog voltage noise will be superimposed on the dc voltage provided by the supply. If the op-amp circuit does not reject this noise, the noise will appear at its output, reducing the signal/noise ratio and the dynamic range. Second, if switching regulators or dc

voltage multipliers are used, a substantial amount of high-frequency switching noise will be present on the supply line. Finally, the clock signals of the various circuits fed from the same line will also usually appear superimposed on the supply voltage with reduced but nonzero amplitude.

In a sampled-data circuit (such as a switched-capacitor circuit), all signals are sampled periodically. This results in a frequency mixing, and as a result the high-frequency noise will be "aliased" into the frequency band of the signal. Hence *any* noise in the overall frequency range is detrimental if it can enter the signal path.

The most likely path through which supply noise can be coupled to the signal is via the op-amps. Thus a high value of the power supply rejection ratio (PSRR), defined in Section 4.1 as the ratio of the open-loop differential gain A_D and the noise gain A_p from the supply to the op-amp output, is of great importance.

At low frequencies, the supply noise is coupled into the op-amp mostly through the bias circuits and can also enter due to the asymmetries in the differential input stage. At high frequencies, on the other hand, the capacitive branches mostly determine the noise gain. Consider the circuits of Figs. 8.2 and 8.3. At high frequencies, the compensation capacitor C_c becomes nearly a short circuit, and (since the drain–source resistance r_8 of Q_8 is small in the linear region), the gate and drain voltages of Q_6 are nearly equal. As shown in Fig. 8.5, therefore, the *incremental* output voltage due to the supply noise v_n is

$$\Delta v_{\text{out}} = v_n + \Delta v_{GS6} = v_n$$

since $v_{GS6} = I_7/g_{m7}$ is constant. Thus $A_p = 1$, independent of frequency, for this stage. The output impedance of Q_6 driving C_L with this noise voltage is low, around $1/g_{m6}$. Since the open-loop gain A_D decreases with increasing frequency at a rate of -6 dB/octave while A_p stays constant, the PSRR due to noise in V_{DD} decreases at the same rate as A_D, reaching 0 dB near the unity-gain frequency.

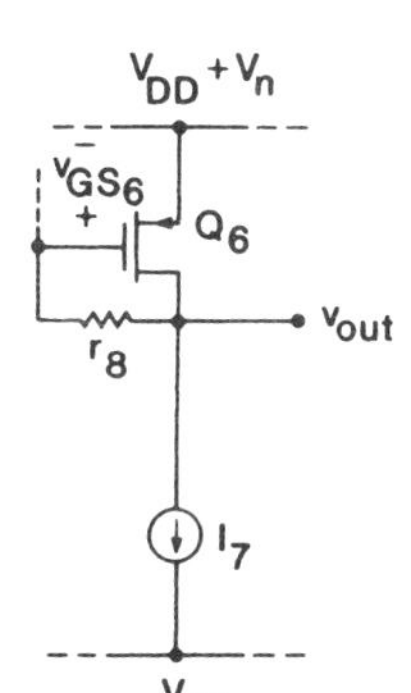

Figure 8.5. High-frequency model of the op-amp output stage.

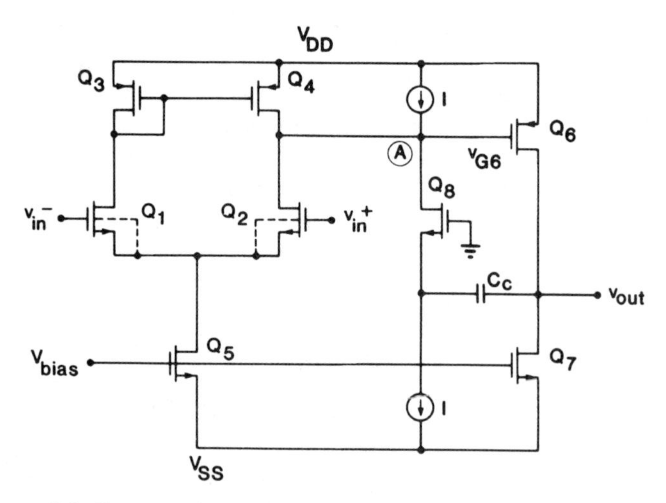

Figure 8.6. Compensation scheme for improving the positive-supply PSRR.

The situation is more favorable with respect to noise on the V_{SS} line. The gain for noise entering via Q_5 is the low common-mode gain. Any noise entering via Q_{10} and Q_7 is added to the signal and attenuated at the same rate (-6 dB/octave) as the signal by the load capacitor C_L since now the noise output impedance is high.

An effective technique [5] for increasing the PSRR for noise coming from the positive supply is illustrated in Fig. 8.6. In contrast to the circuits of Figs. 8.2 and 8.3, where Q_8 operated as a linear resistor for both feedback and feedforward signals, Q_8 is now biased in its saturation region. Hence in the feedback direction (from v_{out} through C_c and Q_8 to the gate of Q_6) the resistance is $1/g_{m8}$, while viewed from node A the drain of Q_8 shows a high impedance r_{d8}. Thus, while the feedback (and also the compensation) remains functional, the feedforward path for noise is interrupted. Specifically, if V_{DD} changes, the source voltage of Q_6 does, too, and (as explained earlier) the gate voltage of v_{g6} must follow. Now, however, the output terminal is not shorted to v_{g6}, and hence v_{out} need not follow v_{G6}. This reduces A_p considerably at high frequencies.

The currents of the two sources I (needed to keep Q_8 in its saturated region) must be carefully matched. This is possible using the strategy explained in connection with Fig. 8.4. Any mismatch will introduce a systematic offset voltage. Also, the impedance at node A is somewhat reduced and therefore so is the gain of the input stage. Finally, due to the added devices, the internally generated noise of the op-amp is increased; however, the effect of increased PSRR usually outweighs this and the overall noise at the op-amp output is reduced.

Another path for power supply noise injection is provided by the stray

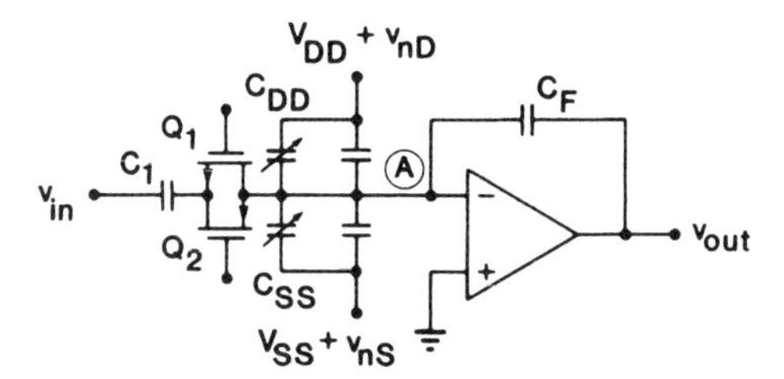

Figure 8.7. Parasitic capacitances affecting the summing node A of an integrator.

capacitances [6,7]. Consider the circuit of Fig. 8.7. It illustrates a typical switched-capacitor integrator in one of its switched states. The two parallel-connected transistors Q_1 and Q_2 constitute the switch. The parasitic junction capacitances coupling the drain of Q_1 and the source of Q_2 to the substrate, as well as the stray capacitances between the lines connected to the inverting input (node A) and the substrate and power lines, are illustrated as C_{DD} and C_{SS}. (As shown, these capacitors contain both linear and nonlinear components.) Consider the effect of v_{nD}. Since node A is a virtual ground, the noise charge entering is $C_{DD}v_{nD}$. This charge flows into C_F and causes an output noise voltage $-(C_{DD}/C_F)v_{nD}$. Thus the noise gain is $-C_{DD}/C_F$. Similarly, the noise gain for v_{nS} is $-C_{SS}/C_F$. Since the signal gain is $v_{\text{out}}/v_{\text{in}} = -C_1/C_F$, the PSRR of the integrator is $C_1/(C_{SS} + C_{DD})$.* Thus the PSRR can be increased by minimizing the stray capacitances and by choosing the values of the capacitors C_1 and C_F sufficiently large—the latter, of course, increases the overall area occupied on the chip by the stage. To reduce the stray capacitances and thereby the influence of the substrate noise, the dimensions of the switches should be chosen as small as possible, and whenever feasible, all lines connected to the input nodes of the op-amps should be shielded by grounded polysilicon or diffusion planes placed between the lines and the substrates.

Parasitic capacitances inside the op-amp also contribute to the power supply noise gain [7]. Consider the equivalent circuit of the op-amp with an external feedback capacitor C_F, as shown in Fig. 8.8. The figure also illustrates the stray capacitances C_{gd} and C_{gs} of the input device Q_1. Noise in the positive line will appear at the drains of Q_3 and Q_4, and from the former will be coupled to the input node A via C_{gd} and from there to the output via C_F. In addition, variations of the bias current I_0 due to the noise in V_{DD} and V_{SS} will change the gate-to-source voltages v_{gs} of Q_1 and Q_2 by $\Delta v_{gs} \simeq (\Delta I_0/2)g_{mi}$. The corresponding change in the source voltages will be coupled to node A by C_{gs}. Similarly, changes in V_{SS} will alter the threshold voltage V_{Tn} of Q_1 and Q_2, unless these devices are placed in an isolated p well. The resulting change in

* This calculation assumes the worst-case condition that v_{nD} and v_{nS} are fully correlated.

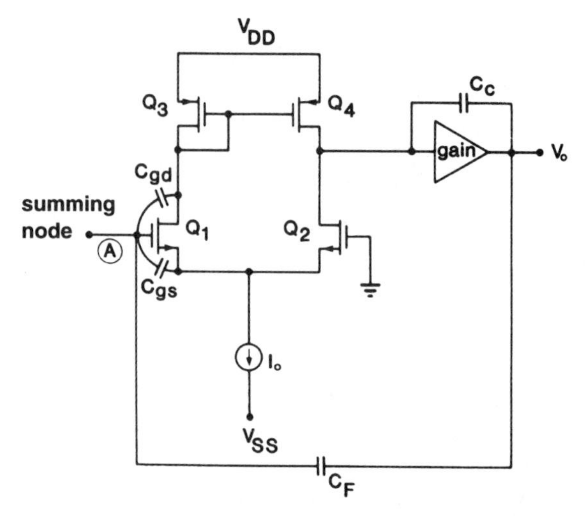

Figure 8.8. Equivalent circuit of the op-amp connected as an integrator.

the source voltages will also be coupled to the input node by C_{gs} and to the output via C_F. Overall, it can be shown [7] that the relations

$$\begin{aligned}\frac{\partial v_{\text{out}}}{\partial V_{SS}} &\simeq \frac{C_{gs}}{C_F}\left(\frac{\partial I_0}{\partial V_{SS}}\frac{1}{2g_{mi}} + \frac{\partial V_{Tn}}{\partial V_{SS}}\right) + \frac{C_{gd}}{C_F}\frac{1}{2g_{ml}}\frac{\partial I_0}{\partial V_{SS}}, \\ \frac{\partial v_{\text{out}}}{\partial V_{DD}} &\simeq \frac{C_{gd}}{C_F}\left(1 - \frac{\partial I_0}{\partial V_{DD}}\frac{1}{2g_{mi}}\right) + \frac{C_{gs}}{C_F}\frac{1}{2g_{mi}}\frac{\partial I_0}{\partial V_{DD}}\end{aligned} \tag{8.27}$$

give the power supply noise gains of the circuit. The terms containing $\partial I_0/\partial V_{SS}$ and $\partial I_0/\partial V_{DD}$ can be eliminated by using supply-independent biasing for the current source I_0. The $\partial V_{Tn}/\partial V_{SS}$ term can be eliminated in CMOS *amplifiers* by using a p well (n well for p-type input devices) for Q_1 and Q_2, connected to their sources.

If all the steps above are taken to reduce the noise gain, the remaining gain is $\partial v_{\text{out}}/\partial V_{DD} \simeq -C_{gd}/C_F$. This can, in principle, be reduced by making Q_1 and Q_2 small and/or C_F large. The former measure, however, results in an increase of internally generated noise, while the latter increases the chip area needed by the stage. A technique that eliminates the problem, at the cost of a slightly reduced common-mode input range, is to use cascode circuitry (Fig. 8.9) in the input stage. The added devices Q_5 and Q_6 then buffer the drains of Q_1 and Q_2 from the variations of V_{DD} provided that V_{bias} is independent of the supply voltages.

8.2. OP-AMP DESIGN TECHNIQUES AND EXAMPLES

The design of MOS op-amps is not an exact scientific process. Typically, the circuit must satisfy many requirements, often conflicting ones. The op-amp performance parameters most often specified are collected in Table 8.1. Other important design

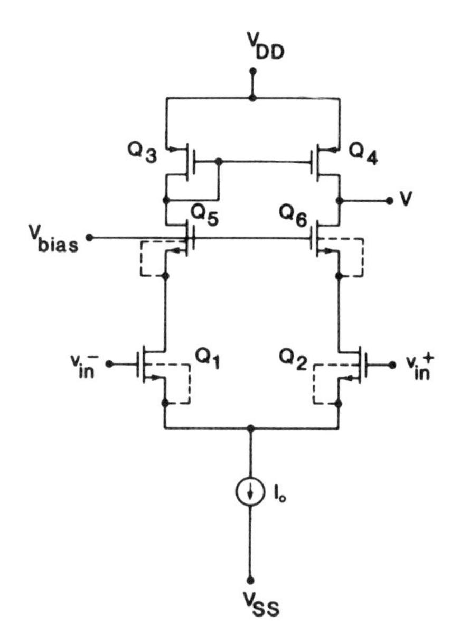

Figure 8.9. Cascode CMOS differential input stage.

TABLE 8.1. Op-Amp Performance Parameters

Design Parameter	Symbol	Relation to Other Parameters[a]	Typical Values
Low-frequency open-loop gain	A_0	$\dfrac{g_{mi}(g_{m6} + g_{m7})}{(g_{dl} + g_{di})(g_{d6} + g_{d7})}$	10^3–10^4
Unity-gain frequency	f_0	$\dfrac{g_{mi}}{2\pi C_c}$	$f_0 = 1$–10 MHz
Slew rate	S_r	$\dfrac{I_0}{C_c}$	2–20 V/μs
Common-mode rejection ratio	CMRR	$2\,\dfrac{g_{mi}g_{ml}}{g_{d5}g_{di}}$	60–80 dB
dc power drain	P_{dc}	$(V_{DD} - V_{SS})(I_0 + I_7 + I_{10})$	0.5–10 mW
Phase margin (open loop)	ϕ_M	$\phi_M > 60°$ for $\lvert s_{p2}\rvert \geq 3\omega_0$	45°–90°
Load impedance	R_L, C_L	None	1–100 kΩ 1–100 pF

[a] The formulas in this column are given for the circuit of Fig. 8.3. Set g_{m7} and I_{10} equal to zero to obtain the relations for the amplifier of Fig. 8.2.

criteria include the noise level, dynamic range, output impedance, and the area occupied on the chip. The specific steps followed in the design depend on the application, the circuit chosen, and the relative importance of the various criteria.

To illustrate the process, the devices in the circuit of Fig. 8.2 will be dimensioned such that the following specifications are satisfied:

Low-frequency gain	$A_0 = 70$ dB
Unity-gain frequency	$f_0 = 2$ MHz
Slew rate	$S_r = 4$ V/μs
Common-mode rejection	CMRR = 80 dB
Phase margin	$\phi_M > 60°$
Load impedance	$C_L = 10$ pF
dc supply voltages	$V_{DD} = -V_{SS} = 5$ V

It will be assumed that the transconductance factor $k' \triangleq \mu C_{ox}/2$ is 30 μA/V^2 for the NMOS devices and 12 μA/V^2 for the PMOS devices. Also, the threshold voltages are assumed to be $V_{Tn} = 1.2$ V and $V_{Tp} = -1$ V.

As mentioned in Section 8.1, it is usual to choose the value of the compensation capacitor C_c equal to C_L. Hence we set

$$C_c = C_L = 10 \text{ pF}. \tag{8.28}$$

For an adequate phase margin ϕ_M, the frequency of the second pole s_{p2} of the open-loop gain should be sufficiently higher than ω_0, the unity-gain frequency. As Fig. 4.22 shows, for $|s_{p2}| \sim 2\omega_0$ the contribution of the factor $j\omega - s_{p2}$ to the phase at $\omega = \omega_0$ is about 30°, and hence the phase margin is 60°. Thus $|s_{p2}| = 3\omega_0$ gives a margin greater than 60°. The value of s_{p2} can be found from the small-signal equivalent circuit of the op-amp of Fig. 8.2, shown in Fig. 8.10. Alternatively, we can use the formula (4.55) derived from the circuit of Fig. 4.28; here, however, we must replace g_{m8} by g_{m6} (since Q_6 is now the output driver) and omit g_{m9} (since Q_7, the

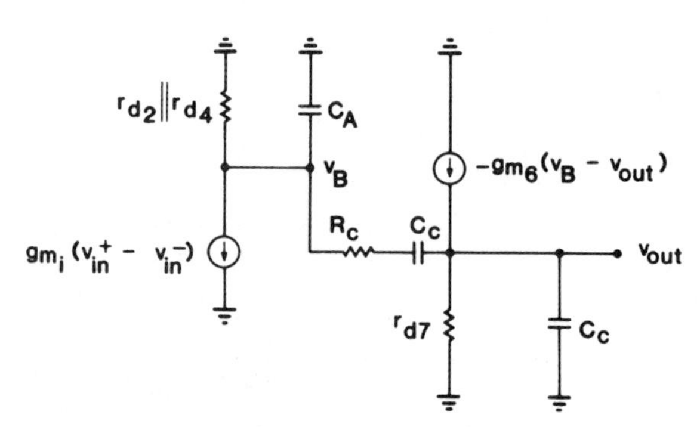

Figure 8.10. Small-signal equivalent circuit of the op-amp of Fig. 8.2.

lower output device, is used merely as a current source). Assuming that $C_A \ll C_L = C_c$ and $g_{m6} \gg g_{d7}$,

$$s_{p2} = \frac{-g_{m6}C_c}{C_A(C_L + C_c) + C_LC_c} \simeq -\frac{g_{m6}}{C_L} \tag{8.29}$$

results. Therefore, as already given in Eq. (8.15),

$$|s_{p2}| = \frac{g_{m6}}{C_L} = 3\omega_0 = \frac{3g_{mi}}{C_c} \tag{8.30}$$

is the design equation. This gives

$$g_{m6} = 3g_{mi} = 3\omega_0 C_L = 3 \times 2\pi \times 2 \times 10^6 \times 10^{-11} \tag{8.31}$$

and hence

$$\begin{aligned} g_{m6} &\simeq 377 \times 10^{-6}\ \text{A/V} = 377\ \mu\text{A/V}, \\ g_{mi} &\simeq 125.7\ \mu\text{A/V}. \end{aligned} \tag{8.32}$$

The specified slew rate requires that the bias current of the input stage satisfy

$$I_0 = S_rC_c \geq 4 \times 10^6 \times 10^{-11} = 40\ \mu\text{A}. \tag{8.33}$$

We can thus choose $I_0 = 40\ \mu$A.

As explained in Section 8.1 [see the discussion preceding Eq. (8.16)] the negative-going slew-rate limitation due to the use of Q_7 as a current source is

$$S_{r0} \leq \frac{I_{\text{bias}}}{C_L}. \tag{8.34}$$

To make this effect small, we can set $S_{r0} = 2.5S_r = 10$ V/μs. Then

$$I_{\text{bias}} = C_LS_{r0} = 10^{-11} \times 10^7 = 100\ \mu\text{A}. \tag{8.35}$$

Such large current also enables the realization of the required large g_{m6} without an excessive aspect ratio $(W/L)_6$.

From Eq. (8.6), to avoid systematic offset voltage, the condition

$$\frac{(W/L)_3}{(W/L)_6} = \frac{(W/L)_4}{(W/L)_6} = \frac{I_0/2}{\text{I}_{\text{bias}}} = \frac{1}{5} \tag{8.36}$$

must hold. Since, by Eq. (2.18) g_m is proportional to $\sqrt{(W/L)i_{D0}}$, Eqs. (8.33) to (8.36) give for the transconductance of the loads,

$$\begin{aligned} g_{ml} = g_{m3} = g_{m4} &= \sqrt{\frac{(W/L)_3I_0/2}{(W/L)_6/I_{\text{bias}}}}\, g_{m6} = \frac{I_0/2}{I_{\text{bias}}}\, g_{m6} = \frac{g_{m6}}{5} \\ &\simeq 75.4\ \mu\text{A/V}. \end{aligned} \tag{8.37}$$

At this point, an estimate of the low-frequency gain A_0 and the common-mode rejection ratio can be found. From Eq. (2.20) the drain conductance of a MOSFET is approximately $g_d \approx \lambda i_D^0$, where λ is the channel-length modulation constant* ($\lambda \simeq 0.03\ \text{V}^{-1}$ for $L \approx 10\ \mu\text{m}$) and i_D^0 is the dc drain current. Hence, from Fig. 8.2 or Table 8.1,

$$\begin{aligned} A_0 &= \frac{g_{mi}g_{m6}}{(g_{dl} + g_{di})(g_{d6} + g_{d7})} \simeq \frac{g_{mi}g_{m6}}{(\lambda I_0)(2\lambda I_{\text{bias}})} \\ &\simeq \frac{125.7 \times 10^{-6} \times 377 \times 10^{-6}}{(0.03 \times 40 \times 10^{-6})(0.06 \times 100 \times 10^{-6})} \simeq 6582, \end{aligned} \tag{8.38}$$

which corresponds to over 76 dB gain. Similarly, from Eq. (3.88), the common-mode rejection can be approximated as follows:

$$\begin{aligned} \text{CMRR} &= 2\frac{g_{mi}g_{ml}}{g_{d5}g_{di}} \simeq \frac{2g_{mi}g_{ml}}{(\lambda I_0)(\lambda I_0/2)} \\ &= \frac{2 \times 125.7 \times 10^{-6} \times 75.4 \times 10^{-6}}{(0.03 \times 40 \times 10^{-6})(0.03 \times 20 \times 10^{-6})} \simeq 26{,}327, \end{aligned} \tag{8.39}$$

which corresponds to about an 88-dB common-mode rejection ratio.

Both values exceed the specifications. If this would not have been the case, the specifications would have been inconsistent. This can be seen by using Eqs. (8.29) to (8.37) to express the parameters entering A_0 and CMRR:

$$\begin{aligned} g_{mi} &= C_c\omega_0, \\ I_0 &= C_cS_r, \\ g_{m6} &= 3\omega_0C_L, \\ I_{\text{bias}} &= S_{r0}C_L, \\ g_{ml} &= \frac{I_0g_{m6}}{2I_{\text{bias}}} = \frac{3C_cS_r\omega_0}{2S_{r0}}. \end{aligned} \tag{8.40}$$

Hence

$$A_0 \simeq \frac{3\omega_0^2}{2\lambda^2S_rS_{r0}}, \tag{8.41}$$

$$\text{CMRR} \simeq \frac{6\omega_0^2}{\lambda^2S_rS_{r0}} = 4A_0. \tag{8.42}$$

* Since, in fact, n- and p-channel devices have slightly different λ values, this calculation gives only a rough estimate of A_0 and the CMRR.

Thus both A_0 and CMRR are fully determined by ω_0 and the slew rates.* They can be increased by choosing g_{mi} and g_{m6} (and hence the realized values of ω_0 and $|s_{p2}|$) larger.

Next, the channel resistance r_8 of Q_8 is found so as to place the zero s_z of $A_v(s)$ at a desirable location. Simple analysis based on Fig. 8.10 shows that for the circuit of Fig. 8.2 the zero is at

$$s_z = \frac{-1}{(R_c - 1/g_{m6})C_6}, \tag{8.43}$$

and hence the required resistance is related to the desired zero location s_z by the formula

$$R_c = \frac{1}{|s_z|C_c} + \frac{1}{g_{m6}}. \tag{8.44}$$

As discussed in Section 4.4, there exist different strategies for choosing s_z. One possibility is to use $s_z = s_{p2}$; another is to shift s_z to ∞. For the former choice, using Eq. (8.29) and $C_c = C_L$, we get

$$R_c = \frac{1}{|s_{p2}C_c|} + \frac{1}{g_{m6}} \simeq \frac{2}{g_{m6}} \simeq 5.3\ \text{k}\Omega. \tag{8.45}$$

For the latter case, $R_c = 1/g_{m6} \simeq 2.65\ \text{k}\Omega$. (Note that an even larger phase margin can be obtained by choosing $|s_z|$ only slightly above ω_0; $|s_z| \sim 1.2\ \omega_0$ is usually a reasonable value.) Here we choose the value given in Eq. (8.45). Note that Q_8 is clearly in its linear region, since its gate is at V_{SS} while its dc drain-to-source voltage is zero. Hence, by Eq. (2.6),

$$\begin{aligned}\frac{1}{R_c} &= \left|\frac{\partial i_D}{\partial v_D}\right| = \mu C_{\text{ox}}\left(\frac{W}{L}\right)_8 |v_{GS8} - V_T| \\ &= 2k_8(|V_{SS} - v_{D8}| - |V_T|).\end{aligned} \tag{8.46}$$

Next, the design of the current sources Q_5 and Q_7 is discussed. The aspect ratios W/L of these transistors should not be too small since otherwise for the given currents (I_0, I_{bias}) the required "excess" gate-to-source voltage ($v_{GS} - V_T$) will be large. This is inconvenient, since the voltages v and v_{out} (Fig. 8.2) are not allowed to drop below $V_{SS} + v_{GS} - V_T$ if Q_5 and Q_7 are to stay in saturation. Hence a large $v_{GS} - V_T$ for Q_5 and Q_7 restricts the voltage swing and thus the dynamic range of the op-amp.

* In single-ended switched-capacitor circuits where the noninverting input is grounded, the CMRR as such is not very important.

On the other hand, the areas of Q_5 and Q_7 should not be too large, either. One reason is that, of course, real estate is very expensive on the chip; the other, that a large area for Q_5 increases the stray capacitance C_w across the current source. This capacitance consists of two parallel-connected reverse-biased junction capacitances; the drain-to-substrate capacitance of Q_5, and the p-well-to-substrate capacitance of Q_1 and Q_2. C_W causes a decrease of the CMRR at high frequencies, since then g_{d5} in Eq. (8.39) is replaced by $g_{d5} + j\omega C_w$. Also, as explained in connection with Figs. 4.42 to 4.44, C_w causes a distortion in the step response of the op-amp. A large stray capacitance across Q_7, caused by a large drain diffusion, will increase C_L and hence reduce the phase margin.

Thus a compromise should be found when Q_5 and Q_7 are dimensioned. From Eq. (2.8) the excess gate-to-source voltages are

$$\begin{aligned} v_{GS5} - V_T &\simeq \sqrt{\frac{I_0}{k_n'(W/L)_5}}, \\ v_{GS7} - V_T &\simeq \sqrt{\frac{I_{\text{bias}}}{k_n'(W/L)_7}}. \end{aligned} \tag{8.47}$$

Assuming again that $k_n' = 30\ \mu\text{A/V}^2$ for the NMOS and $k_p' = 12\ \mu\text{A/V}^2$ for the PMOS devices, and allowing 0.5 V excess voltage for both Q_5 and Q_7, we get

$$\begin{aligned} (W/L)_5 &= \frac{I_0}{k_n'(v_{GS5} - V_T)^2} = \frac{40}{30 \times (0.5)^2} \simeq 5.33, \\ (W/L)_7 &= \frac{I_{\text{bias}}}{k_n'(v_{GS7} - V_T)^2} = \frac{100}{20 \times (0.5)^2} \simeq 13.33. \end{aligned} \tag{8.48}$$

To avoid short-channel effects which occur for $L < 10\ \mu\text{m}$ and which would increase the drain conductance g_d, we choose $L_5 = L_7 = 10\ \mu\text{m}$. Then $W_5 = 54\ \mu\text{m}$ and $W_7 = 133\ \mu\text{m}$ can be used.

We can next calculate the aspect ratios of Q_1 to Q_4 and Q_6 from their transconductances. From Eq. (2.18), assuming that $|\lambda v_{Ds}| \ll 1$, the transconductance is given by

$$g_m \simeq 2\sqrt{k'(W/L)i_D^0}\,. \tag{8.49}$$

Hence the aspect ratios can be found from

$$\begin{aligned} (W/L)_1 &= (W/L)_2 \simeq \frac{g_{mi}^2}{4k_n'I_0/2} = \frac{(125.7)^2}{4 \times 30 \times 20} \simeq 6.58, \\ (W/L)_3 &= (W/L)_4 \simeq \frac{g_{ml}^2}{4k_p'I_0/2} = \frac{(75.4)^2}{4 \times 12 \times 20} \simeq 5.92, \end{aligned} \tag{8.50}$$

and by Eq. (8.36),

$$(W/L)_6 = 5(W/L)_3 \simeq 29.6. \tag{8.51}$$

Again choosing $L = 10\ \mu$m for all transistors, $W_1 = W_2 = 66\ \mu$m, $W_3 = W_4 = 60\ \mu$m, and $W_6 = 300\ \mu$m result. (Note that often noise considerations require that the width of the input devices be chosen much larger, say 200 μm or more!)

Next, we can estimate the (common) dc bias voltages at the drains of Q_1 to Q_4. Since they all carry a dc current $I_0/2$, we have

$$i_{D3} = \frac{I_0}{2} = k'_p(W/L)_3(|v_{GS3}| - |V_{TP}|)^2. \tag{8.52}$$

Since the PMOS threshold voltage is $V_{TP} = -1$ V,

$$|v_{GS3}| = |V_{TP}| + \sqrt{\frac{I_0/2}{k'_p(W/L)_3}} = 1 + \sqrt{\frac{20}{12 \times 6}} \approx 1.527\text{V} \tag{8.53}$$

Hence the drains of Q_1 to Q_4 are at a dc bias voltage $V_{DD} - |v_{Gs3}| \simeq 5 - 1.527 = 3.473$ V.

This is also the drain-and-source bias voltage v_{D8} of Q_8, and hence from Eqs. (8.45) and (8.46),

$$\begin{aligned} 2k'_p(W/L)_8(|-5 - 3.473| - 1) &= \frac{1}{R_c}, \\ (W/L)_8 &= \frac{1}{2 \times 12 \times 10^{-6} \times 5300 \times 7.473} \simeq 1.052. \end{aligned} \tag{8.54}$$

Hence $W_8 = L_8 = 10\ \mu$m can be used.

At this point the dimensions of all devices have been (tentatively) determined, and we also know the values of all currents. The drain voltages of Q_1 and Q_2 have been found; their sources are (for $v_{\text{in}}^- = v_{\text{in}}^+ = 0$) at a voltage v such that

$$k'_n(W/L)_1(-v - V_{Tn})^2 = \frac{I_0}{2}, \tag{8.55}$$

which gives, for $V_{Tn} \simeq 1.2$ V,

$$\begin{aligned} -v &= V_{Tn} + \sqrt{\frac{I_0}{2k'_n(W/L)_1}} \\ &= 1.2 + \sqrt{\frac{40}{2 \times 30 \times 6.6}} = 1.518 \text{ V}, \end{aligned} \tag{8.56}$$

so $v \simeq -1.52$ V.

The only remaining task is to design a bias chain that provides V_{bias}. In Eq. (8.48), Q_5 and Q_7 have been dimensioned such that $v_{GS5} = V_{Tn} + 0.5\text{ V} = 1.7$ V. Thus $V_{\text{bias}} = V_{SS} + v_{GS5} = -3.3$ V. This can be achieved by the simple circuit shown

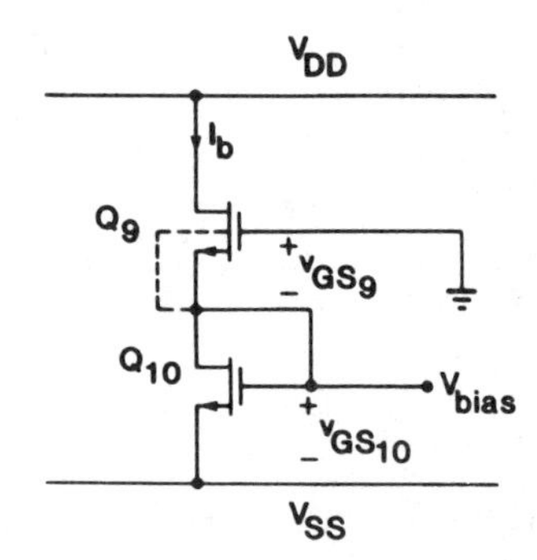

Figure 8.11. Bias circuit for the op-amp of Fig. 8.2.

in Fig. 8.11. Choosing the current of the bias chain $I_b = 20\ \mu\text{A}$, the aspect ratios of Q_9 and Q_{10} are easily found; since $v_{GS9} = 0 - V_{\text{bias}} = 3.3$ V and $v_{GS10} = V_{\text{bias}} - V_{SS} = 1.7$ V are known,

$$(W/L)_9 = \frac{I_b}{k_n'(v_{GS9} - V_{Tn})^2} = \frac{20}{30 \times 2.1^2} \simeq 0.1512, \tag{8.57}$$

$$(W/L)_{10} = \frac{I_b}{k_n'(v_{GS10} - V_{Tn})^2} \simeq 2.667. \tag{8.58}$$

Hence $W_9 = 10\ \mu\text{m}$, $L_9 = 66\ \mu\text{m}$ and $W_{10} = 27\ \mu\text{m}$, $L_{10} = 10\ \mu\text{m}$ can be used.

It should be noted that V_{bias} and I_b are insensitive to variations of V_{DD} but not to changes in V_{SS}. If $V_{SS} \rightarrow V_{SS}' = V_{SS} + \Delta V_{SS}$, then v_{GS9} and v_{GS10} also change, such that

$$v_{GS9}' + v_{GS10}' = -V_{SS} - \Delta V_{SS} = |V_{SS}'|; \tag{8.59}$$

and I_b changes to

$$I_b' = k_9\,(v_{GS9}' - V_{Tn})^2 = k_{10}\,(v_{GS10}' - V_{Tn})^2. \tag{8.60}$$

Here the prime denotes changed values.

From Eq. (8.60),

$$v_{GS9}' = \sqrt{k_{10}/k_9}\,(v_{GS10}' - V_{Tn}) + V_{Tn}. \tag{8.61}$$

Combining with Eq. (8.59), and solving for v_{GS9}' gives

$$v_{GS10}' = \frac{V_{Tn}(\sqrt{k_{10}/k_9} - 1) - V_{SS} - \Delta V_{SS}}{\sqrt{k_{10}/k_9} + 1}. \tag{8.62}$$

Thus a positive change of $+0.1$ V in V_{SS} will change v_{GS10} by

$$\Delta v_{GS10} = \frac{-\Delta V_{SS}}{\sqrt{k_{10}/k_9} + 1} \simeq -0.01915 \text{ V}. \tag{8.63}$$

Hence $V_{\text{bias}} = V_{SS} + v_{GS10}$ changes by $0.1 - 0.01915 \simeq 0.081$ V. The corresponding changes in I_0 and I_{bias} can be found approximately from

$$\begin{aligned} \Delta I_0 &\simeq g_{m5}\,\Delta v_{GS5} = g_{m5}\,\Delta v_{GS10} \\ &\simeq 2\sqrt{k_5 I_0}\,\Delta v_{GS10} = -2\sqrt{30 \times 5.4 \times 40} \times 0.01915 \\ &\simeq -3.1\ \mu\text{A}, \end{aligned} \tag{8.64}$$

$$\begin{aligned} \Delta I_{\text{bias}} &\simeq g_{m7}\Delta v_{GS10} \simeq -2\sqrt{30 \times 13.3 \times 100} \times 0.01915 \\ &= -7.7\ \mu\text{A} \end{aligned} \tag{8.65}$$

If these changes are not acceptable, the bias-independent circuit of Fig. 4.4*a* can be used to provide $v_{o1} = V_{\text{bias}}$.

To verify the accuracy of the design, the overall circuit was analyzed using the popular program SPICE. Figure 8.12 shows the computed gain and phase responses

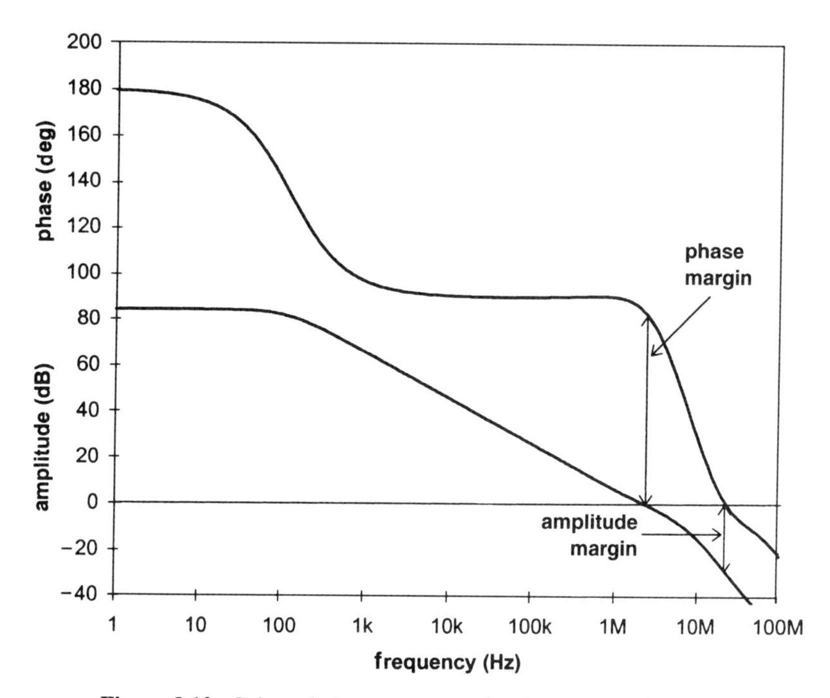

Figure 8.12. Gain and phase responses for the op-amp of Fig. 8.2.

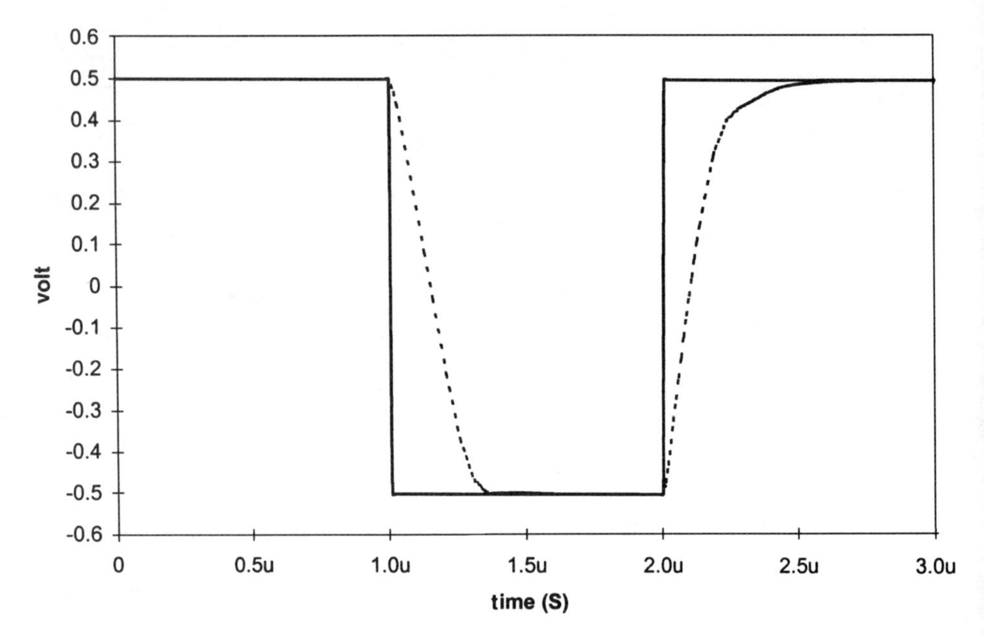

Figure 8.13. Step response of the op-amp for a negative and a positive input step.

of the circuit under open-circuit conditions. The unity-gain bandwidth is about 2.5 MHz, the dc gain 84 dB, and the phase margin about 85°. Thus these specifications are met. The closed-loop slew-rate (S_r) performance of the unity-gain-connected op-amp for a negative and positive input step is illustrated in Fig. 8.13. The maximum slope gives S_r; it is over 5 V/μs in both directions. Hence this requirement is also satisfied. Figure 8.14 illustrates the common-mode gain response. The required 80-dB CMRR is clearly obtained across the full dc-to-unity-gain frequency range.

The systematic input offset voltage can be estimated from the output voltage v_{out}^0 for $v_{\text{in}} = 0$ (Fig. 8.15) as $V_{\text{off}} \approx 0.2$ mV. This is very low and likely to be negligible compared to the random offset. For illustration, the gains for noise entering via the positive (V_{DD}) (Fig. 8.16) as well as the negative (V_{SS}) supplies (Fig. 8.17) are also shown. As predicted in Section 8.1, the supply rejection becomes a problem at higher frequencies; at ω_0 the PSRR is near 0 dB.

The layout geometry of the op-amp has important effects on its rise time, overshoot, and high-frequency response as well as on its sensitivity to process variations and its offset voltage. To achieve good step response and high-frequency response, the components of the amplifier must be arranged so as to minimize the line lengths, especially for lines connecting high-impedance nodes. The devices and lines of the input and output stages should be well separated, to avoid spurious feedback effects. If the nominal design of the op-amp is done carefully, so that no systematic voltage offset exists, some random offset may still occur due to errors in the ratios of (nominally) matched components. For example, as discussed in Section 8.1 [see Eq. (8.11)], a relative error ϵ in the matching of the aspect ratios of the input devices

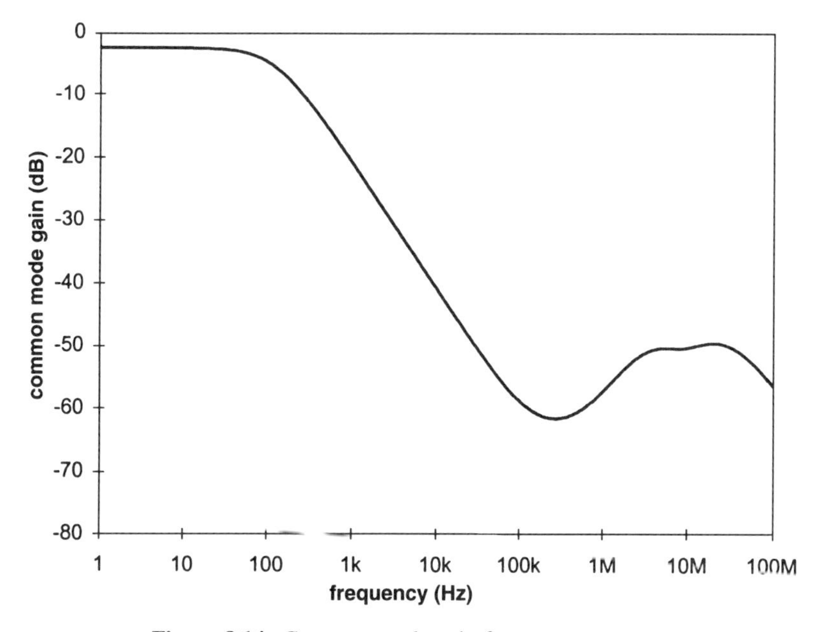

Figure 8.14. Common-mode gain frequency response.

results in an input-referred offset voltage $v_{\text{off}} = \epsilon\,(v_{GS} - V_T)/2$, where v_{GS} and V_T are the gate-to-source and threshold voltages of the input devices, respectively. For example, if $v_{GS} - V_T = 1$ V and $\epsilon = 1\%$, an offset voltage of 5 mV results from this single imperfection. This may be unacceptable in some circuits. Similarly, a matching error ΔV_T between the threshold voltages of the input devices results in an offset voltage equal to ΔV_T. Again, this may give an impractically large offset. To minimize these errors, both the input devices and their loads should be placed side by side, with identical geometries, including all connecting lines. Matched devices require sharing the same well. Only straight-line channels should be used, since the geometry of corners of bent channels is poorly controlled. If high accuracy and a low offset voltage are required, it is advisable to split both input and load devices into two or more unit transistors connected in parallel and arrange them in a common centroid geometry similar to that used for accurate capacitance matching. If a heat source (e.g., high-current output stage) is near the matched elements, the latter should be located symmetrically with respect to it, to ensure matched temperatures.

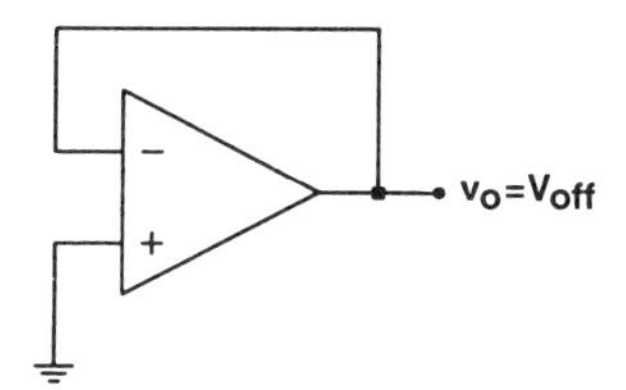

Figure 8.15. Circuit to estimate input dc offset voltage.

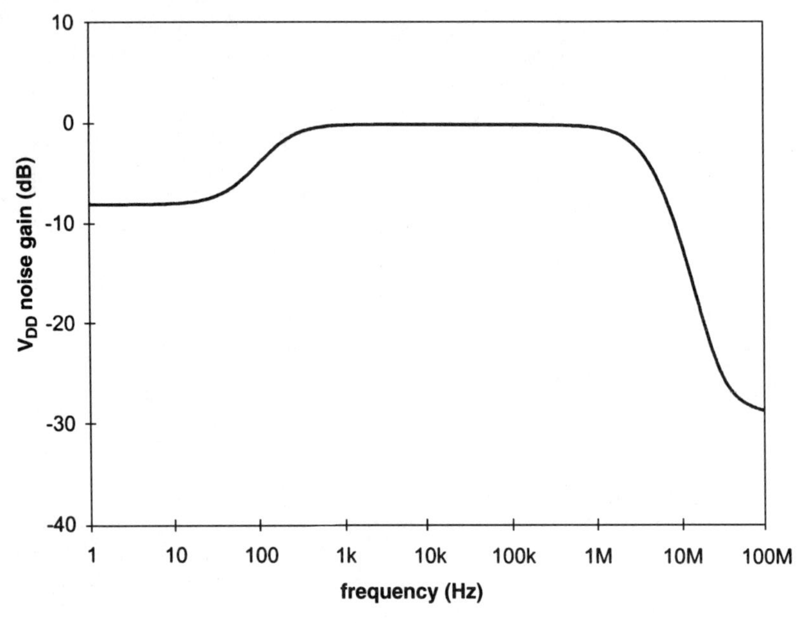

Figure 8.16. V_{dd} noise gain response for the op-amp of Fig. 8.2.

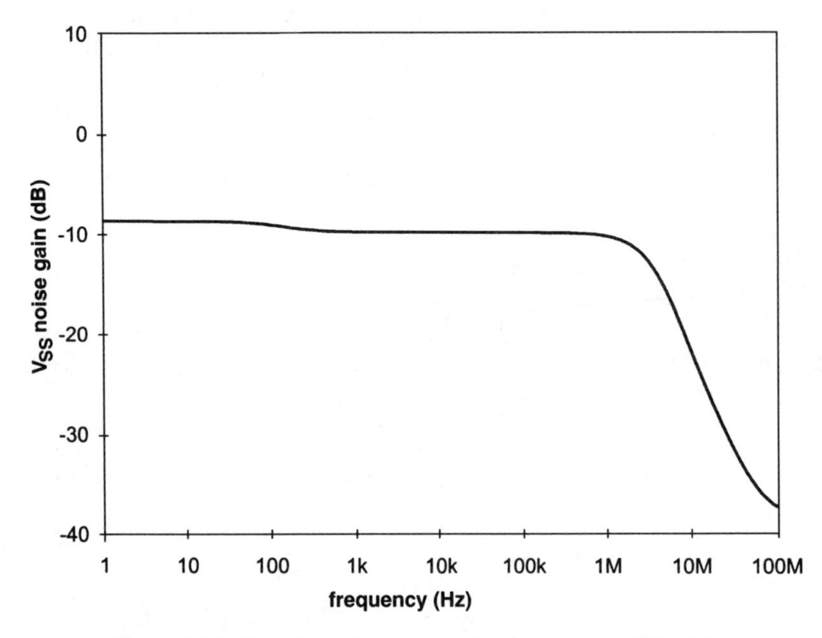

Figure 8.17. V_{ss} noise gain response for the op-amp of Fig. 8.2.

The circuit geometry also has some effect on the internally generated noise. Thus, as already mentioned, the $1/f$ noise may be reduced by using larger input devices in the op-amps; choosing larger capacitors can reduce the wideband and aliased noise. Also, it was observed that very short devices introduce "hot electron" noise if operated at large voltages [8]. This should be avoided by using increased channel lengths for such devices.

As an illustration of efficient layout procedure, the layout of the simple two-stage compensated CMOS op-amp of Fig. 8.2 is illustrated in Fig. 8.18 for a self-aligned silicon-gate p-well process. Note the common centroid layout and the symmetry of the matched devices (Q_1–Q_2 and Q_3–Q_4), the short connecting lines, the separation of the input and output lines, and the compact arrangement of the overall structure, resulting in a small total op-amp area.

Next, we repeat the design for the same specifications but using the circuit of Fig. 8.3 with its class AB output stage. Again, we select $C_c = C_L = 10$ pF, and since $\omega_0 = g_{mi}/C_c$, g_{mi} is given by Eq. (8.32), as before. Also, I (determined by the slew rate and C_c) remains the same, as given by Eq. (8.33). By the argument leading earlier to Eq. (8.48), $(W/L)_5 = 5.4$ can again be used. To obtain the specified CMRR, from Eqs. (3.88) and (2.20),

$$g_{ml} \geq (\text{CMRR}) \frac{g_{d5}g_{di}}{2g_{mi}} \simeq 10^4 \frac{(\lambda I_0)(\lambda I_0/2)}{2g_{mi}}$$

$$\simeq 10^4 \frac{(0.03)^2(40 \times 10^{-6})^2}{2 \times 2 \times 125.7 \times 10^{-6}} \simeq 28.6\ \mu\text{A/V}. \tag{8.66}$$

Hence the values given in Eq. (8.37) remain suitable. In conclusion, the dimensions of the input stage can remain unchanged for the new circuit, since they are determined by the (unchanged) requirements on ω_0, S_r, C_c, and CMRR.

Using Eq. (4.56) and changing subscripts appropriately, the second pole

$$s_{p2} \simeq -\frac{g_{m6} + g_{m7}}{C_L}. \tag{8.67}$$

Hence, for $|s_{p2}| = 3\omega_0$, now the relation

$$g_{m6} + g_{m7} = 3g_{mi} \simeq 377\ \mu\text{A/V} \tag{8.68}$$

must be satisfied. To determine g_{m6} and g_{m7} individually, we note that $i_6 = i_7$ and hence the bias voltages must satisfy

$$k_6(|v_{GS6}| - |V_{Tp}|)^2 = k_7(v_{GS7} - V_{Tn})^2. \tag{8.69}$$

Furthermore,

$$v_{GS6} = v_{DS4} = v_{GS3} \simeq -1.527, \tag{8.70}$$

as given in Eq. (8.53), since the input stage remained unchanged. Also, as suggested

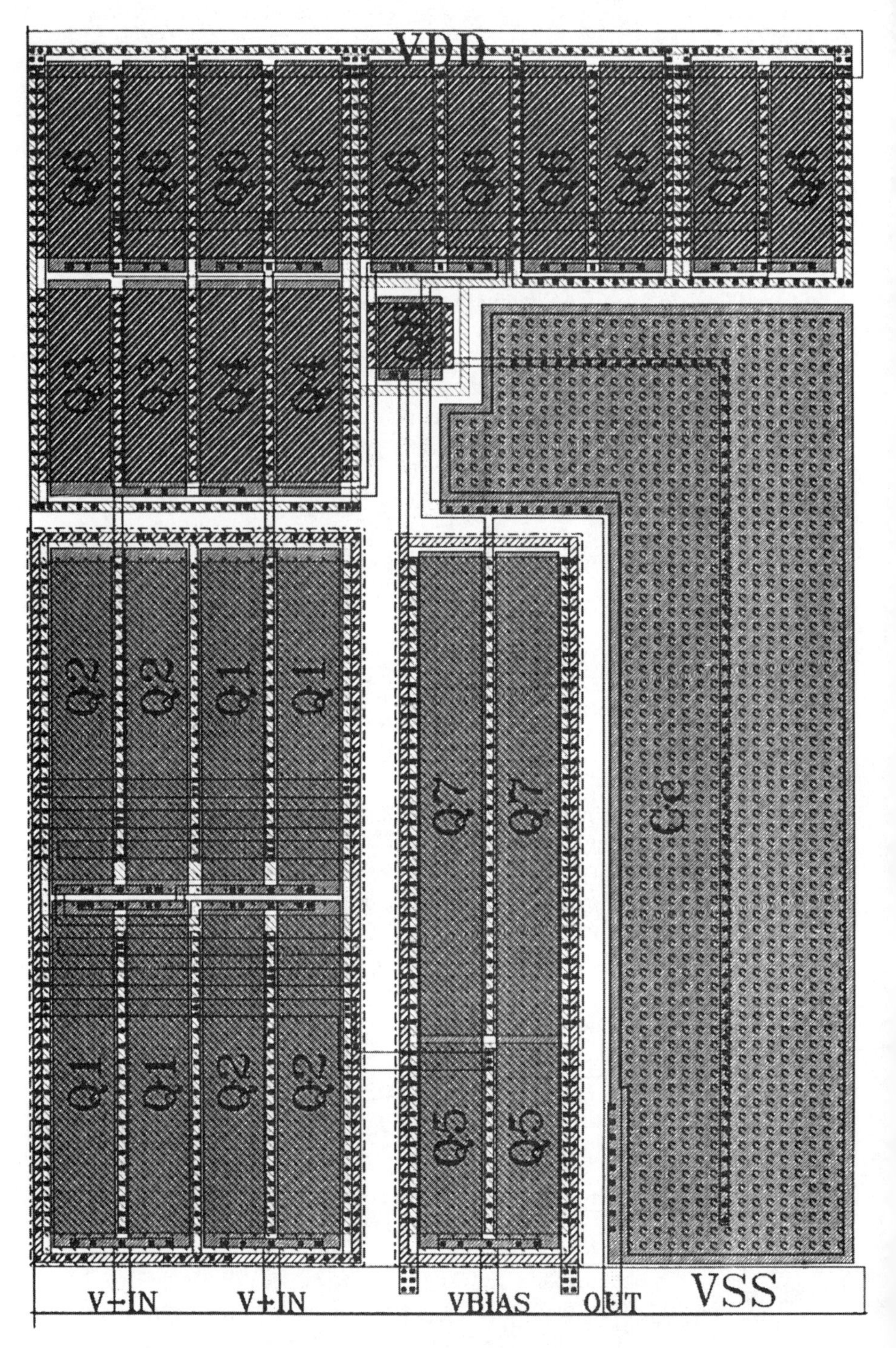

Figure 8.18. Layout of the two-stage CMOS op-amp of Fig. 8.2.

in Eq. (8.23), the bias voltages and currents can be made insensitive to process variations if $v_{GS7} = v_{GS5} = v_{GS10} = V_{\text{bias}} - V_{SS} = -3.3 + 5 + 1.7$ V is chosen. Hence, using Eq. (2.18) and (8.69),

$$\frac{g_{m6}}{g_{m7}} \simeq \frac{2\sqrt{k_6 i_6^0}}{2\sqrt{k_7 i_7^0}} = \sqrt{\frac{k_6}{k_7}} = \frac{v_{GS7} - V_{Tn}}{|v_{GS6}| - |V_{Tp}|}$$

$$= \frac{1.7 - 1.2}{1.527 - 1} \simeq 0.9488. \tag{8.71}$$

Combining Eqs. (8.68) and (8.71), we obtain

$$\begin{aligned} g_{m6} &= 183.5\ \mu\text{A/V}, \\ g_{m7} &= 193.5\ \mu\text{A/V}. \end{aligned} \tag{8.72}$$

To avoid systematic offset, as discussed in connection with Eq. (8.6), we must have

$$\frac{(W/L)_3}{(W/L)_6} = \frac{k_3}{k_6} = \frac{i_3^0}{i_6^0}. \tag{8.73}$$

Hence also

$$\frac{g_{m3}}{g_{m6}} = \frac{2\sqrt{k_3 i_3^0}}{2\sqrt{k_6 i_6^0}} = \frac{k_3}{k_6} = \frac{i_3^0}{i_6^0} = \frac{I_0/2}{i_6^0}. \tag{8.74}$$

Since, from Eqs. (8.37) and (8.72),

$$\frac{g_{m3}}{g_{m6}} \simeq \frac{75.4}{183.5} \simeq 0.4108 \tag{8.75}$$

and $I_0/2 = 20\ \mu$A, we find that $i_6^0 = 49\ \mu$A. Also, from $(W/L)_3 = 6$, $(W/L)_6 \simeq 14.6$. Hence $L_6 = 10\ \mu$m and $W_6 = 146\ \mu$m can be used.

Next, since $i_7^0 = i_6^0 = 49\ \mu$A,

$$(W/L)_7 = \frac{g_{m7}^2}{4k_7' i_7^0} = \frac{(193.5)^2}{4 \times 30 \times 49} \simeq 6.37. \tag{8.76}$$

Therefore, we can choose $L_7 = 10\ \mu$m and $W_7 = 64\ \mu$m.

Finally, the transistors Q_9 and Q_{10} of the level shifter will be dimensioned. As before, $i_9^0 = i_{10}^0$ leads to

$$k_9(v_{GS9} - V_{Tn})^2 = k_{10}\,(v_{GS10} - V_{Tn})^2. \tag{8.77}$$

Here, as Fig. 8.3 shows, $v_{GS9} = V_{DD} - V_{SS} + v_{DS4} - v_{GS7} = 10 - 1.527 - 1.7 \simeq 6.77$ V and $v_{GS10} = v_{GS7} = 1.7$ V. Hence

$$\frac{k_9}{k_{10}} = \left(\frac{v_{GS10} - V_{Tn}}{v_{GS9} - V_{Tn}}\right)^2 \simeq 8.058 \times 10^{-3}. \tag{8.78}$$

The transconductance g_{m9} of Q_9 can be found from the phase shift introduced by the pole–zero pair due to the stray capacitances loading the source terminal of Q_9. Since the source follower Q_9–Q_{10} drives these, the pole and zero are located near $s_{p3} \simeq -g_{m9}/C_p$. Estimating (pessimistically) $C_p = 0.5$ pF, and requiring $|s_{p3}| = 3\omega_0$ to make the contribution of this pole to the phase at ω_0 small, we obtain

$$\begin{aligned} g_{m9} &= 3\omega_0 C_p = 3 \times 2\pi \times 2 \times 10^6 \times 0.5 \times 10^{-12} \\ &\simeq 19\ \mu\text{A/V}. \end{aligned} \tag{8.79}$$

From

$$g_{m9} \simeq 2\sqrt{k_9 i_9^0} = 2k_n'(W/L)_9(v_{GS9} - V_{Tn}), \tag{8.80}$$

we find that

$$\begin{aligned} (W/L)_9 &= \frac{g_{m9}}{2k_n'(v_{GS9} - V_{Tn})} = \frac{19}{2 \times 30 \times (6.77 - 1.2)} \\ &\simeq 0.05685. \end{aligned} \tag{8.81}$$

Hence $W_9 = 10\ \mu$m and $L_9 = 176\ \mu$m can be used. From Eq. (8.78),

$$(W/L)_{10} = \frac{(W/L)_9}{k_9/k_{10}} \simeq 7.06. \tag{8.82}$$

Therefore, $L_{10} = 10\ \mu$m and $W_{10} = 71\ \mu$m can be chosen. The common current of Q_9 and Q_{10} is then

$$\begin{aligned} i_9^0 = i_{10}^0 &= k_n'(W/L)_{10}(v_{GS10} - V_{Tn})^2 \\ &\simeq 53\ \mu\text{A}. \end{aligned} \tag{8.83}$$

As explained in Section 8.1, the bias chain circuit of Fig. 8.4 can bias this circuit. The design formulas for the aspect ratios have been derived in Section 8.1 and given by Eqs. (8.19) to (8.25). Choosing $I_{\text{ref}} = I_0/2 = 20\ \mu$A, we get

$$\begin{aligned} (W/L)_{13} &= (W/L)_5(I_{\text{ref}}/I_0) = \frac{(W/L)_5}{2} = 2.7, \\ (W/L)_{11} &= (W/L)_3(2I_{\text{ref}}/I_0) = (W/L)_3 = 6, \\ (W/L)_{12} &= (W/L)_9(I_{\text{ref}}/I_{10}) \simeq 0.05685 \times \frac{20}{53} \\ &\simeq 0.02145. \end{aligned} \tag{8.84}$$

Hence we can use $L_{11} = 10$, $W_{11} = 60$, $L_{12} = 466$, $W_{12} = 10$, and $L_{13} = 10$, $W_{13} = 27$, all in μm.

The compensation branch of the circuit remains unchanged if we again choose $s_z = s_{p2}$. This is because s_{p2} remained at $-3\omega_0$, and the dc potential of the drain of Q_8 also remained the same. Hence we can once again use $W_8 = L_8 = 10\ \mu$m.

SPICE analysis of the class AB amplifier with the aspect ratios calculated above indicates that the dc bias voltage of the output terminal is unsuitable for proper operation. Its value is too low (about -4.5 V) to allow Q_7 to saturate. This occurs only under open-circuit conditions and is a consequence of the simplifying assumptions, primarily the neglect of the channel-length modulation factor $(1 + \lambda v_{DS})$, made in the calculations. The problem is an artificial one, since the circuit never functions without a dc load and/or feedback. Adding a 1-MΩ load resistor between the output terminal and ground, or a feedback resistor of (say) 100 MΩ between the inverting input terminal (v_{in}^-) and the output terminal of the op-amp, the output voltage returns to a value sufficient to keep both Q_6 and Q_7 in saturation.

As an exercise in design, however, as well as a way to show how to reduce systematic offset, we are next going to redesign the output stage so as to obtain a satisfactory dc bias value for v_{out} even under open-circuit conditions. The SPICE analysis for the circuit gave $v_{GS6} = -1.572$ V, $v_{GS7} = 1.821$ V, and $i_6^0 = i_7^0 = 74\ \mu$A. To achieve $v_{out} = 0$ V while retaining the values of v_{GS6}, v_{GS7}, and $i_6^0 = i_7^0$, we must have, by Eq. (2.11),

$$k_p' (W/L)_6(v_{GS6} - V_{Tp})^2(1 - \lambda V_{SS}) = k_n'(W/L)_7(v_{GS7} - V_{Tn})^2(1 + \lambda V_{DD}) = i_6^0. \quad (8.85)$$

This relation now includes the channel-length modulation effect and is hence more accurate.

Substituting $\lambda = 0.03\ \text{V}^{-1}$, as well as the given values, $(W/L)_6 \simeq 16.4$ and $(W/L)_7 \simeq 50/9$ results. Thus $W_6 = 164\ \mu$m, $L_6 = 10\ \mu$m, $W_7 = 50\ \mu$m, and $L_7 = 9\ \mu$m can be used. The resulting output bias voltage is only 0.04 V.

Figure 8.19 shows the gain and phase responses of the redesigned circuit under open-circuited output conditions. The unity-gain bandwidth is again near 2.5 MHz, while the dc gain is over 90 dB. The phase margin is over 83°; this is more than adequate and indicates that the bias current of Q_6, Q_7, Q_9, and Q_{10} (and thus their transconductances) can be reduced and still adequate stability maintained. This was not attempted. The common-mode gain response is shown in Fig. 8.20, the CMRR is over 80 dB across the full frequency range 0 to ω_0. The systematic input-referred offset voltage is negligible.

The slew-rate performance of the op-amp for positive and negative voltage input steps was computed and was found to be around 5 V/μs for both polarities. The gain response for V_{DD} and V_{SS} noise are illustrated in Figs. 8.21 and 8.22, respectively. The layout of the op-amp of Fig. 8.3 is illustrated in Fig. 8.23.

The design of a 3-V CMOS single-stage folded-cascade op-amp is presented next. The design will be based on a 0.6-μm silicon-gate bulk CMOS n-well process.

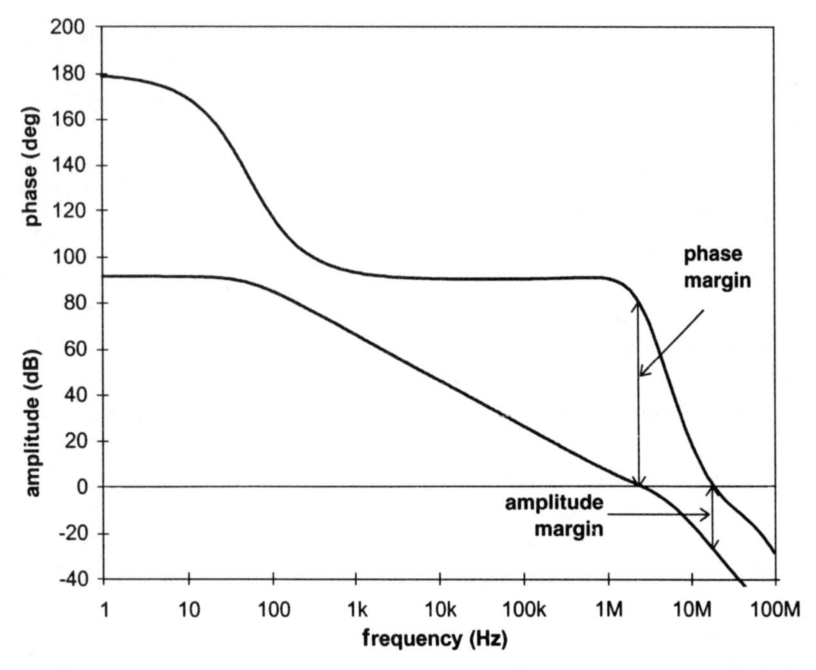

Figure 8.19. Gain and phase response of the op-amp of Fig. 8.3.

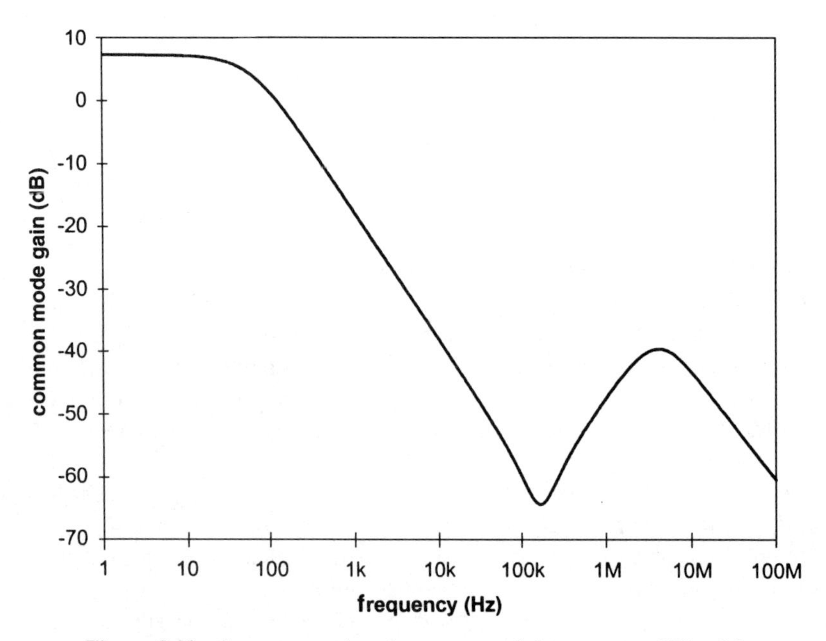

Figure 8.20. Common-mode gain response of the op-amp of Fig. 8.3.

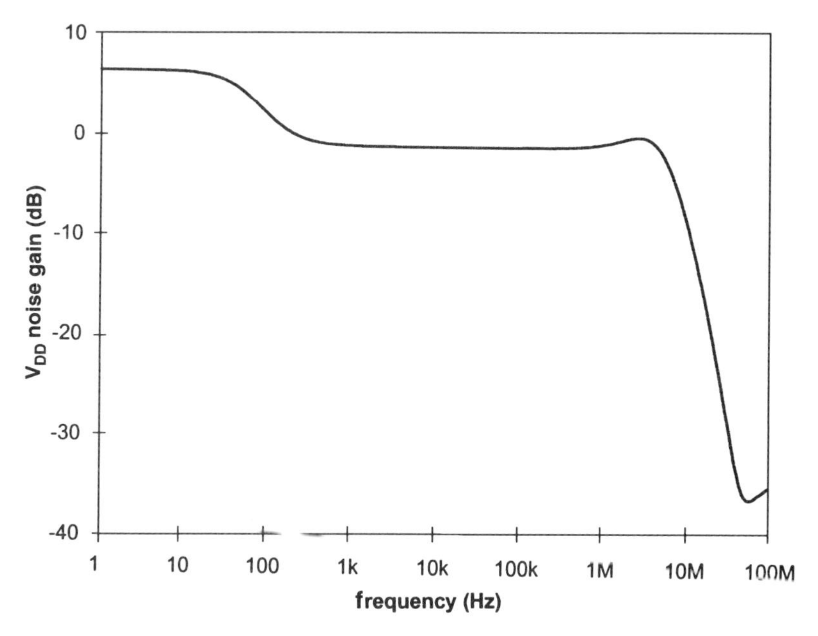

Figure 8.21. V_{DD} noise gain response of the op-amp of Fig. 8.3.

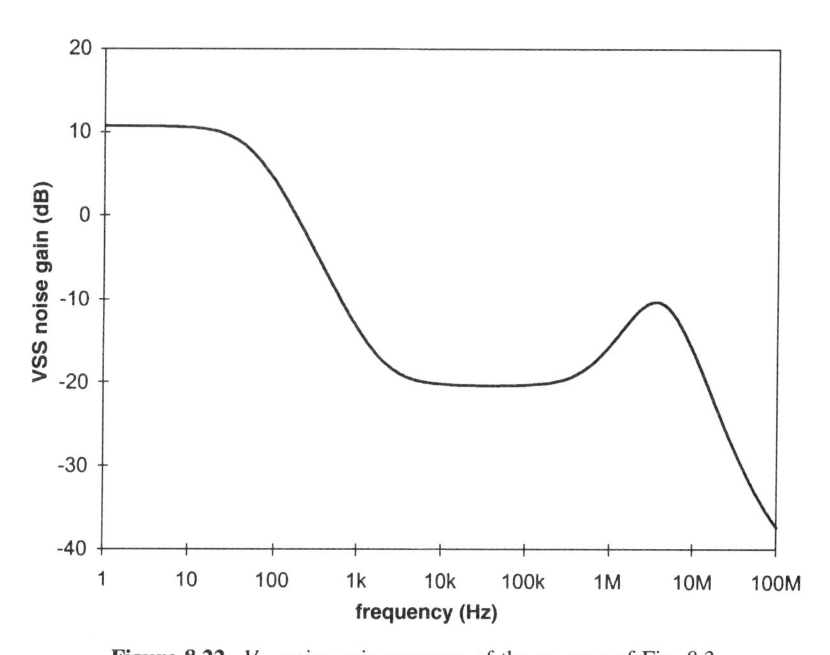

Figure 8.22. V_{SS} noise gain response of the op-amp of Fig. 8.3.

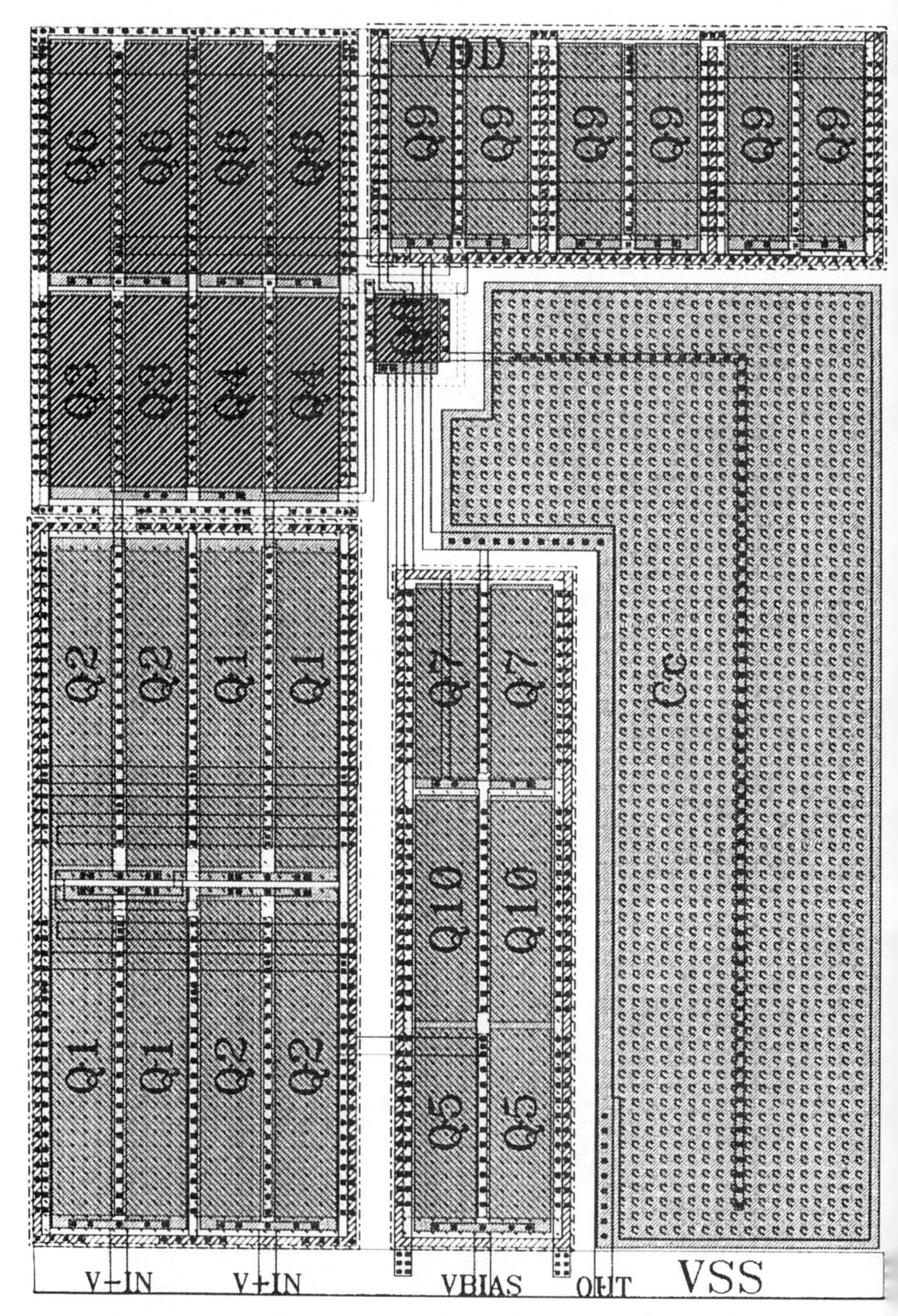

Figure 8.23. Layout of the two-stage class AB CMOS op-amp of Fig. 8.3.

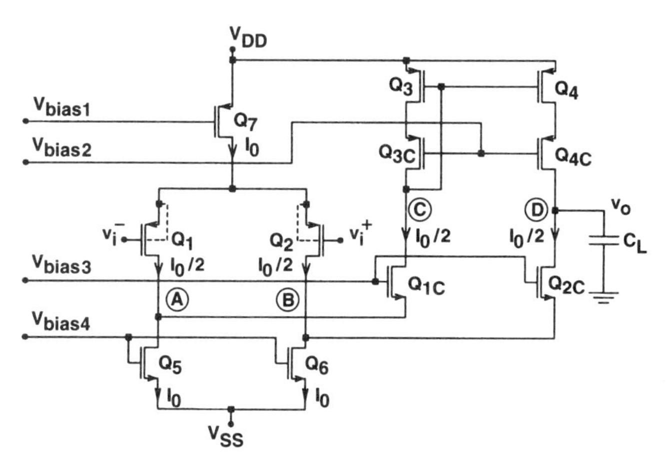

Figure 8.24. Folded-cascode single-stage op-amp.

The single-stage folded-cascade op-amp was discussed earlier (cf. Fig. 4.10) and is reproduced in Fig. 8.24. The devices in the circuit of Fig. 8.24 are dimensioned such that the following specifications are satisfied:

Low-frequency gain	$A_0 \geq 65$ dB
Unity-gain frequency	$f_0 \geq 5$ MHz
Slew rate	$S_r \geq 10$ V/μs
Phase margin	$\phi_M > 60°$
Load impedance	$C_L = 10$ pF
dc supply voltages	$V_{DD} = 3$ V, $V_{SS} = 0$ V

It will be assumed that the transconductance factor $k' \triangleq \mu C_{ox}/2$ is 55 μA/V^2 for NMOS and 22 μA/V^2 for the PMOS transistors. Also, the threshold voltages are assumed to be $V_{Tn} = 0.8$ V and $V_{Tp} = -0.9$ V. The op-amp performance parameters are summarized in Table 8.2.

For the specified slew rate it is necessary that the bias current of the input stage satisfy

$$I_0 = S_r C_L \geq 10 \times 10^6 \times 10 \times 10^{-12} = 100 \ \mu\text{A}. \tag{8.86}$$

We can then choose $I_0 = 150$ μA.

TABLE 8.2. Single-Stage Folded-Cascode Op-Amp Performance Parameters

Low-frequency open-loop gain	$A_0 = g_{m1}(g_{m2c}\, r_{d2c}\, r_{d2} \parallel g_{m4c}\, r_{d4c}\, r_{d4})$
Unity-gain frequency	$f_0 = \dfrac{g_{m1}}{2\pi C_L}$
Slew rate	$S_r = \dfrac{I_0}{C_L}$

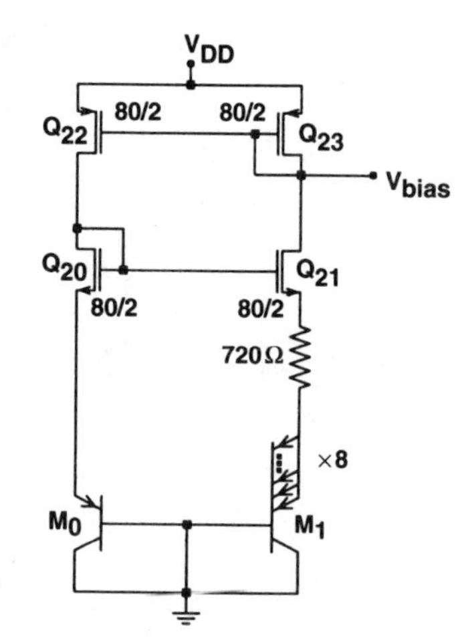

Figure 8.25. ΔV_{BE}-based bias generator for the folded-cascode op-amp.

For a unity-gain frequency of 5 MHz we should choose g_{m1} from

$$g_{m1} = f_0 \times 2\pi C_L = 5 \times 10^6 \times 2\pi \times 10^{-12} \times 10 \simeq 314\ \mu\text{A/V}. \tag{8.87}$$

The transconductance of the input devices is given by

$$g_{m1} \simeq 2\sqrt{k_p'\left(\frac{W}{L}\right)I_0/2}. \tag{8.88}$$

Hence the aspect ratios of the input devices can be found from

$$\left(\frac{W}{L}\right)_1 = \left(\frac{W}{L}\right)_2 \simeq \frac{g_{m1}^2}{4k_p'(I_0/2)} = \frac{(314)^2}{4 \times 22 \times 75} \simeq 15, \tag{8.89}$$

choosing $L = 2\ \mu$m for the transistors; $W_1 = W_2 = 30\ \mu$m results. Note that noise considerations require the width of the input devices to be chosen larger, say $W_1 = W_2 = 200\ \mu$m or more.

Based on the values above, a ΔV_{BE}-based supply-independent bias circuit will be designed to operate the op-amp with maximum output voltage swing. Since the power supply is 3 V, there is not enough headroom to use the high-performance cascode-load bias generator circuit shown in Fig. 3.6*b*. Therefore, the supply-independent bias generator of Fig. 3.6*a* will be used instead. The bias circuit is shown in Fig. 8.25, where $M = 8$. If we set the bias standby current I_{bias}, to 75 μA, from Eq. (3.13) we have

$$R = \frac{\Delta V_{BE}}{I_{\text{bias}}} = \frac{V_T \ln(M)}{I_{\text{bias}}} = \frac{26 \times 10^{-3} \times \ln(8)}{75 \times 10^{-6}} \simeq 720\ \Omega. \tag{8.90}$$

For transistors Q_{20} to Q_{23} in the bias generator, we choose $L = 2$ μm for the channel lengths and $W = 80$ μm for the channel widths.

The complete op-amp and the bias generator are shown in Fig. 8.26. Transistors Q_{24} and Q_{28} form current mirrors and are used to bias all devices in the op-amp. The currents in Q_{25} and Q_{26} are 75 μA. If we select $(W/L)_{26} = 80/2$, then for Q_5 and Q_6 to carry 150 μA we have $(W/L)_5 = (W/L)_6 = 160/2$. For ease of physical layout we should also choose $(W/L)_{1c} = (W/L)_{2c} = 160/2$. To maximize the output voltage swing the W/L ratio of Q_{25} can be calculated from Eq. (4.15) and is given by

$$\left(\frac{W}{L}\right)_{25} = \frac{1}{3 + 2\sqrt{2}} \times \left(\frac{W}{L}\right)_5 = 0.1716 \times \frac{160}{2} \simeq \frac{27}{2}. \tag{8.91}$$

In order to bias Q_5 and Q_6 slightly above the saturation voltage, we will select $(W/L)_{25} = 25/2$. If the W/L ratios of Q_3, Q_4, Q_{3c}, and Q_{4c} are also selected as 160/2, then from Eq. (4.14) we have

$$\left(\frac{W}{L}\right)_{28} = \frac{1}{4} \times \left(\frac{W}{L}\right)_3 = \frac{160}{2} \times \frac{1}{4} = \frac{40}{2}. \tag{8.92}$$

Once again to ensure that Q_3 and Q_4 both operate well in the saturation region, the W/L ratio of Q_{28} will be selected as $(W/L)_{28} = 38/2$. Finally, the W/L ratio of Q_7 will be selected as 160/2, which will set the op-amp tail current to 150 μA.

As discussed earlier, the capacitive load acts as the compensation capacitor for the op-amp. The larger the capacitive load, the greater the phase margin and narrower the unity gain frequency of the op-amp.

To verify the accuracy of the design, the overall circuit was analyzed using SPICE. The results of the analysis showed that $(V_{DS})_{Q5} = (V_{DS})_{Q6} = 157$ mV and $(V_{DS})_{Q4} = (V_{DS})_{Q3} = -205$ mV, while $(V_{D\text{sat}})_{Q5} = (V_{D\text{sat}})_{Q6} = 144$ mV and $(V_{D\text{sat}})_{Q4} = (V_{D\text{sat}})_{Q3} = -199$ mV. Thus Q_3, Q_4, Q_5, and Q_6 were all biased on the edge of the saturation region. Also since $(V_{D\text{sat}})_{Q4C} = -196$ mV and $(V_{D\text{sat}})_{Q2C} = 93$ mV, the output voltage range is given by

$$(v_{\text{out}})_{\text{range}} = V_{DD} - (V_{d\text{sat}})_{Q6} - (V_{d\text{sat}})_{Q2c} + (V_{d\text{sat}})_{Q4} + (V_{d\text{sat}})_{Q4c} = 2.362 \text{ V} \tag{8.93}$$

or

$$V_{DD} + (V_{D\text{sat}})_{Q4} + (V_{D\text{sat}})_{Q4c} \leq v_{\text{out}} \leq (V_{D\text{sat}})_{Q6} + (V_{D\text{sat}})_{Q2c}, \tag{9.94}$$

which simplifies to

$$2.599 \text{ V} \leq v_{\text{out}} \leq 0.237 \text{ V}. \tag{8.95}$$

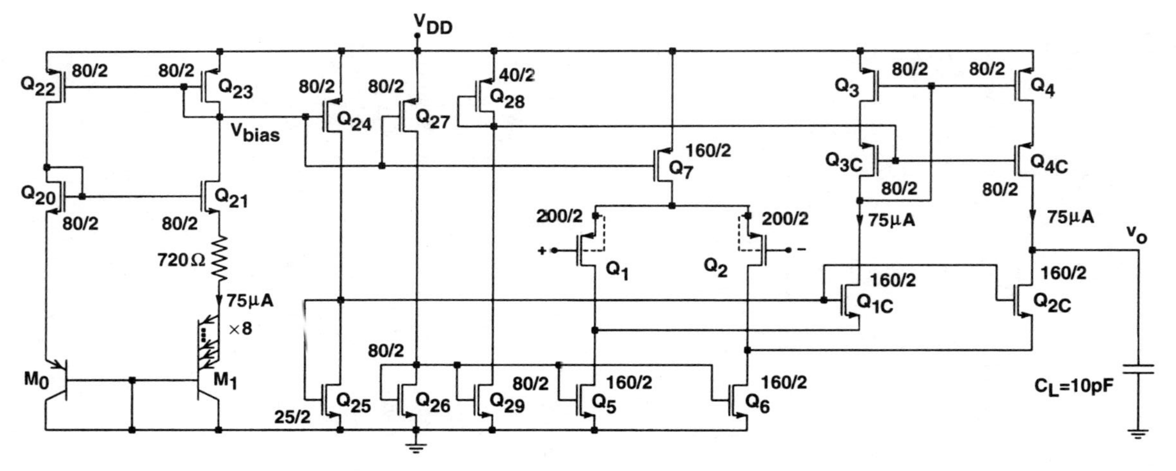

Figure 8.26. Complete CMOS folded-cascode op-amp with supply independent ΔV_{BE}-based bias circuit.

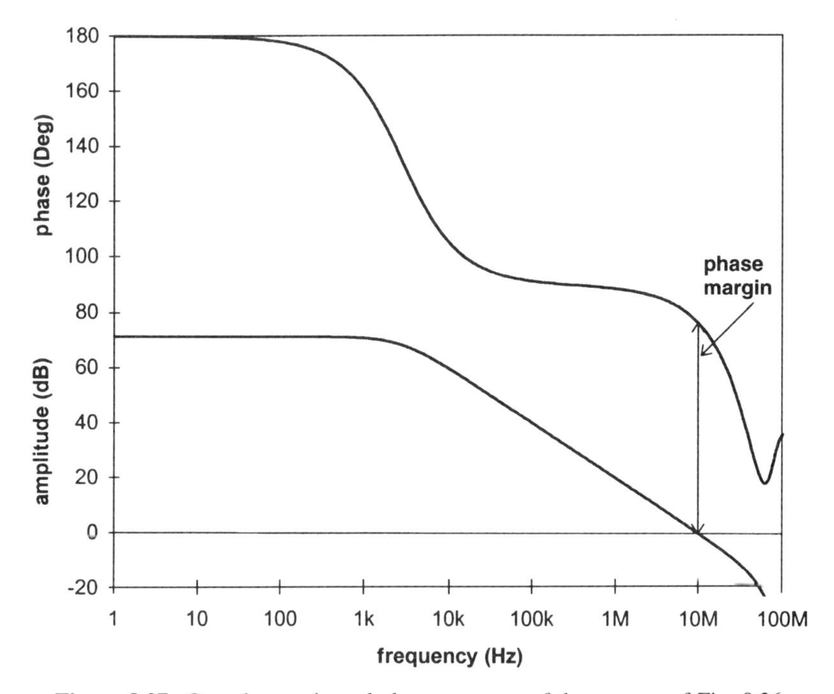

Figure 8.27. Open-loop gain and phase response of the op-amp of Fig. 8.26.

Utilizing the improved biasing scheme provides maximum signal swing at the op-amp output.

Figure 8.27 shows the computed gain and phase responses of the circuit under open-loop conditions. The unity-gain bandwidth is about 10 MHz, the dc gain 70 dB, and the phase margin 79°. So the target specifications are met. The slew-rate (S_r) performance is illustrated in Fig. 8.28. The maximum slew rate for negative and positive inputs steps are 11.5 V/μs. Hence this requirement is also satisfied. To estimate the systematic dc offset voltage, the op-amps was connected as a unity-gain buffer with the positive input connected to 1.5 V; the output was estimated at 1.5003 V, which corresponds to a 0.3-mV systematic dc offset voltage. This is very low and likely to be negligible compared to the random offset. The ΔV_{BE}-based bias generator of Fig. 8.25 needs a startup circuit, which is not shown. In the actual implementation a startup circuit similar to the one shown in Fig. 3.5 is highly recommended. Figure 8.29 shows the physical layout of the single-stage folded-cascode op-amp of Fig. 8.26.

The next example discusses a buffered CMOS op-amp that is able to drive a small resistive load on the order of 50 Ω. This example consists of the combination of the unbuffered single-stage folded-cascode op-amp of Fig. 8.26 and the output stage of Fig. 4.72. The design will be based on the same performance specification of the previous example with the added requirement that the op-amp will be driving a 50-Ω resistive lead with a signal swing of 2 V for a 3-V ($\pm$1.5-V) dc supply voltage.

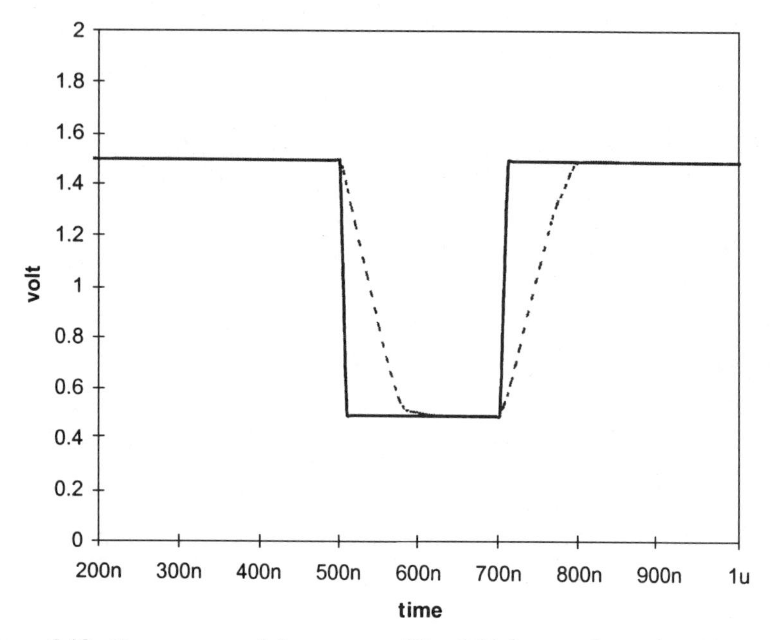

Figure 8.28. Step response of the op-amp of Fig. 8.26 for negative and positive steps.

The schematic of the buffered op-amp is shown in Fig. 8.30*a*, where devices Q_8 to Q_{11} have been added to the circuit of Fig. 8.28 to form the output stage. The bias circuit is shown in Fig. 8.30*b*. It has also been modified by adding devices Q_{30} to Q_{35} to provide bias voltages for the output stage. Next, the devices in the output stage and bias circuit will be dimensioned to satisfy the output drive requirements.

In steady state the current through transistor Q_4 and Q_{4c} ($I = 75\ \mu\text{A}$) is equally divided between Q_8 and Q_9, so that each carries a current equal to 37.5 μA. From Eqs. (4.127) and (4.128) we have

$$\left(\frac{W}{L}\right)_8 = \left(\frac{W}{L}\right)_{31}, \tag{8.96}$$

$$\left(\frac{W}{L}\right)_9 = \left(\frac{W}{L}\right)_{34} \tag{8.97}$$

by choosing

$$\left(\frac{W}{L}\right)_{30} = \frac{1}{2}\left(\frac{W}{L}\right)_{26}, \tag{8.98}$$

$$\left(\frac{W}{L}\right)_{33} = \frac{1}{2}\left(\frac{W}{L}\right)_{23}. \tag{8.99}$$

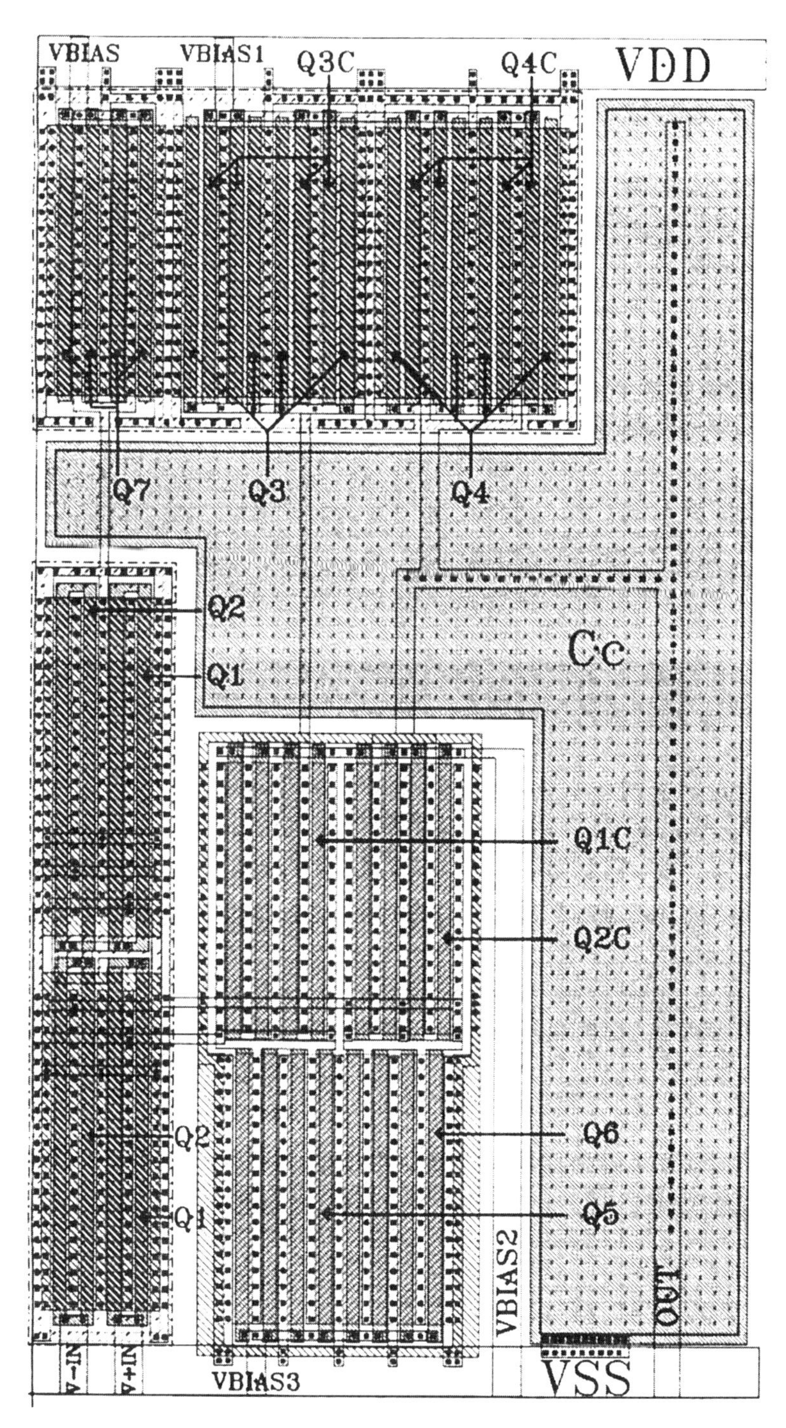

Figure 8.29. Layout of the single-stage folded-cascode op-amp of Fig. 8.26.

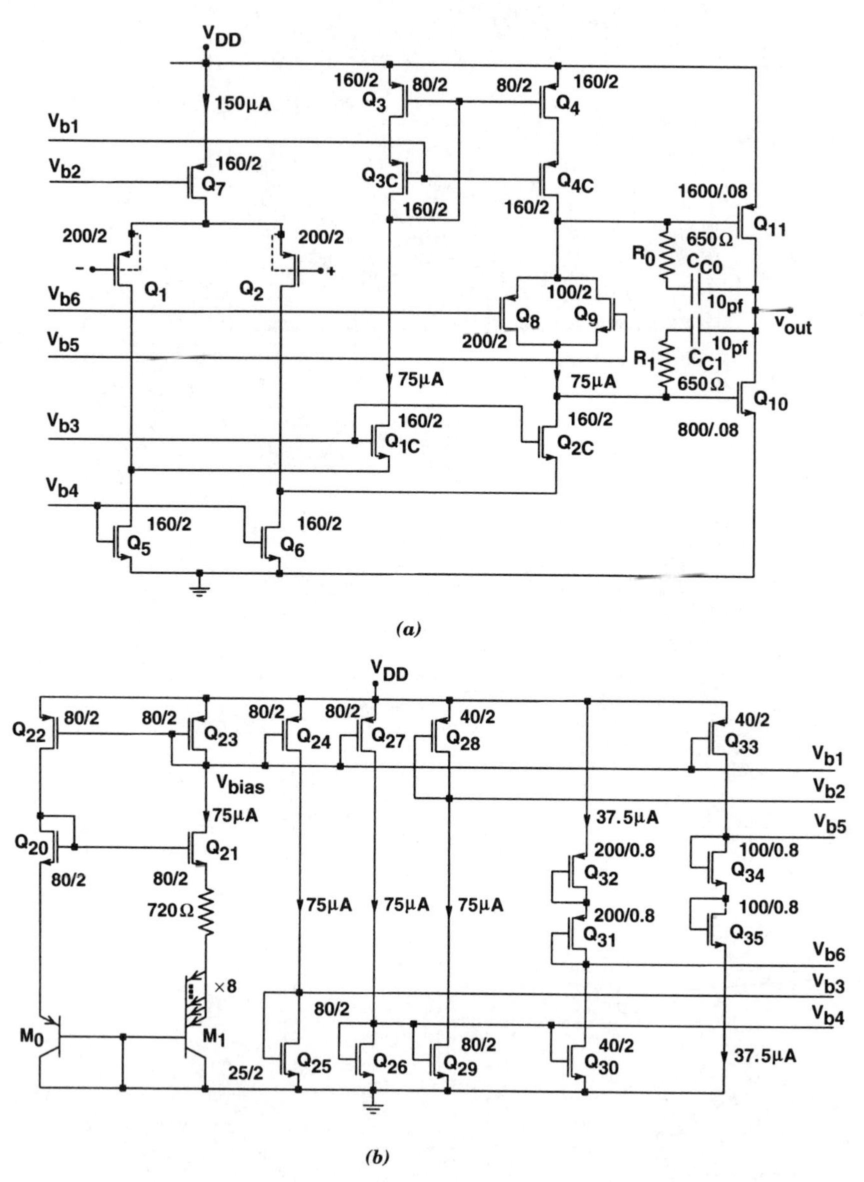

Figure 8.30. (*a*) CMOS buffer op-amp. (*b*) ΔV_{BE}-based bias circuit.

Hence Q_{31} and Q_8 and Q_{34} and Q_9, will have the same gate-to-source voltage and will carry the same drain currents, equal to $I_b = 37.5\ \mu A$. Therefore, as described in Section 4.9 and given by Eq. (4.129), the standby current of the output devices Q_{10} and Q_{11} is given by

$$I_0 = \frac{(W/L)_{11}}{(W/L)_{32}} \times 37.5\ \mu A = \frac{(W/L)_{10}}{(W/L)_{35}} \times 37.5\ \mu A. \tag{8.100}$$

The external load resistance R_L determines the load current and thus the dimensions of the output devices Q_{10} and Q_{11}. Assume that the resistance of the load R_L that is ac coupled to the output of the op-amp is 50 Ω. Then for an output signal with 2-V peak-to-peak voltage swings, the op-amp output varies from 0.5 to 2.5 V. The maximum source and sink currents may be calculated from

$$I_{\text{sink}} = I_{\text{source}} = \frac{1\ \text{V}}{50\ \Omega} = 20\ \text{mA}. \tag{8.101}$$

During the negative half-cycle of the output voltage swing, the drain of transistor Q_{10} pulls to 0.5 V and its gate is pulled up toward the positive supply V_{DD}. To keep the gain-stage output devices in the saturation region, the gate voltage positive swing should be limited to $V_{DD} + 2(V_{DSat})_{Q4} + (V_{DS})_{Q8} \simeq 2.5$ V. In this range the NMOS transistor Q_{10} will operate in the linear region and its current can be calculated from Eq. (2.10) as

$$(I_D)_{Q10} = 20 \times 10^{-3} = 55 \times 10^{-6} \times \left(\frac{W}{L}\right)_{Q10} \left(2.5 + 0.8 - \frac{0.5}{2}\right) \times 0.5 \tag{8.102}$$

or

$$(I_D)_{Q10} = 20 \times 10^{-3} = 3.9875 \times 10^{-5} \times \left(\frac{W}{L}\right)_{Q10}. \tag{8.103}$$

The W/L ratio of Q_{10} can now be calculated from Eq. (8.103) and is given by

$$\left(\frac{W}{L}\right)_{Q10} = 502. \tag{8.104}$$

For $L_{10} = 0.8\ \mu m$ from Eq. (8.104) we have $W_{10} \approx 400\ \mu m$.

The dimensions of the p-channel output device can be calculated using the same procedure. Again, using Eq. (2.10), we have

$$(I_D)_{Q11} = 20\times 10^{-3} = 22 \times 10^{-6}\times\left(\frac{W}{L}\right)_{Q11}\left(-2.5 + 0.9 + \frac{0.5}{2}\right)\times(-0.5) \tag{8.105}$$

or

$$(I_D)_{Q11} = 20 \times 10^{-3} = 1.485 \times 10^{-5} \times \left(\frac{W}{L}\right)_{Q11}. \tag{8.106}$$

So for $L_{11} = 0.8\ \mu\text{m}$ from Eq. (8.106) we have $W_{11} \approx 1100\ \mu\text{m}$.

Assuming a standby current of 300 μA in the output stage, the aspect ratios of Q_{32} and Q_{35} can be calculated from Eq. (8.100):

$$(W/L)_{32} = (W/L)_{11} \times \frac{37.5}{300} = \frac{1}{8} \times (W/L)_{11}, \tag{8.107}$$

$$(W/L)_{35} = (W/L)_{10} \times \frac{37.5}{300} = \frac{1}{8} \times (W/L)_{10}. \tag{8.108}$$

Hence $L_{32} = L_{35} = 0.8\ \mu\text{m}$, $W_{35} = 50\ \mu\text{m}$, and $W_{32} = 137.5\ \mu\text{m}$ can be used.

The only remaining task is to design the bias chain consisting of devices Q_{30}, Q_{31}, Q_{32} and Q_{33}, Q_{34}, Q_{35}. The dimensions of Q_{31} and Q_{34} should be selected such that devices Q_{30} and Q_{32} remain in saturation. To meet this condition the following two equations should be satisfied:

$$V_{DD} - (V_{GS})_{Q34} - (V_{GS})_{Q35} \geq -(V_{D\text{Sat}})_{Q33}, \tag{8.109}$$

$$V_{DD} + (V_{GS})_{Q31} + (V_{GS})_{Q32} \geq (V_{D\text{Sat}})_{Q30}. \tag{8.110}$$

The physical layout of the op-amp can be considerably simplified if the dimensions of Q_{31} and Q_{34} are selected the same as Q_{32} and Q_{35}. This assumption will be validated later.

Figure 8.30*a* shows the buffered op-amp with the two feedback branches (R_0, C_{c0} and R_1, C_{c1}) added for frequency compensation. As explained in Section 8.1, the unity-gain bandwidth of the op-amp is given by $\omega_0 = g_{mi}/C_c$, where $C_c = C_{c0} + C_{c1}$ and g_{mi} is the transconductance of the input device pair. The value of the compensation capacitance C_c can be determined from knowledge of the op-amp unity-gain bandwidth and the transconductance of the input device pair. For $\omega_0 = 2\pi \times 5 \times 10^6$ (5 MHz bandwidth) and

$$g_{mi} = 2\sqrt{\frac{k'_p(W/L)I_0}{2}} = 2\sqrt{22 \times 10^{-6} \times \frac{200}{2} \times 75 \times 10^{-6}}$$
$$= 8.12 \times 10^{-4} \text{ A/V},$$

the compensation capacitor can be calculated as $C_c = 8.12 \times 10^{-4}/(2\pi \times 5 \times 10^6) \approx 26$ pF. Assuming that the two compensation capacitors are equal, we have $C_{c0} = C_{c1} = 13$ pF.

At this point the dimensions of all devices have been (tentatively) determined, and we also know the value of all currents. To verify the accuracy of the design, the overall circuit was analyzed using SPICE. Figure 8.31 shows plots of the gain and phase responses under open-loop conditions. The dc gain without load resistance is 107.9 dB, the unity-gain bandwidth is 8 MHz, and the phase margin about 90°. The final device aspect ratios are shown in Fig. 8.30*a*. Based on a worst-case analysis, in order to satisfy the minimum load current requirements, the widths of the output devices Q_{10} and Q_{11} were increased to $W_{10} = 800$ μm and $W_{11} = 1100$ μm, respectively. Additionally, the dimensions of Q_{35} and Q_{32} were modified based on Eqs. (8.107) and (8.108), to $W_{35} = 100$ μm and $W_{32} = 200$ μm. Finally, the choices

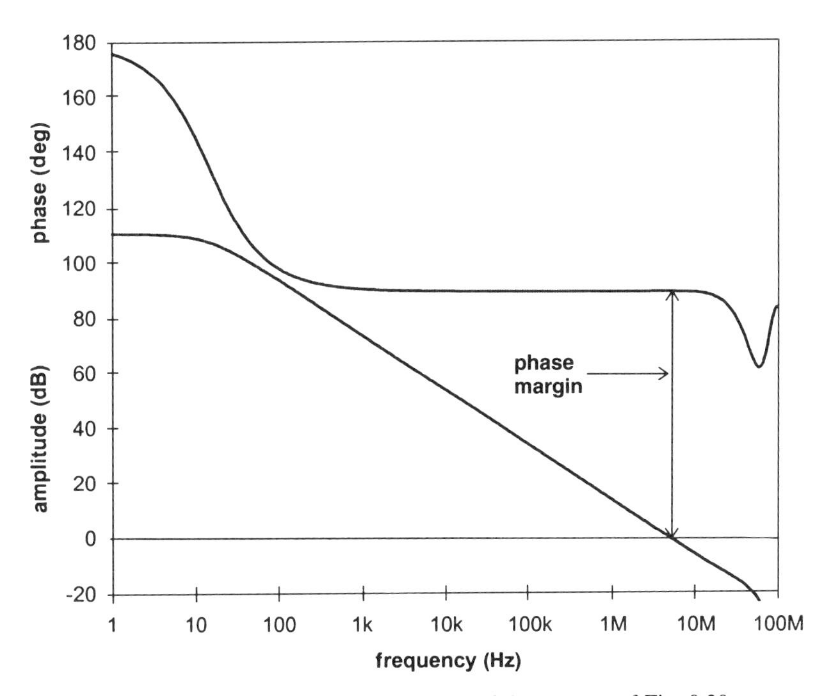

Figure 8.31. Gain and phase responses of the op-amp of Fig. 8.30*a*.

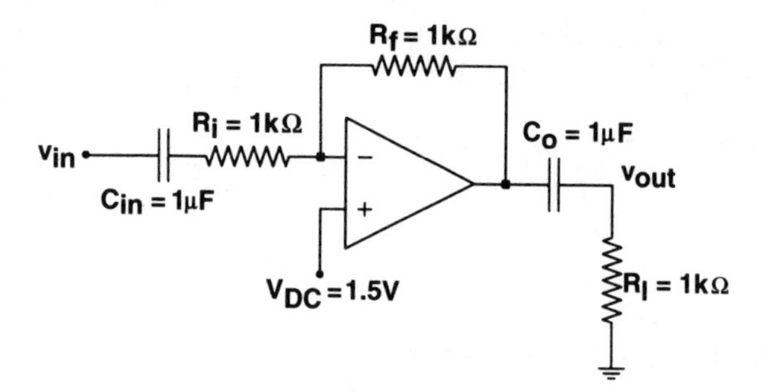

Figure 8.32. Test circuit to compute the output drive performance.

$W_8 = W_{31} = W_{32} = 200\ \mu\text{m}$ and $W_9 = W_{34} = W_{35} = 100\ \mu\text{m}$ were verified to satisfy the conditions given by Eqs. (8.109) and (8.110). The output drive performance of the op-amp was computed by using the inverting-gain test circuit illustrated in Fig. 8.32. The output of the test circuit for a sine-wave input signal with 2-V peak-to-peak amplitude is shown in Fig. 8.33.

In the examples above, the assumption is made that the currents in two transistors with equal gate-to-source and drain-to-source voltages have the same ratio as their

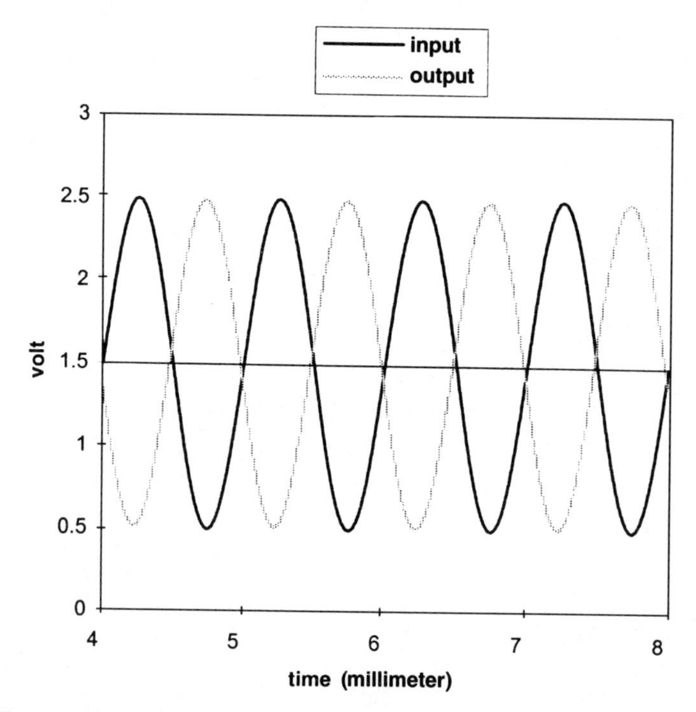

Figure 8.33. Output response of the inverting gain test circuit of Fig. 8.32.

dimensions. One such example is the current in transistor Q_{10}, which is designed to be eight times the current of Q_{35}. In practice due to mask, photolithographic, and etch variations there are differences in the drawn values of W and L. In other words, it is not guaranteed that a transistor with an aspect ratio of $W/L = 1600/0.8$ will carry eight times the current of another transistor with an aspect ratio of $W/L = 200/0.8$ even if they both have the same gate-to-source and drain-to-source voltages. For improved performance, the proper layout technique is to implement the 1:8 aspect ratio using eight duplicates of a unit transistor. This way tolerances of L and W will not affect the current ratio. This strategy should be used in all current mirrors used to multiply current and function as a current amplifier.

8.3. COMPARATOR DESIGN TECHNIQUES AND EXAMPLES

In this section, design procedures for two types of comparators are discussed. Once again the design will be based on the 0.6 μm n-well CMOS process of Section 8.2. In the first example the task is to design the three-stage direct-coupled comparator of Fig. 5.19 to switch state with a minimum overdrive voltage on the order of 250 μV in less than 200 ns. In the worst case the input will change from a level that drives the comparator to some initial saturated condition, to an opposite level barely in excess of 250 μV. The schematic of the three-stage direct-coupled comparator with a V_{BE}-based bias generator is shown in Fig. 8.34. It consists of two direct-coupled resistive-load differential gain stages followed by a two-stage differential comparator.

After the comparison phase, the output swing of the comparator should be equal to the full power supply voltage, $V_{DD} = 3$ V. As a result, the overall gain of the comparator must be at least (3 V/250 μV) = 12,000 V/V. If the third comparator stage provides a minimum gain of 60, the required combined gain from the first two stages will be 200, or $\sqrt{200} \approx 14.2$ for each of the first two stages. The response time of the third-stage differential comparator is largely determined by its parasitic capacitances, the currents supplied by transistors Q_{17} and Q_{19}, and the amount of the overdrive voltage available at the input of the differential stage. The bias current is given by

$$I_B = \frac{V_{BE}}{R_B} = \frac{0.7}{7000} = 100\ \mu\text{A}. \tag{8.111}$$

Assuming that $(W/L)_{17} = (W/L)_{19} = (W/L)_6$, the currents of transistors Q_{17} and Q_{19} will be 100 μA. We will choose the dimensions of the input-stage source-coupled transistors as $L_{13} = L_{14} = 1$ μm *and* $W_{13} = W_{14} = 200$ μm, and $L_{15} = L_{16} = L_{18} = 2$ μm and $W_{15} = W_{16} = W_{18} = 40$ μm for transistors Q_{15}, Q_{16}, and Q_{18}. Since the overall gain of the first two stages is approximately $A_d \approx 200$, so a 250-μV differential signal at the input of the first stage will appear as a 50-mV overdrive signal at the input of the third stage. It was verified through SPICE simula-

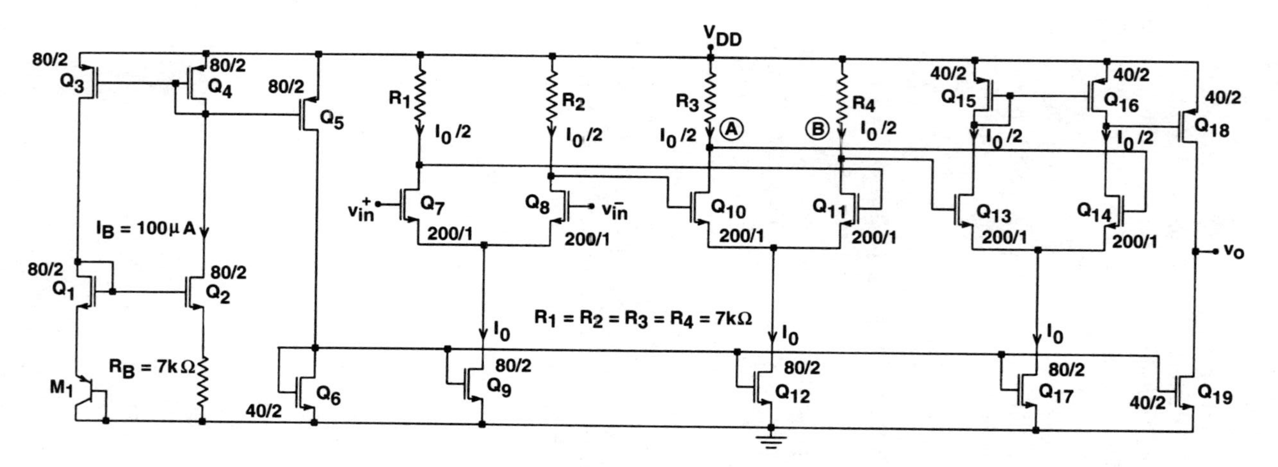

Figure 8.34. Direct-coupled three-stage comparator circuit.

tion that a differential overdrive signal of 50 mV will cause the third-stage comparator to switch in less than 50 ns.

The next step in the design process is to select component and current values for the two resistive-load differential amplifiers. Using $(W/L)_9 = (W/L)_{12} = 2(W/L)_6$, the differential stage tail currents will be $I_0 = 2 \times 100 = 200\ \mu\text{A}$. Then if we choose $R_1 = R_2 = R_3 = R_4 = R_B = 7000\Omega$, the voltage drop across the loads of the differential amplifiers will be $V_{RL} = 7000 \times (200 \times 10^{-6})/2 = 0.7$ V. This results in an output bias voltage of $V_o = 3 - 0.7 = 2.3$ V, which is an adequate value for a 3-V power supply voltage. In addition, the maximum output voltage swing of the differential stage is a well-controlled value given by $\Delta V_o = 200 \times 10^{-6} \times 7000 = 1.4$ V. The gain of the resistive-load differential amplifier was derived earlier and is given by Eq. (5.12) as

$$A_d = g_{mi} R_L, \tag{8.112}$$

where g_{mi} is the transconductance of the NMOS input devices. For the target gain of $A_d = 14.2$ the transconductance can be calculated from Eq. (8.112) as

$$g_{mi} = \frac{14.2}{7000} = 2.03 \times 10^{-3} \text{ mhos.} \tag{8.113}$$

Using the transconductance factor for the NMOS transistor given in Section 8.2 ($k_n' = 55 \times 10^{-6}\ \mu\text{A/V}^2$), the aspect ratios of the input devices can be calculated from the following relationship:

$$g_{mi} = 2\sqrt{k_n'(W/L)_i I_i}, \tag{8.114}$$

where $I_i = I_0/2 = 100\ \mu\text{A}$. Substituting the value of $g_{mi} = 2.03 \times 10^{-3}$ mhos into Eq. (8.114), the W/L ratio can be calculated as $(W/L)_i \approx 182$. Once again, using a channel length of $L_i = 1\ \mu\text{m}$ for all input-stage source-coupled devices results in $W_7 = W_8 = W_{10} = W_{11} = 182\ \mu\text{m}$ for the channel widths. For simplicity and added margin, the channel widths were increased to 200 μm.

The response time of the resistive-load differential amplifier is determined mainly by the resistance of the load and the parasitic capacitances connected to the output nodes. The parasitics include the junction capacitances of the drain diffusion and the gate-to-source (C_{Gs}) capacitances of the devices that load the output node. For the comparator of Fig. 8.34 an estimate of the total capacitances at the output of the differential stage is 1.5 pF. Hence the transient response time constant of each differential amplifier is on the order of $\tau_0 = 1.5 \times 10^{-12} \times 7000 = 10.5$ ns, which is well within the limits of the required 200-ns response time.

At this point the dimensions of all devices have been determined and the schematic of the three-stage comparator and the corresponding device sizes are shown in Fig. 8.34. To verify the accuracy of the design a SPICE simulation was carried out to analyze the circuit. A single power supply voltage of 3 V was used and one input

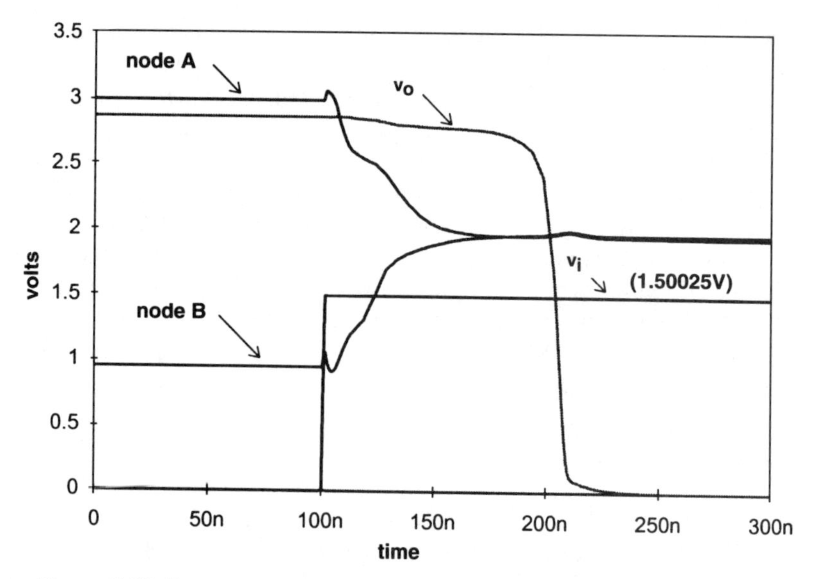

Figure 8.35. Response of the comparator of Fig. 8.34 to a 250-μV overdrive.

of the comparator was connected to 1.5 V, which was considered to be analog ground. For a worst-case condition, the input overdrive was varied from 0 V (heavily saturated condition) to 1.50025 V (250 μV above the 1.5-V reference ground voltage). The actual result of the SPICE simulation is illustrated in Fig. 8.35. As the figure shows, the response time is on the order of 105 ns, which is clearly well within the required 200-ns design limit.

The last example of this section deals with the design of a comparator with hysteresis. The hysteresis is a characteristic of the comparator which changes the input threshold voltage as a function of the output level. Comparators with hysteresis were discussed in Section 5.4. The purpose of this example is to design a comparator with 76-mV hysteresis based on the regenerative differential amplifier of Fig. 5.23. The complete schematic of the comparator with the output stage and a V_{BE}-based bias generator is shown in Fig. 8.36. The bias current was calculated in the preceding example as $I_B = 100$ μA. Assuming that $(W/L)_0 = 2\,(W/L)_{16}$, the tail current of the regenerative differential amplifier will be $I_0 = 200$ μA. We next choose the dimensions of the input source-coupled device pairs as $W_1 = W_2 = 100$ μm and $L_1 = L_2 = 2$ μm. For simplicity we assume that plus (+1.5 V) and minus (−1.5 V) supplies are used and that the gate of Q_2 is tied to analog ground (0 V). Thus the purpose of this example is to calculate the dimensions of the remaining devices so that the comparator exhibits positive and negative threshold points of +38 mV and −38 mV, respectively.

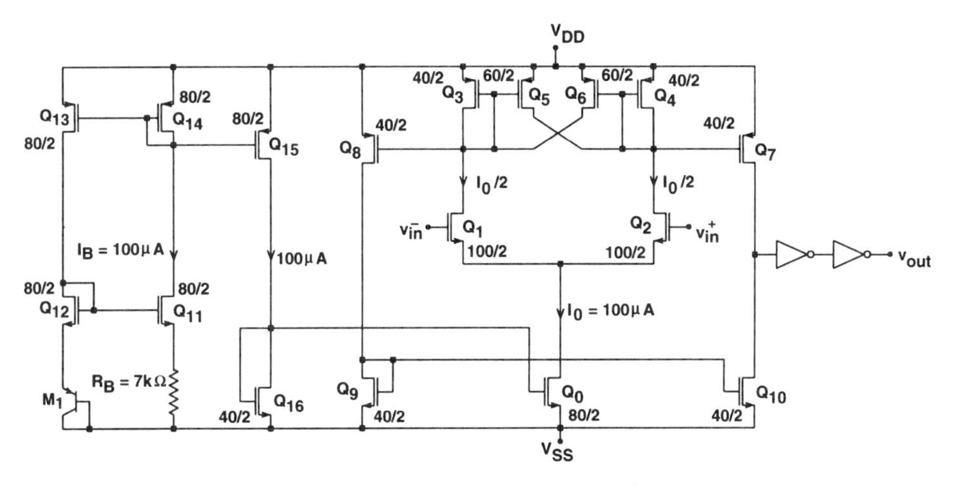

Figure 8.36. Comparator with hysteresis.

The trigger points of the regenerative comparator of Fig. 8.36 were calculated in Chapter 5 as Eqs. (5.28) and (5.29) and is repeated here for convenience:

$$V_{\text{trig}}^{+} = -V_{\text{trig}}^{-} = \sqrt{\frac{I_0}{k_n'(W/L)_i}}\,\frac{\sqrt{\alpha} - 1}{\sqrt{1 + \alpha}}, \tag{8.115}$$

where the $(W/L)_i$ represent the aspect ratio of the input devices and $\alpha = (W/L)_5/(W/L)_3 = (W/L)_6/(W/L)_4$. Using the transistor device parameters given in Section 8.2 ($k'_n = 55 \times 10^{-6}$ A/V^2), we have

$$0.038 = \sqrt{\frac{200 \times 10^{-6}}{55 \times 10^{-6} \times (100/2)}}\,\frac{\sqrt{\alpha} - 1}{\sqrt{1 + \alpha}}. \tag{8.116}$$

Simplifying Eq. (8.116) and solving for $\sqrt{\alpha}$, we get

$$\alpha - 2.04051\sqrt{\alpha} + 1 = 0. \tag{8.117}$$

Solving Eq. (8.117) yields $\alpha \simeq 1.5$. Using this value of α, the aspect ratio of the load transistors can be calculated from the following relationship:

$$\alpha = \frac{(W/L)_5}{(W/L)_3} = \frac{(W/L)_6}{(W/L)_4} = 1.5. \tag{8.118}$$

Using $L = 2\ \mu$m for all transistors in Eq. (8.118) results in the following relationship between the channel widths:

$$W_5 = W_6 = 1.5W_3 = 1.5W_4. \tag{8.119}$$

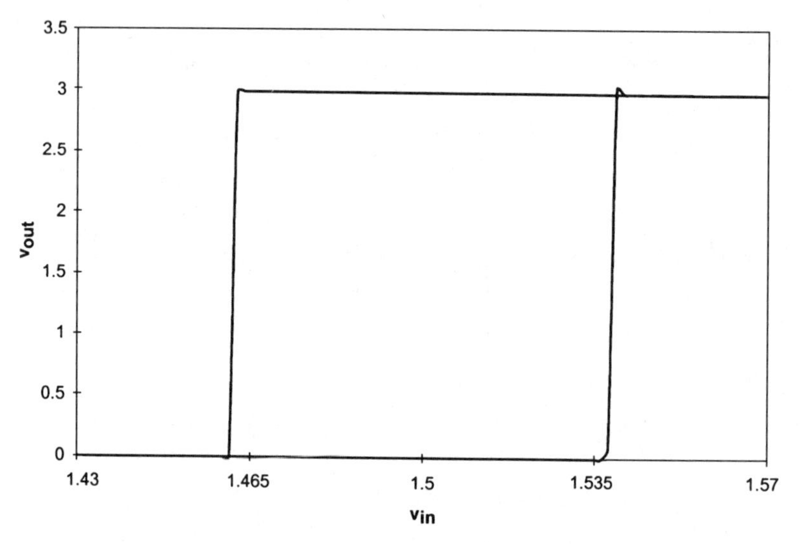

Figure 8.37. Simulation results of the comparator of Fig. 8.36 with hysteresis.

Next, using $W_3 = W_4 = 40\ \mu\text{m}$, the widths of transistors Q_5 and Q_6 can be calculated from Eq. (8.119) as $W_5 = W_6 = 60\ \mu\text{m}$. Finally, a reasonable choice for the widths of the transistors in the differential to single-ended converter is

$$\begin{aligned} W_8 &= W_3 = W_4 = W_7 = 40\ \mu\text{m}, \\ W_9 &= W_{10} = 40\ \mu\text{m}. \end{aligned} \tag{8.120}$$

where, once again, $L = 2\ \mu\text{m}$ has been assumed for the lengths of all transistors in the output stage.

In this example the aspect ratios of the devices were selected in such a way that Eq. (8.118) was satisfied. The choices of the actual device dimensions, however, were somewhat arbitrary. Another criterion that was ignored in this analysis is the comparator transient response time. The sizes of the devices determine the values of the parasitic capacitances at the critical nodes, which affect the transient response. Imposing a requirement on the minimum response time sets an additional condition that can be used for the optimum selection of the device dimensions.

The circuit of Fig. 8.36 was simulated on SPICE and the simulation results are illustrated in Fig. 8.37. Notice that as the input starts from a negative value and goes positive, the output does not make a transition until the input reaches the positive threshold, 38 mV. Once the output switches state, the effective threshold is changed to -38 mV. When the input returns in the direction of its original negative value, the output does not change until the input reaches the negative trip point of -38 mV. The amount of the hysteresis should normally be equal to or greater than the largest expected noise amplitude.

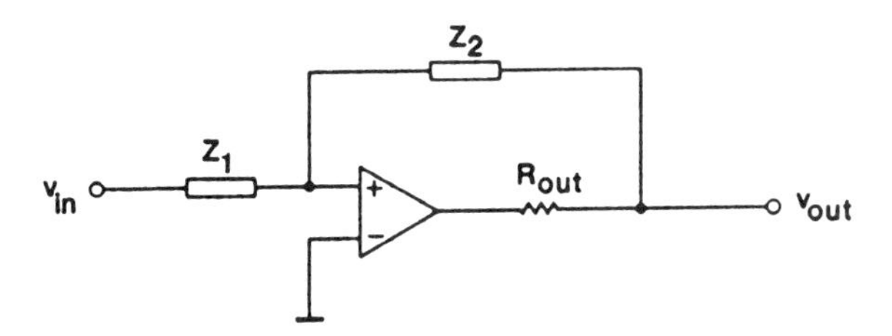

Figure 8.38. Op-amp with nonzero output impedance (Problem 8.4).

PROBLEMS

8.1. In the input stage of the circuit of Fig. 8.2, assume that Q_1 and Q_2 are imperfectly matched such that $g_{m1} = (1 + \epsilon)g_{m2}$. Show that if a common-mode voltage v_c exists at the input, the corresponding CMRR is $(1/\epsilon)g_{mi}/g_{di}$.

8.2. Prove Eq. (8.17) for the circuit of Fig. 8.1.

8.3. Find the dc offset voltage of the op-amp of Fig. 8.6 if the currents of the two sources shown as I are mismatched by ϵ.

8.4. Show that in the circuit of Fig. 8.38, the effective output impedance is $R_{\text{out}} = (-A_c + 1)/A$, where $A_c \cong -Z_2/Z_1$ is the closed-loop gain of the stage. Assume that $A \gg 1$ and $R_{\text{out}}/A \ll |Z_1|$ and $|Z_2|$.

REFERENCES

1. P. R. Gray and R. G. Meyer, *Analysis and Design of Analog Integrated Circuits,* Wiley, New York, 1977.
2. O. H. Schade, Jr., *IEEE J. Solid-State Circuits, SC-3*(6), 791–798 (1978).
3. G. C. Temes, *IEEE J. Solid-State Circuits, SC-15*(3), 358–361 (1980).
4. K. Martin and A. S. Sedra, *IEEE Trans. Circuits Syst., CAS-28*(8), 822–829 (1981).
5. P. R. Gray and R. G. Meyer, *IEEE J. Solid-State Circuits, SC-17*(6), 969–982 (1982)
6. W. C. Black, D. J. Allstat, and R. A. Reid, *IEEE J. Solid-State Circuits, SC-15*(6) 929–938 (1980).
7. K. Ohara, P. R. Gray, W. M. Baxter, C. F. Rahim, and J. L. McCreary, *IEEE J. Solid-State Circuits, SC-15*(6), 1005–1013 (1980).
8. P. R. Gray, R. W. Brodersen, D. A. Hodges, T. C. Chai, R. Kameshiro, and K. C. Hsieh, *Proc. Int. Symp. Circuits Syst.,* pp. 419–422, 1982.

INDEX